C0-ABV-529

Chromosomes in Evolution of Eukaryotic Groups

Volume I

Editors

Arun Kumar Sharma, D.Sc., F.N.A., F.A.Sc.
Ghosh Professor and Program Coordinator
Center of Advanced Study
Department of Botany
University of Calcutta
Calcutta

Archana Sharma, Ph.D., D.Sc., F.N.A., F.A.Sc.
Professor of Genetics and Head
Department of Botany
University of Calcutta
Calcutta

CRC Press, Inc.
Boca Raton, Florida

Library of Congress Cataloging in Publication Data

Main entry under title:

Chromosomes in evolution of eukaryotic groups.

Bibliography: p.
Includes index.
1. Chromosomes—Evolution. 2. Eukaryotic cells—Evolution. 3. Evolution. I. Sharma, Arun Kumar, 1923- II. Sharma, Archana Mookerjea.
QH371.C5 1983 574.87'322 82-9449
ISBN 0-8493-6496-5 (v. 1) AACR2
ISBN 0-8493-6497-3 (v. 2)

Direct all inquiries to CRC Press, Inc., 2000 Corporate Blvd., N.W., Boca Raton, Florida, 33431.

International Standard Book Number 0-8493-6496-5 (Volume I)
International Standard Book Number 0-8493-6497-3 (Volume II)

Library of Congress Card Number 82-9449
Printed in the United States

FOREWORD

Chromosomes of higher organisms exhibit myriads of forms of wide diversity, but their mechanism of origin is still a controversial issue. Their prokaryotic ancestry, so far considered as undisputed, has run into bad weather, due to the discovery of intervening sequences associated with the property of splicing. This theory, envisaging a gradual evolution of complexity in the chromosome from the genophore, is facing replacement by the concept of independent origin of eukaryotic chromosomes parallel to that of the prokaryota.

The chromosomes of higher organisms have undergone diverse modifications both during and after evolution. Several structural features are indeed of phylogenetic and evolutionary significance. At one end of the broad spectrum is the peculiar chromosome of the Dinophyceae with very little histone and absence of functional differentiation of segments. The other extreme is represented by the complex chromosome structure of primates, having reverse-banded segments associated with even special genetic attributes.

Structural features of chromosomes with high phylogenetic potential include, among others, the centromere — diffuse, polycentric, or localized; number and nature of nucleolar constrictions; sex chromosomes with special reference to their multiple mechanisms; early and late replication, heterochromatin, repeated DNA and banding pattern. The nature and position of repeats in introns and even exons, as brought out lately, may all reveal facts of fundamental significance to a evolutionist.

Aspects of chromosome behavior often stressed in the study of evolution are heteroploidy, fragmentation, and translocation. The association of such factors with the evolution of species is undeniable. However, all three mechanisms, excepting the relative infrequency of polyploids in animal systems, are widespread in eukaryota, though not restricted to any particular level of taxonomic hierarchy. In fact, in plant systems, such changes occur with equal prominence within and between species, as well as between genera and families. As such, their universal incidence at all levels of taxonomic hierarchy disqualifies them for consideration as specific parameters of phylogenetic significance. This understanding is of supreme importance in the application of chromosome science to the study of evolution.

The amount of data on the importance of chromosome structural analysis in the study of evolution is indeed enormous. The analysis of such a compendium of information is a task of tremendous magnitude. This exercise is also complicated as the paths of evolution are divergent. A definite pattern of chromosome evolution in the eukaryotic system is yet to be established, even in structural features of chromosomes of established phylogenetic significance. As a result it is difficult to draw a single undisputed phylogenetic tree for a plant or animal system as a whole, based solely on chromosomal characteristics.

It was initially desired to present the pathways of evolution of chromosomes in different groups of eukaryota according to their sequence in an established progression of complexity. Such a desire, however pious it may be, could not be fulfilled due to various reasons beyond our control for which we offer our apologies in advance. Limitations of space and time also did not permit a discussion of all the groups which may have to be taken up later in detail.

THE EDITORS

Arun Kumar Sharma, D.Sc., F.N.A., F.A.Sc., F.N.A.Sc., is Sir Rashbehari Ghosh Professor and Programme Coordinator (Director), Centre of Advanced Study on Cell and Chromosome Research, Department of Botany, University of Calcutta and the President, Indian National Science Academy, New Delhi (1983 to 1984). He obtained his M.Sc. (1945) and D.Sc. (1955) degrees from the University of Calcutta and was Head of the Department of Botany from 1969 to 1980. He has made significant contributions on different aspects of chromosome research; and he has built up one of the largest centers of chromosome research in the world. His works cover cytotaxonomic studies on angiosperms, particularly monocotyledons, speciation in asexual plants, and development of techniques of chromosome analysis from both meristematic and differentiated nuclei. Among his more than 350 papers, recent works include theory of chromosome dynamism, demonstration of the variability of chemical components of the chromosome during organogensis, additional genetic elements in chromosomes, dynamic DNA, and study of the chromosome involving in vitro mutagenasis.

Professor Sharma has been the General President of the Indian Science Congress Association (1980), Vice-President of the Indian National Science Academy, Chairman of the Indian National Committee of IUBS and Co-Chairman, Global Seminar on Role of Scientific Societies (AAAS/INSA/ISCA). He has been a member of the councils of all three Academies of India and President of several societies, including the Indian Botanical Society, Indian Society of Cell Biology, Genetic Association of India, Society of Cytologists and Geneticists, and others. His numerous Awards include the Shanti Swarup Bhatnagar Award of the Council of Scientific and Industrial Research, the J. C. Bose Award of the University Grants Commission, the Silver Jubilee Medal of the Indian National Science Academy, the Birbal Sahni Medal of the Indian Botanical Society, and the Jawaharlal Nehru Fellowship.

He has been visiting lecturer to different centers of the world and led the Indian delegation several times to international conferences including the International Genetics Congress, the Hague (1963); the International Botanical Congress, Leningrad (1975); the International Cell Biology Congress, Berlin (1980); and the IUBS General Assembly, Bangalore (1975), and Ottawa (1982). He is the founding Editor of the international journal, *The Nucleus,* and member of the editorial board of several journals. He is co-author, with Archana Sharma of *Chromosome Techniques—Theory and Practice,* a standard reference and textbook.

Archana Sharma, Ph.D., D.Sc., F.N.A., F.A.Sc., F.N.A.Sc., is Professor of Genetics and Head of the Department of Botany, University of Calcutta (1980 to 1982). She obtained her M.Sc. (1951), Ph.D. (1955), and D.Sc. (1960) degrees from the University of Calcutta and specialized in cytogenetics and human genetics. She has made outstanding contributions to cytotaxonomy, the cause of polyteny in differentiated nuclei, and the development of new techniques for the study of chromosome structure. Her group is actively engaged in the study of chromosomal and genetic polymorphisms in normal and pathological human populations, differentiating patterns in the human fibroblast, and genetic polymorphisms in relation to environmental mutagensis and genetic diseases. Other significant research includes studies of the effect of metallic pollutants on genetic systems, both antagonistic and synergistic. She has more than 150 papers and several books to her credit.

Professor Sharma is a Fellow of all three Academies of India, member of the council of the Indian National Science Academy, member of the Science and Engineering Research Council of the Government of India, and General Secretary of the Indian Science Congress Association, with which she has been involved for nearly two decades. As official delegate of the Government of India, she has participated in several international conferences, including the IUBS General Assembly Session at Helsinki, and the International Cell Biology Congress at Berlin, and has been the Visiting Scientist in the U.S.S.R. under the Government of India exchange program and a member of the delegation from the Academy to the People's Republic of China. She is the recipient of the Shanti Swarup Bhatnagar Award of the Council of Scientific and Industrial Research (1978) and the J. C. Bose Award of the University Grants Commission, and National Lecturer, University Grants Commission. She is the Editor of *The Nucleus*, and a member of the Editorial Board of a number of other journals.

CONTRIBUTORS

Bernard John, Ph.D., D.Sc.
Director
Research School of Biological Sciences
Australian National University
Canberra
Australia

Y.S.R.K. Sarma, Ph.D.
Professor of Botany
Center of Advanced Study in Botany
Department of Botany
Banaras Hindu University
Varanasi
India

Yoshio Ojima
Professor of Biology
Department of Biology
Kwansei Gakuin University
Nishinomiya
Japan

A. J. E. Smith, D.Phil., D.Sc.
Reader in Botany
School of Plant Biology
University College of North Wales
Bangor
Gwynedd
Wales

Tosihide H. Yosida, D.Sc.
Head
Department of Cytogenetics
National Institute of Genetics
Misima, Sizuoka-ken
Japan

TABLE OF CONTENTS

Volume I

Chapter 1

THE ROLE OF CHROMOSOME CHANGE IN THE EVOLUTION OF ORTHOPTEROID INSECTS

Bernard John

TABLE OF CONTENTS

I. INTRODUCTION

The chromosome changes that occur in the cell lineages of all eukaryotes as a mutational undercurrent are of three different kinds:

1. Structural mutations leading to either intra (inversion) or inter (translocation) chromosome rearrangements
2. Numerical mutations leading to either aneuploid or polyploid changes in members of the regular chromosome complement
3. Changes in the content of heterochromatin involving either the addition of supernumerary segments to normal members of the chromosome complement or else the addition of novel supernumerary chromosomes

The contribution which the chromosomes of orthopteroids have made to the evolution of the group can, therefore, only be gauged from an examination of the chromosome variation present within and between populations of extant species and kinds of differences demonstrated by such an examination.

These differences are of three kinds:

1. Intrapopulation polymorphisms in which distinct karyomorphs differing in respect of the structure, number, or heterochromatin content of their chromosomes coexist within individual populations of a given species.
2. Interpopulation polytypisms in which fixed karyomorphs are present within individual populations of a given species but where different populations differ in respect of the nature of these fixed differences. Most commonly, one or more forms of chromosome change present in a homozygous state in one population, or one series of populations, are absent from others in different geographical areas. In some such cases restricted polymorphic populations are sometimes present in hybrid zones of geographical overlap. The one unusual form of polytypism is that where a particular category of chromosome change reaches fixation as a permanent heterozygote in one sex but not the other. Here the fixed differences that obtain between different populations are concerned only with the form of permanent heterozygosity present in one of the sexes.
3. Interspecies differences in which closely related species differ in respect of chromosome structure, chromosome number, and/or heterochromatin content. Here, too, one or more chromosome changes present in a homozygous state in one taxon is absent in the other.

Before embarking on an in-depth study of each of these three categories, it is necessary to define the range of organisms which the term orthopteroid encompasses (Table 1). The group includes five orders of insects, all of which are well known under the common names of roaches, mantids, termites, stick insects, crickets, and grasshoppers. In presenting the chromosome differences which occur within and between populations and species of these orders, I will concentrate on the relatively few well-worked examples rather than attempt to mention the very many more cases which have been noted but not analyzed in detail. An interested reader will find them well summarized in the recent reviews of White[1] and Hewitt.[2]

II. CHROMOSOME POLYMORPHISMS

In orthopteroid insects two categories of chromosome mutation, in particular, have been involved in the development of polymorphic populations:

Table 1
A CLASSIFICATION OF THE GENERA OF ORTHOPTEROID INSECTS REFERRED TO IN THIS ARTICLE

SUPERORDER ORTHOPTEROIDEA

Order 1 — Blattodea	
Family 1 — Blattidae	
Subfamily Blattinae	*Periplaneta*
Family 2 — Blattellidae	
Subfamily Blatellinae	*Blatella*
Family 3 — Blaberidae	
Subfamily 1 — Blaberinae	*Blaberus*
Subfamily 2 — Pycnoscelinae	*Pycnoscelus*
Order 2 — Mantodea	
Family 1 — Amorphoscelidae	*Amorphoscelus, Cliomantis, Glabromantis*
Family 2 — Eremiaphilidae	*Humbertiella*
Family 3 — Hymenopodidae	*Harpagomantis*
Family 4 — Mantidae	
Subfamily 1 — Amelinae	*Holaptilon*
Subfamily 2 — Caliridinae	*Leptomantis*
Subfamily 3 — Iridiopteryginae	*Bolbe, Halwania, Ima, Kongobatha*
Subfamily 4 — Mantinae	*Callimantis*
Subfamily 5 — Thespinae	*Promiopteryx, Pseudomiopteryx*
Subfamily 6 — Photininae	*Brunneria*
Order 3 — Isoptera	
Family 1 — Kalotermitidae	*Incisitermes, Kalotermes*
Family 2 — Rhinotermitidae	*Reticulitermes*
Family 3 — Termitidae	
Subfamily 1 — Apicotermitinae	*Acidnotermes, Microcerotermes*
Subfamily 2 — Termitinae	*Cubitermes, Crenetermes, Noditermes, Ophiotermes, Pericapritermes, Procubitermes, Thoracotermes, Tuberculitermes*
Subfamily 3 — Macrotermitinae	*Macrotermes, Odontotermes, Protermes, Pseudocanthotermes*
Subfamily 4 — Nasulitermitinae	*Nasutitermes*
Order 4 — Phasmatodea	
Family 1 — Phylliidae	
Subfamily Bacillinae	*Bacillus, Clitumnus, Clonopsis, Epibacillus, Leptynia*
Family 2 — Phasmidae	
Subfamily 1 — Lonchodinae	*Carausius*
Subfamily 2 — Necrosciinae	*Parasipyloidea*
Subfamily 3 — Phasminae	*Baculum, Ctenomorpha, Phobaeticus*
Subfamily 4 — Podocanthinae	*Didymuria, Extatosoma*
Order 5 — Orthoptera	
Suborder 1 — Ensifera	
Superfamily 1 — Gryllodea	
Family 1 — Gryllidae	
Subfamily 1 — Gryllinae	*Scapsipedus, Teleogryllus*
Subfamily 2 — Nemobiinae	*Allonemobius, Eunemobius, Neonemobius, Nemobius*
Subfamily 3 — Trigonidiinae	*Anaxipha*
Family 2 — Gryllotalpidae	*Gryllotalpa*
Superfamily 2 — Tettigoniodea	
Family 1 — Tettigoniidae	
Subfamily 1 — Meconematinae	*Xiphidiopsis*

Table 1 (continued)
A CLASSIFICATION OF THE GENERA OF ORTHOPTEROID INSECTS REFERRED TO IN THIS ARTICLE

Subfamily 2 — Phaneropterinae	*Odontura, Poecilimon*
Subfamily 3 — Saginae	*Saga*
Subfamily 4 — Tettigoniinae	*Decticus, Metrioptera* (and see Table 53)
Suborder 2 — Caelifera	
Superfamily 1 — Acridoidea	
Family 1 — Acrididae	
Subfamily 1 — Acridinae	*Acrida, Austroicetes, Caledia, Chortoicetes, Cryptobothrus*
Subfamily 2 — Calliptaminae	*Calliptamus*
Subfamily 3 — Catantopinae	*Buforania, Cuparessa, Eurenephilus, Gonista, Leiotettix, Macrotona, Peakesia, Percassa, Pezotettix, Phaulacridium, Tolgadia*
Subfamily 4 — Cyrtacanthacridinae	*Patanga, Schistocerca*
Subfamily 5 — Eyprepocnemidinae	*Eyprepocnemis, Heteracris*
Subfamily 6 — Gomophocerinae	*Acyptera, Bootettix, Chloealtis, Chorthippus, Chrysochroan, Euthystira, Myrmeleotettix, Neopodismopsis, Omocestus, Stauroderus, Stenobothrus, Stethophyma*
Subfamily 7 — Melanoplinae	*Boonacris, Dichroplus, Melanoplus, Micropodisma, Miramella, Oedaleonotus, Podisma*
Subfamily 8 — Oedipodinae	*Acrotylus, Aerochoreutes, Camnulla, Chimerocephala, Circotettix, Conozoa, Derotemema, Encotophus, Gastrimargus, Locusta, Oedaleus, Oedipoda, Parapleurus, Trimerotropis*
Subfamily 9 — Romaleinae	*Spaniacris*
Family 2 — Eumastacidae	
Subfamily Morabinae	*Culmacris, Keyacris, Moraba,* P24, P25, P45, P52, P53, P85, P169, P196, *Vandiemenella, Warramaba*
Family 3 — Lentulidae	*Karruacris*
Family 4 — Pyrgomorphidae	*Atractomorpha*
Superfamily 2 — Tetrigoidea	
Family Tetrigidae	*Tetrigidea*

1. Spontaneous exchanges either within or between chromosomes leading to the production of structural rearrangements
2. Spontaneous amplification of particular chromosome regions resulting in the formation of supernumerary segments

These two events are also sometimes coupled so that amplification occurs at or following exchange events. In this way a small centric fragment, produced as a by-product of an exchange event, may enlarge into a supernumerary chromosome.

A. Inversion Systems

In chromosome systems with localized centromeres, which is true of all orthopteroids, inversions of chromosome material can be conveniently considered in two groups according to whether the centromere itself lies within (pericentric) or outside (paracentric) the inverted region. Paracentric inversions do not alter arm lengths or, therefore, arm ratios. Pericentric inversions, on the other hand, by producing a change in centromere position, may affect both arm length and arm ratio.

1. Paracentrics

There is only one case on record in orthopteroids of a polymorphism involving a

Table 2
STRUCTURE OF THE MARY'S PEAK POPULATION OF *BOONACRIS ALTICOLA*[3]

	No. of males		
Year	Heterozygous for In(2)MP-1	Homozygous for bivalent #2	Total sample
1975	9 (60%)	6 (40%)	15
1976	25 (62%)	15 (38%)	40

Table 3
MEIOTIC BEHAVIOR OF HETEROZYGOUS IN(2)MP-1 BIVALENTS OF *BOONACRIS ALTICOLA*[3]

	Year		
Meiotic stage	1975, $n = 9$	1976, $n = 10$	Totals
Pachytene			
Straight pairing	144 (60%)	201 (57%)	
Incomplete pairing	84 (35%)	129 (37%)	
Reverse loop pairing	11 (5%)	19 (62%)	
Totals	239	350	589
First anaphase			
Normal segregation	450 (95%)	470 (94%)	
Dicentric and acentric	15 (3%)	16 (3.2%)	
Dicentric only	7 (1.5%)	11 (2.2%)	
Acentric only	3 (0.6%)	3 (0.6%)	
Totals	475	500	975

paracentric inversion, that is, in the grasshopper *Boonacris alticola* (Acrididae, Melanoplinae), a species which has a chromosome complement of 21 rods and in which, as is most usual in orthopteroids, the autosomes are numbered from 1 to 10 in decreasing order of size.[3] A single population of this flightless form from Mary's Peak, Ore., U.S. proved to be polymorphic for a long paracentric inversion in autosome 2 designated as In(2)MP-1.

In both years over which the population was sampled there was an excess of individuals heterozygous for the inversion, which comprised some 60% of the entire population (Table 2). Since the basic and the inverted homozygote classes could not be distinguished, it was not possible to partition the homozygotes more precisely. The data relating to the meiotic behavior of the inversion heterozygotes were remarkably consistent over both years of sampling (Table 3). Of the 589 pachytene cells from heterozygous individuals, 59% showed straight pairing of the relatively inverted homologs. A further 36% showed asynapsis within the relatively inverted region, and only 5% formed a reverse loop of the kind expected from homologous pairing. In keeping with these figures, some 95% of the 975 anaphase-I cells examined showed normal segregation. Dicentrics and/or acentrics were present in the remaining 5%.

The straight pairing observed in this case must, of course, be nonhomologous and cannot lead to crossing over. The interrupted pairing (partial asynapsis) in the inverted region has the same effect. Hence, the contraction of the linkage group, the most

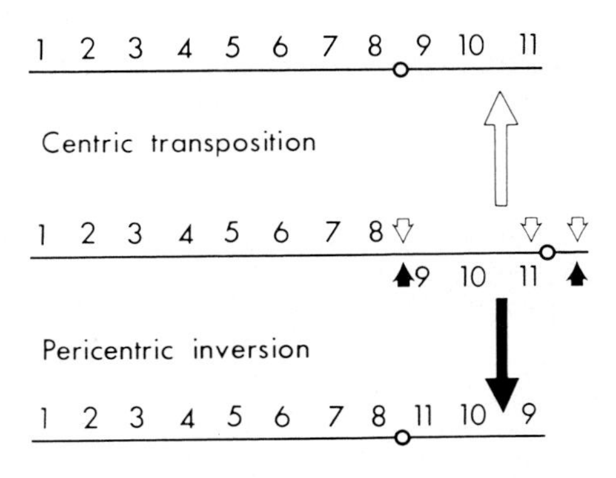

FIGURE 1. Two modes of pericentric rearrangement. In heterozygotes for a centric transposition, straight pairing at pachytene will, by definition, be homologous in character.

$$\frac{9.10.11}{9.10.11}$$

In a pericentric inversion heterozygote straight pairing will be nonhomologous.

$$\frac{9.10.11}{11.10.9}$$

obvious effect of inversion hybridity, is achieved in this case without any serious accompanying sterility of the kind expected following regular crossing over within reverse pairing loops.

Essentially the same result was obtained[4] in a study of a single mutant individual of the grasshopper *Camnulla pellucida* collected at Olema, Marin County, Calif., U.S. which was heterozygous for a paracentric inversion occupying approximately 10% of the length of one of the two longest autosomes. Here reverse loops were present in not more than 4% of the 297 pachytene cells analyzed with a further 8% of the cells showing partial asynapsis and the remainder, 88%, with straight pairing. An analysis of 603 cells in anaphase I and II, scored for the presence of dicentric bridges and acentric fragments, indicated that crossing over within the inversion could not have occurred in more than 8% of the meiocytes.

2. Pericentrics

By contrast with the apparent rarity of paracentric inversion polymorphisms, pericentric polymorphisms are much more common in orthopteroids. It has been suggested by several authors[2,5] that these polymorphisms are better described simply as centric shifts, since in all of them there is no reverse looping of the kind conventionally assumed to be diagnostic for inversions. Consequently, they might also be explained as a result of a three-break centric transposition,[6] which would, of course, be expected to produce straight pairing at pachytene (Figure 1).

The behavior of paracentric inversion heterozygotes in orthopteroids, outlined in Section II.A.1, shows that the absence of reverse loop pairing does not militate against an inversion explanation in the case of pericentric rearrangements. The important difference between the two hypotheses rests on the precise nature of pairing. In a centric transposition system the straight pairing is homologous in nature. As such, it does not prohibit crossing over and such crossing over would, of course, give rise to a dicentric and an acentric chromatid at first anaphase, as would also arise following crossing over within a reverse loop in the case of a heterozygous pericentric inversion. In a pericentric inversion system the straight pairing is nonhomologous and so prohibits crossing over. The fact that in the pericentric rearrangements of orthopteroids there is

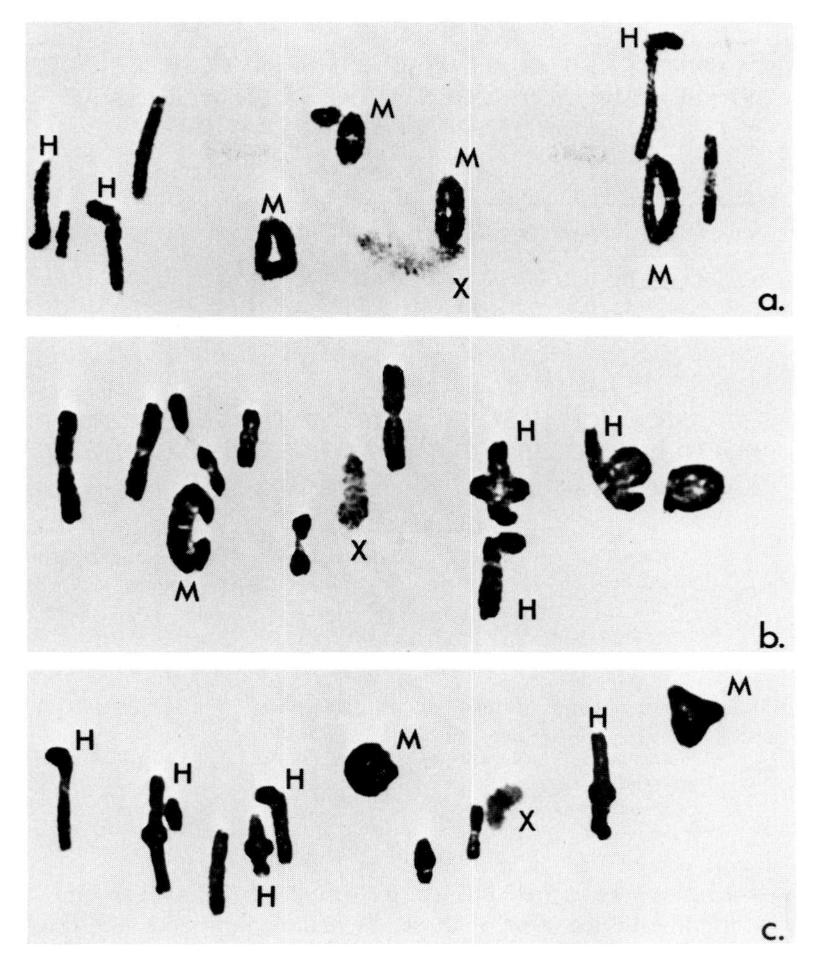

FIGURE 2. Heterozygosity for pericentric inversion in trimerotropine grasshoppers. (a) *Trimerotropis thalassica*, (b) and (c) *T. pseudofasciata*, both with $2n = 23♂$, XO. M denotes a metacentric bivalent, homozygous for inversion, while H identifies a bivalent heterozygous for an inversion.

neither reverse looping at pachytene nor dicentrics and acentrics at anaphase I is only explicable if the straight pairing is indeed nonhomologous. Added to this, in at least one case, that of the grasshopper *Trimerotropis helferi*, reverse loops were seen in two out of several hundred pachytene cells.[7]

Pericentric inversion polymorphisms are known in many grasshoppers. In America, for example, such polymorphisms occur in at least 20 species of the predominantly western genera *Trimerotropis, Circotettix,* and *Aerochoreutes* (Acrididae, Oedipodinae), while in Australia several species of morabine grasshoppers are similarly polymorphic.[8] There are, however, only six studies which deal in some detail with the distribution of the polymorphisms within individual species and we shall confine our attention to these. All of them suffer from some disadvantage or other. Thus, most of them are relatively short-term studies, so we have little accurate information concerning their stability. Undoubtedly the most complete studies to date are in the trimerotropines. Here the number of chromosomes involved in the inversion system varies considerably. Only 1 pair is polymorphic in *T. helferi, C. coconino,* and *C. crotalum*, whereas in *T. sparsa* populations are polymorphic for pericentric inversions in at least 7 of the 11 pairs of autosomes, and in *T. thalassica* up to 10 of the 11 autosome pairs may be polymorphic (Figure 2).

Table 4
FREQUENCIES OF THE 13 DIFFERENT M-INVERSION MORPHS PRESENT IN TWO ARGENTINIAN POPULATIONS OF *TRIMEROTROPIS PALLIDIPENNIS* [9]

	M4		M5		M6			M7			M8		
Population	Acro	Telo	Telo	Meta	Telo	Submeta	Meta	Telo	Submeta	Meta	Telo	Submeta	Meta
Choele-Choel	0.11	0.89	0.97	0.03	0.44	0.00	0.66	0.03	0.63	0.14	0.14	0.13	0.73
Sierra de la Ventana	0.62	0.38	0.95	0.05	0.52	0.17	0.31	0.45	0.36	0.33	0.33	0.12	0.55

Table 5
CYTOLOGICAL CHARACTERISTICS OF THREE ARGENTINIAN POPULATIONS OF *TRIMEROTROPIS PALLIDIPENNIS* [9]

		M4—M8 autosomes			
Population	Climatic characteristics	Inversions per male	Heterozygous bivalents per male	Mean cell Xa frequency	Mean no. interstitial Xta per cell
Laguna Blanca	Coldest climate, precordillera	Nil	Nil	20.42 ± 1.18	7.34 ± 1.29
Choele-Choel	Typical desert area	5.16 (4—6)	1.16 (0—4)	17.33 ± 1.08	1.23 ± 0.60
Sierra de la Ventana	Mildest climate, most rainfall, at eastern border of species range	4.10 (2—6)	2.43 (0—4)	15.60 ± 0.89	1.49 ± 0.71

Most trimerotropine species are confined to North America. Two species have, however, invaded South America. One of these, *T. pallidipennis*, extends from southern Canada over the whole of western U.S. and Mexico but is absent from Central America. It is, however, found southward all along the Andes in Peru, Ecuador, Bolivia, Chile, and Argentina. From here, it has extended to lower altitudes by invading adjacent arid or semiarid regions. Consequently, in South America it exhibits an amazing plasticity in its adaptation to ecological and altitudinal situations.

If the 11 autosomes of the haploid set are numbered in decreasing size order, the three largest chromosomes (L1 to 3) are homozygous metacentrics throughout the entire species range. Similarly, the three smallest pairs (S9 to 11) are invariably homozygous rods. In certain of the southern populations, however, the five medium-sized autosomes (M4 to 8) are polymorphic for a series of pericentric inversions[9] (Table 4), a condition which is never found in North America. Of the three populations sampled in Argentina (Table 5), that from Laguna Blanca, near the Andes, shows the same structural characteristics as the North American populations. In the two polymorphic populations there are significantly more homozygous M inversions per male in Choele-Choel and significantly more heterozygous M inversions per male in Sierra de la Ventana. Added to this, the total number of M inversions per male is significantly greater in Choele-Choel.

The two polymorphic populations have a reduced mean cell chiasma frequency compared to that of Laguna Blanca. The mean frequency at Choele-Choel is, however, significantly higher than that at Sierra de la Ventana. The reduction is, thus, most pronounced in the population with the highest number of heterozygous inversions. The number of interstitial chiasmata is not significantly different in the two poly-

Table 6
CHIASMA DISTRIBUTION DATA FOR 18 INDIVIDUALS OF *TRIMEROTROPIS PSEUDOFASCIATA* FROM SANTA CRUZ ISLAND AND FOR 10 INDIVIDUALS FROM SAN NICOLAS ISLAND[5]

Population	Karyomorph	Distal Xta (%)
Santa Cruz Island,	L_2 Metacentric	81
sheep area	L_1L_3	
	Basic homozygote	23
	Inversion heterozygote	34
	Inversion homozygote	81
	M_8—S_{11} Telocentric	82
	M_{4-7}	
	Basic homozygote	53
	Inversion heterozygote	99
	Inversion homozygote	82
San Nicolas Island,	L_2 Metacentric	76
basic island	L_1L_3 Telocentric	29
karyotype	M_{4-8} Telocentric	56
	S_{9-11} Telocentric	83

Note: In the basic island karyotype referred to, all chromosomes other than the L2 and the X (both fixed metacentrics) are telocentric. The L1, 3, M5, 6, and 7 are polymorphic on the islands.

morphic populations, though in both there is a marked and highly significant reduction compared to the population at Laguna Blanca. Thus, apart from the virtual abolition of crossing over within the inverted regions themselves, there is a marked movement of the chiasmata distally in the M bivalents in both polymorphic populations.

A somewhat similar situation involving patterned chromosome variation with respect to pericentric inversion has been described in *T. pseudofasciata* in North America[5] on comparison of mainland populations from Jacalitos and Bakersfield (California) and from Sand Mountain and Hawthorne (Nevada) with those of five of the Californian Channel Islands, namely Santa Rosa, Santa Cruz, San Nicolas, Santa Catalina, and San Clemente. The mainland populations were fixed for inversions in the L_1, L_2, L_3, and M_5 chromosomes but polymorphic for the M_6, M_7, M_8, and S_9. Only minor differences existed between the different mainland populations with reference both to the kind and the frequency of inversions present. The average percentage of the autosomal complement that was structurally heterozygous was, however, low compared to island populations with an equivalent number of inverted elements.

In the island populations the L_2 was the only autosome fixed as an inversion homozygote, although the L_1, L_3, and M_{5-7} chromosomes were all polymorphic. On any one island the densest populations had more different chromosomes polymorphic and a higher average number of inversions per individual. Similarly, between islands, there was a direct relationship between average population density and the number of different chromosomes carrying inversions. Thus, in *T. pseudofasciata* inversion heterozygosity does not fall off at geographic margins. Only areas that are ecologically marginal, as gauged by population density, show reduced chromosomal polymorphism.

Chiasma comparisons, carried out by Weissman, indicated that:

Table 7
CHIASMA CHARACTERISTICS IN NINE POPULATIONS OF *TRIMEROTROPIS PSEUDOFASCIATA*[5]

Population	Site	*n*	Mean cell Xa frequency	Mean cell interstitial Xta	Average inversion het. (%)
San Nicolas Is.	Base	11	17.31 ± 1.04	6.47 ± 0.76	0.27
Santa Catalina Is.	Middle Ranch	20	14.35 ± 0.70	5.41 ± 1.42	1.35
	Isthmus	20	14.07 ± 0.78	5.11 ± 1.10	1.37
	Ben Weston	21	13.93 ± 0.79	4.72 ± 1.70	2.46
Mainland	Bakersfield	25	15.57 ± 1.08	3.56 ± 1.40	2.70
San Clemente Is.	Mosquito Cove	19	14.26 ± 0.95	5.57 ± 1.76	3.77
	O.P. Gate	20	14.47 ± 1.07	5.67 ± 1.38	4.46
Santa Cruz Is.	Field Station	19	15.15 ± 1.10	5.01 ± 1.59	7.22
	Sheep Area	18	14.45 ± 0.82	4.63 ± 1.55	7.85

Table 8
CYTOLOGICAL CHARACTERISTICS OF 24 POPULATIONS OF *TRIMEROTROPIS SPARSA*[10,11]

Population	Sample size	No. het. bivalents per individual								Mean no. het. bivalents per individual	Mean no. metacentrics per individual
		0	1	2	3	4	5	6	7		
Eastern populations											
$2n = 23♂$											
Cedar Creek	31	6	16	7	2	—	—	—	—	1.16 ± .15	11.36
Olathe	16	5	7	2	2	—	—	—	—	1.06 ± .25	11.75
Delta	5	—	3	2	—	—	—	—	—	1.40 ± .25	11.40
Whitewater	82	37	29	14	1	—	—	—	1	0.93	19.05
Mack	54	12	22	14	6	—	—	—	—	1.26 ± .13	11.81
Pojaque	143	7	49	71	45	15	1	—	—	2.08	11.05
De Besque	6	1	3	2	—	—	—	—	—	1.17 ± .31	16.17
Rifle I	8	—	1	2	3	1	1	—	—	2.88 ± .44	12.12
Rifle II	6	1	3	1	1	—	—	—	—	1.33 ± .42	13.33
Piceance Ck.	5	3	2	—	—	—	—	—	—	0.40 ± .24	8.40
Cortez	16	3	7	5	1	—	—	—	—	1.25 ± .21	11.81
Jensen	35	4	21	8	2	—	—	—	—	1.23 ± .12	11.57
Duchesne	26	3	6	10	4	3	—	—	—	1.92 ± .21	11.12
$2n = 21♂$ (Fixed fusion)											
Craig	71	21	31	19	—	—	—	—	—	1.03	9.44
Meeker	11	2	6	3	—	—	—	—	—	1.25	9.42
Great Basin populations, all $2n = 23♂$											
Delle	21	1	3	6	4	6	1	—	—	2.67 ± .29	10.09
Hinkley	23	—	3	10	5	4	1	—	—	2.57 ± .23	7.96
White Valley	11	—	6	3	1	—	1	—	—	1.82 ± .38	8.55
Robinson's Ranch	7	—	—	3	2	2	—	—	—	2.86 ± .34	10.29
Cherry Ck.	58	—	16	16	21	3	2	—	—	2.29 ± .14	5.33
Railroad Valley	22	—	6	6	4	5	1	—	—	2.50 ± .27	10.05
Lovelock	5	—	—	2	1	2	—	—	—	3.00 ± .45	9.00
Rye Patch Dam	14	—	2	5	5	2	—	—	—	2.50 ± .20	10.64
Battle Mt.	29	1	6	10	5	5	2	—	—	2.45 ± .24	8.66

Note: The Duchesne population is polymorphic for a fusion but contains predominantly $2n = 21$ individuals.

Table 9
THE INFLUENCE OF THE SEVIER (SEV) INVERSION ON THE CHIASMA CHARACTERISTICS OF THE L1 CHROMOSOME OF *TRIMEROTROPIS SPARSA*[12]

Bivalent type	No. of individuals analyzed	Total bivalents	Prox. Xta 0 / Distal Xta 1	0 / 2	0 / 3	1 / 0	1 / 1	1 / 2	2 / 0	2 / 1	Mean proximal Xta	Mean distal Xta	Mean bivalent Xa frequency
St/St	14	1118	74	96	0	24	800	60	1	3	0.875	1.105	1.98
St/Sev	20	2234	1441	784	6	0	3	0	0	0	0.001	1.374	1.37
Sev/Sev	9	473	54	35	0	3	326	55	0	0	0.798	1.108	1.91

Table 10
MEAN CELL CHIASMA FREQUENCIES FOR 45 INDIVIDUALS OF *TRIMEROTROPIS SUFFUSA* FROM TRUCKEE, CALIF.[12]

No. heterozygous bivalents	No. individuals analyzed	Mean cell Xa frequency
0	3	14.2 ± 0.56
1	12	14.4 ± 0.28
2	13	14.6 ± 0.26
3	10	14.2 ± 0.30
4	6	13.9 ± 0.39
5	1	13.8 ± 0.98

1. While there was a clear tendency for chiasmata to be distributed distally in all karyomorphs, the proportion of distal chiasmata increases in those bivalents which carry inversions whether as heterozygotes or homozygotes.
2. As far as the M_{4-7} chromosomes are concerned inversion heterozygotes and inversion homozygotes have similar effects on chiasma distribution (Table 6). In all the L_{1-3} chromosomes, on the other hand, there are many more distal chiasmata in the inversion homozygotes.
3. Populations with the lowest inversion frequency have the highest average number of interstitial chiasmata and the lowest population densities (Table 7).

A very different situation has been described in *T. sparsa,* a species found in arid regions of western North America. It has a wide distribution ranging from northern New Mexico and Arizona to Alberta and from Colorado, western Nebraska, and the Dakotas across the Great Basin to northern Nevada and California. Despite such a wide species range, populations are restricted to rather specialized habitats and individuals have only a small home range.

White,[10,11] in 24 populations of *T. sparsa* in Utah, Nevada, and western Colorado, observed (Table 8) an extremely complex system which is broken up into a number of microgeographical races, each with its own distinctive pattern of cytological polymorphism. Moreover, quite striking differences can exist between populations, like those at Delta and Whitewater, which are less than 20 mi apart, populations which incidentally are not separated by any obvious geographical or ecological barriers. The data also indicate that, in general, the Great Basin samples collected in Utah and northern Nevada differ from those of western Colorado in having a much greater degree of structural heterozygosity. Added to this, microgeographical races are much more evi-

Table 11
CHROMOSOME AND KARYOTYPE FREQUENCIES IN SIX POPULATIONS OF *TRIMEROTROPIS HELFERI*.[7]

Population	Sample size	Karyotype frequencies			Chromosome frequencies		Inversion frequencies
		8^a8^a	8^a8^m	8^m8^m	8^a	8^m	
Bandon	10♂	0	0	10	0.0000	1.0000	1.00
Orick	96♂	13	44	39	0.3646	0.6354	0.64
Arcata	159♂	25	72	62	0.3836	0.6164	0.62
Petrolia	95♂	31	47	17	0.5737	0.4263	0.43
Cleone							
1966	87♂	24	53	10	0.5805	0.4195	0.42
1967	103♂ 80♀	38	96	49	0.4660 0.4750	0.5340 0.5250	0.53
Point Arena							
1966 1967	100♂ 100♀	6	62	132	0.1800 0.1850	0.8200 0.8150	0.82

Note: The populations are arranged in sequence from north to south.

dent in western Colorado and eastern Utah compared to the Great Basin where the species is confined to the very uniform environment of *Sarcobatus* flats. Despite this, the amount of structural heterozygosity is much greater in the Great Basin populations where, however, the species appears to be a relative newcomer. Unfortunately, no detailed chiasma analysis of the kind carried out in *T. pallidipennis* and *T. pseudofasciata* is available for *T. sparsa.* White and Morley,[12] in studying the chiasma conditions for the inversion in chromosome-1, find that the chiasma frequency in the region distal to the rearrangement increases in inversion heterozygotes, an effect which is, however, not maintained in the inversion homozygote (Table 9). No scores are given for the mean chiasma frequency per individual for any of the populations, though they record that in a population of *T. suffusa* carrying from zero to six structurally heterozygous bivalents per individual, the frequency of inversion heterozygosity has little influence on total chiasma frequency (Table 10). This implies the operation of an interchromosomal effect which compensates for the reduction in chiasma frequency within a bivalent caused by the presence of a heterozygous inversion.

Yet a further distinctive system exists in *T. helferi,* a species which is confined to northern California and southern Oregon and over a very restricted range extending from Point Arena, Calif. to Coos Bay, Ore. Here it inhabits coastal beaches and the first line of sand dunes in localities where sand verbena is abundant. Schroeter[7] analyzed 830 individuals from six sites covering the entire range of this species (Table 11). *T. helferi* is monomorphic for the three long metacentrics (L1 to 3) but polymorphic for a single pericentric inversion in the M_8 which, thus exists in two morphs, one metacentric (8^m) and one acrocentric (8^a).

The data indicate that, in this species, peripheral populations are characterized by higher inversion frequencies, whereas lower frequencies are found in the populations occupying the central part of the range. Variation is, thus, essentially clinal. Thus, there are clear differences between the two collections taken at Cleone. Although these were only 50 to 75 yd apart, they were separated by a well-used logging road running along the bluff between the two.

The presence of inversion again has a marked effect on chiasma frequency. Individuals which are either heterozygous or homozygous for the inversion have a mean cell chiasma frequency which is significantly higher than that of the basic homozygote

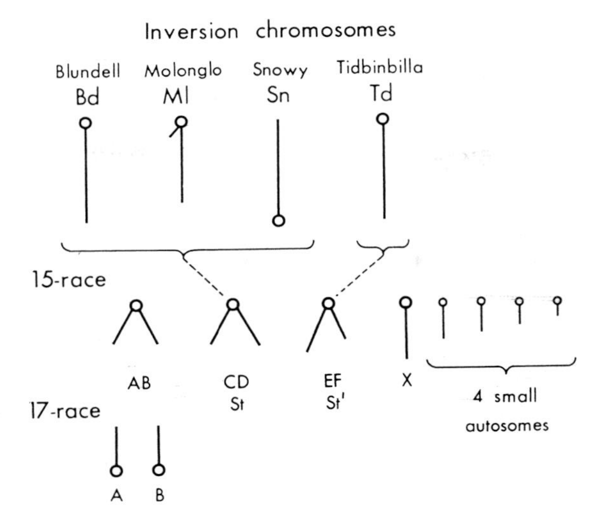

FIGURE 3. Karyotype variation in the morabine grasshopper *Keyacris scurra.*

Table 12
A COMPARISON OF THE MEAN CELL CHIASMA FREQUENCIES IN FOUR POPULATIONS OF *TRIMEROTROPIS HELFERI*[7]

Karyomorph	Mean cell Xa frequency				Overall means
	Pt. Arena	Cleone	Arcata	Orick	
8^a8^a	Not scored	20.68	21.95	22.18	21.62
8^a8^m	20.98	22.57	23.84	24.11	23.51
8^m8^m	21.28	23.15	24.33	24.07	23.85
Population mean	21.13	22.13	23.34	23.45	

(Table 12). Moreover, this difference holds for both L and M chromosomes. The S chromosomes, by comparison, tend to form only single chiasmata, so that here there is little or no effect.

In Australia the morabine grasshoppers constitute a wholly endemic subfamily of the Eumastacidae, a family with a widespread distribution mainly in the tropics and subtropics of both the old and the new world. All the species are completely apterous and the habitat of many of them has been fragmented and largely destroyed by the development of the land and especially by grazing. As a result, many of them are reduced to small relict populations. Of the 180 known species of morabines, 17 are polymorphic for pericentric inversions in from one to three pairs of chromosomes,[8] but in only one of these, *Keyacris scurra,* has there been any attempt to analyze the situation in detail.

This species exists as two chromosome races, one with 15 and the other with 17 chromosomes[13,14] (Figure 3). In the former, there is a single large AB metacentric which in the 17-race exists as two rod chromosomes, A and B, one of which is equivalent to each of the two arms of the AB metacentric. The inversion series affects the two next largest pairs, designated as CD and EF, which exist as a series of morphs in both races. There are four of these in the CD, namely Standard (St), Blundell (Bl), Molonglo

Table 13
KARYOMORPH FREQUENCIES OF THE BLUNDELL (B) AND TIDBINBILLA (T) INVERSIONS IN TWO $2n = 15$ POPULATIONS OF *KEYACRIS SCURRA* OVER SUCCESSIVE YEARS OF SAMPLING[17]

			Karyomorph combinations									Total
Population	Year	Sex	SSS′S′	SBS′S′	BBS′S′	SSS′T	SBS′T	BBS′T	SSTT	SBTT	BBTT	sample size
Murrumbateman Cemetery	1961	♂	39	194	132	10	47	54	0	1	3	480
		♀	62	196	191	12	57	61	0	3	5	587
	1962	♂	71	242	230	22	103	70	2	2	4	746
		♀	33	172	137	11	62	68	0	5	2	490
	1963	♂	18	88	64	3	22	21	0	3	1	220
		♀	20	81	50	7	30	18	0	0	1	207
	1964	♂	31	110	85	8	30	21	0	2	1	288
		♀	29	121	85	7	25	17	0	1	0	285
	1965	♂	22	65	50	4	13	14	0	0	0	168
		♀	23	61	49	2	12	10	0	0	1	158
	1966M	♂	26	85	66	1	12	22	0	2	1	215
		♀	31	104	68	8	32	30	0	3	1	277
	1966A	♂	32	124	81	9	14	33	0	2	6	301
		♀	25	102	89	3	26	17	0	0	3	265
Tarago Swamp	1955	♂	1	24	80	1	13	60	0	0	21	200
	1956	♂	0	9	92	0	15	70	0	2	12	200
	1960	♂	2	56	314	0	32	220	0	10	35	669
	1961	♂	2	16	57	0	9	58	0	3	5	150
	1962	♂	2	18	125	2	22	91	0	4	19	288

Note: The noninverted standard equivalents of these two inversions are designated as S and S′, respectively. Note the two 1966 samples from Murrumbateman Cemetery were taken from two segregated burial areas, (M) Methodist and (A) Anglican.

Table 14
VARIATION IN THE STANDARD CD AND EF CHROMOSOMES FROM TWO $2n = 15$ POPULATIONS OF *KEYACRIS SCURRA* BETWEEN SEX AND TIME[17]

Year	Standard CD frequency		Standard EF frequency	
	Male	Female	Male	Female
Murrumbateman population				
1961	0.354	0.344	0.876	0.875
1962	0.360	0.334	0.859	0.842
1963	0.352	0.399	0.877	0.862
1964	0.382	0.384	0.887	0.910
1965	0.387	0.389	0.908	0.918
1966 M	0.356	0.392	0.905	0.859
1966 A	0.368	0.347	0.880	0.902
Tarago Swamp population				
1955	0.103	—	0.710	—
1956	0.065	—	0.718	—
1960	0.076	—	0.744	—
1961	0.107	—	0.723	—
1962	0.092	—	0.716	—

Note: The two 1966 samples from Murrumbateman Cemetery were taken from segregated burial areas, (M) Methodist and (A) Anglican.

(Mol), and Snowy (Sn), though the Mol inversion has a restricted distribution and the Sn inversion is known only from a single locality. The EF system is simpler; here there are only two morphs, Standard (St′) and Tidbinbilla (Td).

The most common category of population and the only one studied in depth is that which includes the St, Bl, St′, and Td morphs. Here nine combinations are possible, namely:

Cd	EF	CD	EF	CD	EF
(1) St St	St′ St′	(4) St St	St′ Td	(7) St St	Td Td
(2) St Bl	St′ St′	(5) St Bl	St′ Td	(8) St Bl	Td Td
(3) Bl Bl	St′ St′	(6) Bl Bl	St′ Td	(9) Bl Bl	Td Td

The frequency of these morphs varies considerably in different populations (Table 13), though the last three karyomorphs are almost always rare, and combinations 2 and 3 the most common.

Lewontin and White[15] and later White et al.[16] concluded that within any one deme the different karyomorphs exist in a stable equilibrium which remains unchanged from year to year. On this premise, they proceeded to calculate the relative viabilities of the different morphs. Colgan and Cheney[17] in an analysis of data collected over six successive years at Murrumbateman (Table 14) have, however, provided evidence that there are significant interactions between year and genotype frequencies, on the one hand, and sex and genotype frequencies on the other. Thus, contrary to the earlier assumptions, there is certainly no stability of inversion frequencies at Murrumbateman which, without question, represents the most complete data base currently available. The con-

Table 15
THE INFLUENCE OF THE B1 INVERSION OF THE CD CHROMOSOME OF *KEYACRIS SCURRA* ON CHIASMA DISTRIBUTION[12]

Karyomorph	No. individuals analyzed	No. bivalents analyzed	Prox. Xta / Distal Xta	0 / 1	1 / 0	1 / 1	Mean prox. Xta	Mean distal Xta	Total bivalent Xa frequency
StSt	4	700		112	97	491	0.811	0.841	1.65
StB1	4	800		800	—	—	0.000	1.000	1.00
B1B1	7	1196		278	421	497	0.762	0.630	1.39

Note: The data is taken from the Blundell population (15-chromosome race) in which the B1 inversion was first discovered.

clusions reached by White and colleagues concerning the interactive effects of the CD and EF inversions on the viability of males are, thus, open to criticism.

White and Andrew[18] claimed that the standard forms of both CD and EF were size increasing while the Bl and Td arrangements were size decreasing. This claim was, however, complicated by the fact that size shows considerable geographical variation in this species, so that Bl Bl Td Td individuals at one locality may actually be larger than St St St′ St′ individuals at another. Moreover, while there was a correlation between karyomorph type and size in the 17-chromosome populations at Wombat and Wallendbeen and the 15-chromosome populations at Murrumbateman, Royalla, and Tarago Swamp, there was no such correlation in the 15-chromosome population at Michelago. Added to this, it is not known whether the size correlation holds for any of the other populations sampled.

Even so, White has concluded that there has been a differential accumulation of size increasing and size decreasing alleles in the mutually inverted sequences of *K. scurra* and that the biometrical effects produced by these alleles is additive. He thus assumes that the CD and EF polymorphisms are functionally related, since each locks up unique polygenic combinations affecting body weight. In the light of the findings of Colgan and Cheney,[17] these conclusions, too, must be viewed with reserve.

Finally, in keeping with the effects of inversions seen in the trimerotropines, there is a pronounced effect of the CD and EF inversions on chiasma distribution in *K. scurra*, though this has been quantified only for the CD chromosome (Table 15).

The genus *Austroicetes*, which is also endemic to Australia, includes seven distinct morphological species and two sibling species, *A. pulsilla* and *A. interioris*.[19,20] The latter is the only member of this genus to show inversion polymorphism (Figure 4). Three pairs of medium-sized chromosomes are polymorphic for pericentric inversions throughout the species range, and the individual inversion morphs are known as Trangie (chromosome 2), Quorn (chromosome 4), and Flinders (chromosome 6). The X chromosome is also polymorphic for the Buronga inversion. Nankivell,[20] in four populations of this species (Table 16), found no evidence for an excess of any particular karyomorph in any of the populations sampled with the exception of Quorn. Nevertheless, inversion karyomorph frequencies were not uniform for the various sites analyzed and there were some significant differences between populations with respect to autosomes 4 and 6 and the X-chromosome polymorphism.

Only one inversion, that of 4, showed any apparent influence on mean body weight. Karyomorphs with 4^s4^s tended to be heavier, while those with 4^Q4^Q tended to be lighter than the average for the population, but even then, only in one of the two Euston collections. Nevertheless, it was suggested that an interaction of the different inversions influenced the variability of body weight in the Euston population. The effect of this influence was claimed to make for an increased variance around the mean in

(a) St St 2, 4, 6

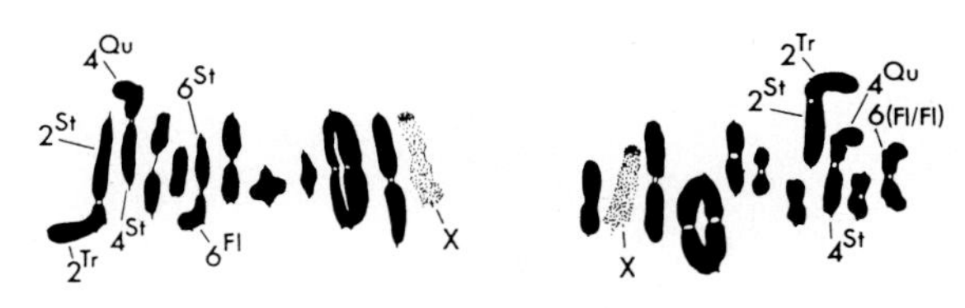

(b) St Tr 2, St Qu 4, St Fl 6 (c) St Tr 2, St Qu 4, Fl Fl 6

FIGURE 4. Karyotype variation in the grasshopper *Austroicetes interioris.* The three male meiocytes are metaphase-I preparations from individuals, respectively, (a) standard (St St) for all chromosomes; (b) heterozygous for the Trangie (St Tr), Quorn (St Qu), and Flinders (St Fl) inversions in chromosomes 2, 4, and 6; and (c) heterozygous St Tr and St Qu but homozygous Fl Fl. (After Nankivell, R. N., *Chromosoma*, 22, 42, 1967. With permission.)

Table 16
KARYOMORPH SCORES IN FIVE POPULATION SAMPLES OF *AUSTROICETES INTERIORIS*[20]

Population	Sample size	Collection time	Autosome 2			Autosome 4			Autosome 6			X chromosome	
			SS	TS	TT	SS	QS	QQ	SS	FS	FF	S	B
Euston	336	Summer 63/64	238	95	3	110	174	52	45	165	126	55	27
	372	Summer 64/65	269	100	3	119	198	58	48	181	143	162	71
Trangie A	90	17.1.66	65	24	1	16	44[a]	30	1	22	67	58	32
Trangie B	37	18.1.66	28	9	0	14	16[a]	7	2	7	28	25	12
Quorn (Pichi-Richi Pass)	113	5.2.67	95	18	0	41	64[b]	8	18	57	38	57	56
Balranald	91♂	Summer 67/68	66	23	2	24	53	14	6[c]	45[c]	40[c]	71	20
	64♀		42	21	1	11	36	17	6[c]	18[c]	40[c]	55	55
												BS	28
												SS	4

G Test for Homogeneity (5df)

Trangie (T)	Quorn (Q)	Flinders (F)	Buronga (B)
8.850	30.608	43.666	19.34
Not significant	Significant at $p = 0.001$		

Note: T, Q, F, and B represent the inversion chromosomes and S their standard equivalents.

[a] Not homogeneous, therefore regarded by Nankivell as a separate population, although less than 5 km apart.

[b] Excess heterozygotes.

[c] Significant difference between sexes.

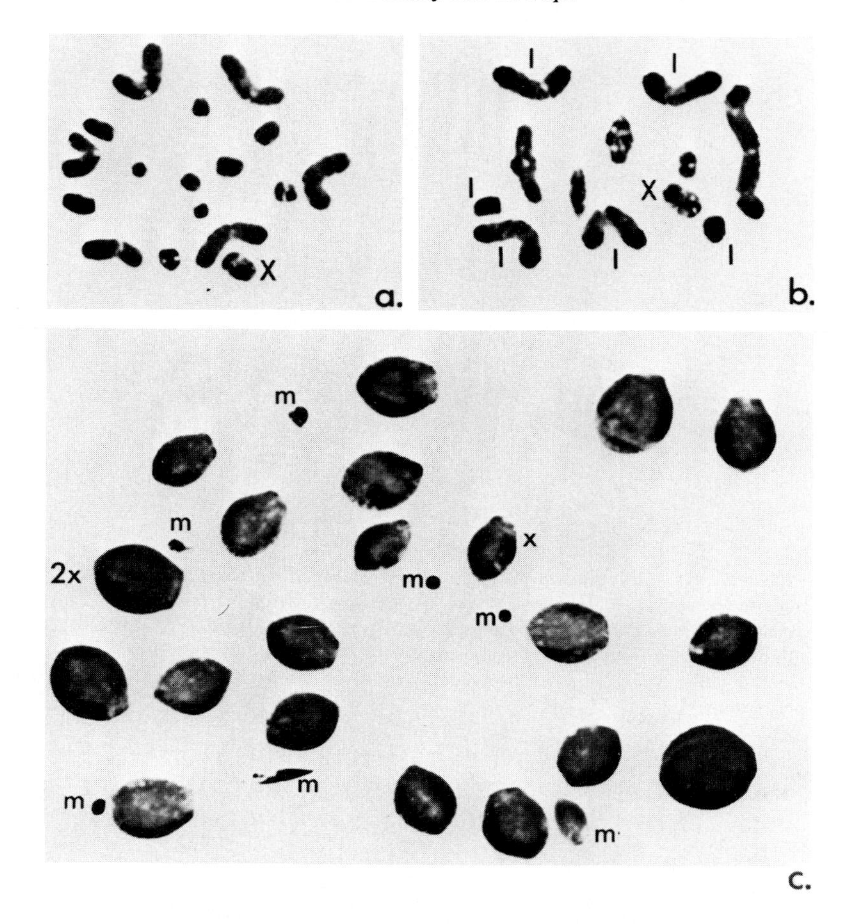

FIGURE 5. Meiosis and its consequences in a spontaneous asynaptic mutant male of the grasshopper *Chorthippus dorsatus* ($2n$ = 17♂,XO). In place of the expected 8 II + X, first metaphase meiocytes of this mutant contained from 6 (b) to 17 (a) univalents (1). As a result of this extensive univalency, spermatids were of many different sizes including microspermatids (m), subhaploid, haploid (x), diploid ($2x$), and tetraploid types.

individuals with a higher degree of chromosome homozygosity when this was expressed as the summation of all classes of homozygotes. The data, however, are not particularly convincing.

Outside of the *Acrididae* the only other recorded instance of pericentric inversion polymorphism is found in the *Gryllidae*. For example, Ohmachi and Ueshima[21] found three cytotypes in populations of *Scapsipedus aspersus*, all with $2n$ = 21 = 10 II + X. One of these has all its autosomes in the forms of rods, the second has one pair of metacentric autosomes, and the third two pairs of metacentric autosomes. In the latter, heterozygous bivalents were sometimes found confirming the inversion origin of the variation in this case.

B. Translocation Systems

Whereas an inversion is confined to a single chromosome, translocations involve two, or sometimes more than two, different chromosomes. As such they fall into two distinct categories. First, there are those that are reciprocal involving an exchange of segments between two nonhomologs. These are referred to as interchanges. They may, or may not, alter chromosome morphology depending on whether they are unequal or

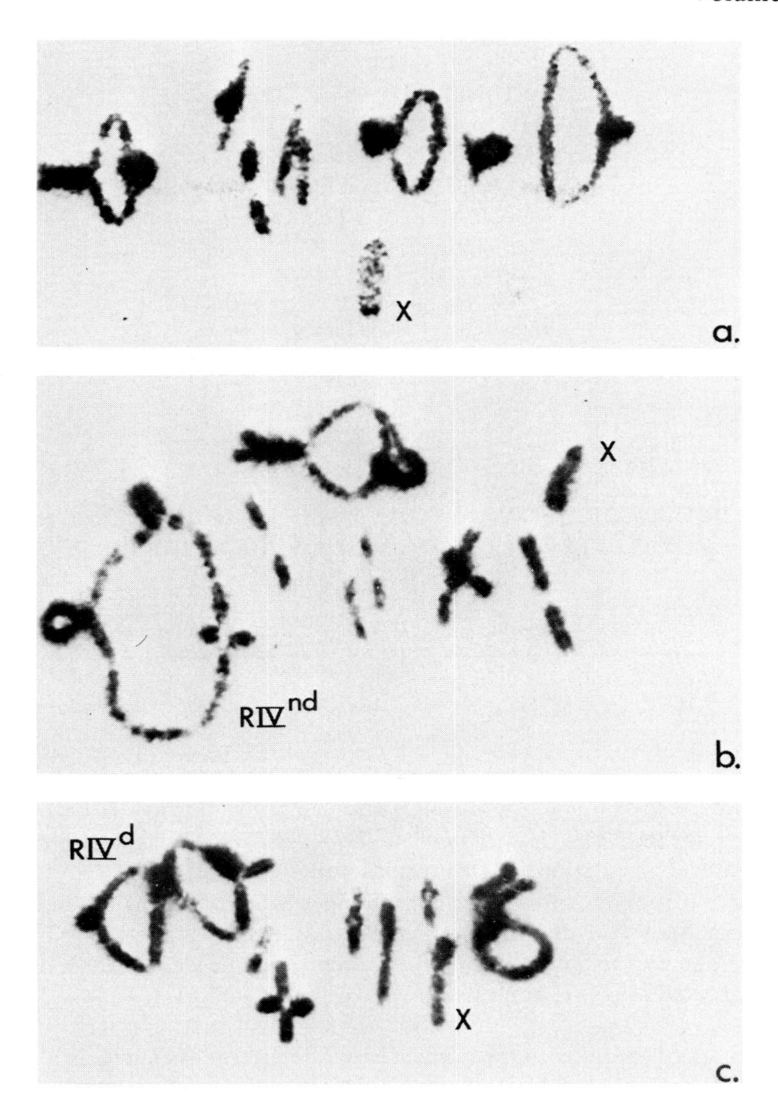

FIGURE 6. A spontaneous interchange heterozygote involving two metacentric chromosomes in the grasshopper *Myrmeleotettix maculatus* ($2n = 17♂$,XO). Conventionally, it forms 8 II + X at male meiosis (a) but the two mutant meiocytes include a ring of four interchange multiple (R IV + 6 II + X). In (b) the ring is in nondisjunctional orientation whereas in (c) it is twisted (⋈) so as to lie in disjunctional orientation.

equal in character but, in either event, they leave chromosome number unaffected. Second, there are those that involve a union of two nonhomologs into a single chromosome and which, therefore, are nonreciprocal. These are known as fusions and they not only change chromosome morphology but, in addition, reduce chromosome number.

1. Interchanges

Polymorphisms for interchanges are found in only one group of orthopteroids, the Blattodea. Even here they are known in natural populations of only one species, *Periplaneta americana*, the American cockroach.[22-24] Their occurrence in laboratory stocks

Table 17
PAIRING BEHAVIOR OF THE MEMBERS OF THE SPONTANEOUS L1-2 INTERCHANGE IN 487 MALE MEIOCYTES OF *MELANOPLUS DIFFERENTIALIS*[25]

Configuration	Frequency	%
C.IV	349	72
R.IV	129	26
2.II	9	2

Table 18
MEIOTIC BEHAVIOR OF TWO SPONTANEOUS MALE INTERCHANGE HETEROZYGOTES IN THE GRASSHOPPER *CHORTHIPPUS BRUNNEUS*[28,29]

	Pairing behavior					Orientation behavior of IVs		
Chromosomes involved	2.II	C.III + I	C.IV	R.IV	Total cells	Disjunctional	Non-disjunctional	Total cells
L3—L4	0	10	176	1	187	49 (29.2%)	116 (70.8%)	165
L1—M5	2	17	170	11	200	5 (6.8%)	68 (93.2%)	73

of a second roach, *Blaberus discoidalis*,[25] however, strongly suggests that they must be present in natural populations of this species, too.

Such a paucity of polymorphism for interchanges is by no means limited to orthopteroids, but is a general feature for all animal populations.

There is only one way to achieve consistent gametic balance and, hence, high fertility in an interchange heterozygote and this depends on three factors:[26]

1. All members of the interchange system must regularly associate as a single multiple configuration, that is, a ring or chain of four chromosomes. Where univalents form they invariably create a problem situation, at least in male orthopteroids. Thus, if they segregate they do so irregularly and give rise to aneuploid meiotic products. If they fail to segregate then they either give rise to microgametes or else, by impeding cytokinesis at first telophase and/or second division, to macrogametes (Figure 5). Neither of these classes are able to fertilize.
2. The multiple must orient at first meiotic metaphase in such a way that the two noninterchanged members of the multiple pass to the same pole at first anaphase with the two interchanged members moving to the other pole. In rings this requires a twist to produce a figure of eight (Figure 6). In chains it involves a zigzag arrangement. Not only do these arrangements produce balanced gametes but they are the only ones that will do so regularly.
3. The chiasmata necessary to maintain the maximum multiple must be excluded from the segments which lie between the centromeres and the points of exchange, the so-called interstitial segments. If a chiasma does occur in one of these segments then no pattern of orientation is capable of producing regular fertility, even when maximum multiples are formed, for such a chiasma automatically leads to the movement of interchanged and noninterchanged chromatids to the

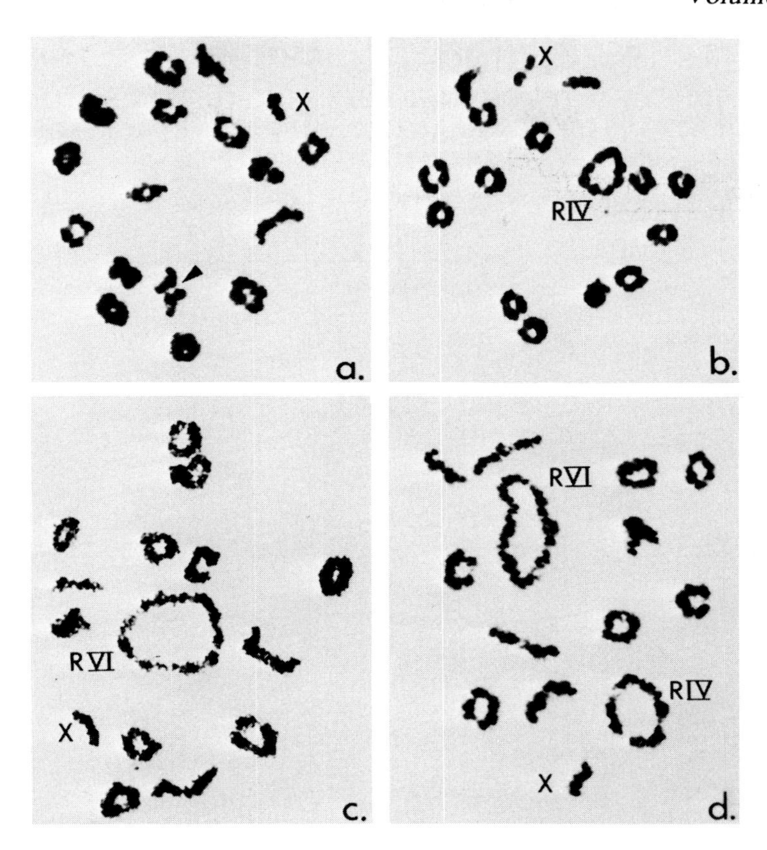

FIGURE 7. Interchange heterozygosity at first meiotic division in males of the American cockroach, *Periplaneta americana* (2*n* = 33♂,XO). The standard complement of 16 II + X is shown in (a), whereas (b), (c), and (d) show, respectively, R IV + 14 II + X, R VI + 13 II + X, and R VI + R IV + 11 II + X. In only one bivalent (►) in one meiocyte (a) is there a clear indication of an interstitial chiasma.

same pole at first anaphase and this, inevitably, produces imbalance in half of the products of meiosis.

These are very demanding requirements and it is clear that a system of interchange hybridity is not easily evolved. Indeed species which have adopted such a system must have been under considerable pressure to have done so. They must also have been preadapted in respect of the properties required for the successful operation of the system at meiosis. Thus, when acro- or telocentric chromosomes are involved in an interchange it follows that crossing over within an interstitial segment is obligatory in order to maintain the maximum multiple. This, in turn, means the system is doomed to fail for it is self-sterilizing. For example, Wise and Rickards[27] described a spontaneous mutant male of the grasshopper *Melanoplus differentialis* heterozygous for an interchange between the two largest members of the complement, both of which were acrocentrics. The behavior of this heterozygote (Table 17) indicates that maximum multiples, in fact, formed in 98% of the meiocytes. Nevertheless, the orientation behavior of the multiple was invariably of the kind expected to lead to infertility. The same was true of two spontaneous interchange mutants observed involving one metacentric and one acrocentric in *Chorthippus brunneus*[28,29] (Table 18).

When we turn to the situation in *P. americana* the contrast in behavior to these

Table 19
PATTERNS OF INTERCHANGE POLYMORPHISM IN NATURAL POPULATIONS OF *PERIPLANETA AMERICANA* FROM SOUTH WALES[22,23] AND PAKISTAN[24]

Population		Homozygotes (16.II)	Heterozygotes 1.IV	2.IV	1.VI	1.VI + 1.IV	1.VIII	Total individuals
South Wales								
Tonypandy	all coal mines	4	4	2	1	—	—	11
Cilfyndd	all coal mines	5	8	1	1	—	—	15
Treharris 1	all coal mines	12	16	10	7	5	—	50
Treharris 2	all coal mines	12	22	9	—	—	—	43
Totals		33	50	22	9	5	—	119
			86					heterozygotes = 71%
Pakistan								
Karachi								
1 — House		4	2	3	1	2	—	
2 — Sewer		7	4	2	1	—	—	
Lahore								
1 — House		3	1	—	2	3	—	
2 — Fruit market		7	1	—	2	—	1	
Totals		21	8	5	6	5	1	46
			25					heterozygotes = 54%

Note: The two Treharris populations, significantly different in composition, are from localities less than a few hundred yards apart but separated by a main roadway which is permanently illuminated and in constant usage. It is not possible to determine whether the homozygous members of the populations were basic or interchange in character because the two classes could not be distinguished morphologically.

spontaneous mutants is striking. All natural populations of this species which have been sampled, whether in Britain (coal mine populations from South Wales) or Pakistan, have proved to be polymorphic for interchanges (Figure 7) with some 71 and 54% of the individuals, respectively, being heterozygotes (Table 19).

There are a number of features which predispose this system to a measure of success. First, the interstitial segments are in all cases very short. Coupled with this the chiasmata show extreme distal localization so that there is really a zero possibility of chiasmata forming in the interstitial segments. Added to this maximum multiple formation was generally high. Thus, in only 1 of the 25 heterozygous individuals from Pakistan was it less than 70%. Likewise, in South Wales, too, 88% of the heterozygous individuals examined formed maximum multiples in more than 70% of their meiocytes. Finally, the frequency of disjunction in the multiples that do form is high (Table 20). Failure of multiple formation was quite pronounced in some individuals, especially those with multiple heterozygosity. White[8] found this "disturbing". This, however, is not the only interchange system where the organism has had to pay the price of infer-

Table 20
ORIENTATION BEHAVIOR OF INTERCHANGE MULTIPLES IN BRITISH POPULATIONS OF *PERIPLANETA AMERICANA*[23]

	Interchange type					
	1.IV		2.IV			Mean disjunction frequency
Orientation	Ring	Chain	Ring	Chain	Totals	
Disjunctional	49	1	35	31	116	0.92
Nondisjunctional	8	—	2	—	10	
Total cells	57	1	37	31	126	
	58		68			

tility to survive. James[30] notes that in the interchange system of the plant *Isotoma petraea* there is a parallel drop in pollen fertility and seed production as the interchange increases in size. Likewise, parthenogenetic forms not uncommonly produce a high percentage of inviable progeny, presumably due to imperfections of the unusual and specialized cytogenetic systems involved.

White[1] has also suggested that the ring and chain multiples in *Periplaneta* are not due to interchange heterozygosity but to the presence of "small distal chromosome segments in the tetrasomic state." The presence of the relevant multiple pachytene pairing crosses which involve all segments of the chromosome arms makes it quite clear that homology is not confined to chromosome ends in the manner which he suggests.

The interchange basis of the multiples in *Periplaneta* is also confirmed by the very considerable evidence that now exists for the experimental induction of interchanges in the German cockroach *Blatella germanica*, ignored by White on the issue. In this species the 12 members of the haploid set have been numbered from 1 to 12 beginning with the smallest X-chromosome. More than 20 interchange stocks have been produced in the laboratory, all except two following exposure to ionizing radiation.[31] A majority of these involve one of the larger chromosomes, 9 through 12, and as in *Periplaneta* typical pachytene crosses are found in all but two of the stocks where the interchanged segments were very short. The validity of the interchange origin of these induced multiples in *Blatella* has been confirmed by linkage data using known genetic markers. Indeed the first indication that interchanges might have resulted from the radiation treatment came from the identification of phenotypic mutants inherited as autosomal semidominant lethals. These experiments also confirm that cockroaches do indeed show a propensity for interchange. Thus, in most cases, the exchange sites occur close to the centromere so that the interstitial segments are short, again as in *Periplaneta*. Moreover, chiasmata, if not invariably localized, are certainly much more frequent in distal regions.

These induced interchanges provide information on five points not readily available from the natural interchange system of *P. americana:*

1. The single exchanges fall into two major categories showing, respectively, random orientation or preferential orientation (>50% disjunction). The double exchanges, on the other hand, all have low disjunction frequencies (Table 21).
2. Of six stocks tested only two, T(7,12) and T(11,12), gave viable interchange homozygotes. Consequently most of the interchange stocks are maintained as heterozygotes by backcrossing to the wild type at each generation.

Table 21
ORIENTATION BEHAVIOR OF NINE INDUCED INTERCHANGES IN MALES OF *BLATELLA GERMANICA*[32-35]

Type	Stock	Chain formation (%)	Multiple orientation		Disjunction (%)
			Disjunctional	Nondisjunctional	
1.IV	T(2,11)[a]	—	306	310	49.7
	T(7,12)[b]	13	243	237	50.6
	T(8,9)[c]	2	470	286	62.2
	T(9,10)[d]	1	1064	611	63.5
	T(11,12)	2	1144	510	69.2
	T(3,12)[e]	7	532	205	72.2
1.VI	T(3,7,12)	13			28—33
	T(4,5,10)	6	No precise data given		33
	T(4,8,10)	15			33

[a] ♀s Are sterile.
[b] ♀ Shows preferential disjunction.
[c] ♀ Shows random disjunction.
[d] ♀ Shows preferential disjunction.
[e] ♀ Shows preferential disjunction.

Table 22
EFFECT OF INTERCHANGE HETEROZYGOSITY ON MEAN BIVALENT CHIASMA FREQUENCY IN *BLATELLA GERMANICA*[35]

Stock	Mean Xa frequency per bivalent
Wild type	1.34 ± 0.16
T(8,9)	1.99 ± 0.01
T(4,5,10)	1.98 ± 0.01

3. Linkage estimates using simple genetic markers suggest either equality or relatively minor sex differences in crossing over. Chiasma frequency is, however, increased in the chromosomes involved in interchange though the extra chiasmata that form are again restricted distally (Table 22).
4. In *Blatella* eggs are enclosed as groups in a discrete egg case, or ootheca, which is carried externally until it is nearly ready to hatch. Hatch reduction in interchange stocks depends on both genetic lethality, due to the production of unbalanced gametes as a result of nondisjunction, and to ancillary sterility effects from embryonic trapping, i.e., the inability of the reduced numbers of viable embryos to force open the egg case at the time of hatching. Fortunately lethal effects are easily distinguished because dead embryos are visible through the semitransparent shell of the egg case and can be scored in the intact ootheca. Embryonic lethality, estimated from observations on mature oothecae, provides a good indication of disjunction frequency.
5. The disjunction pattern of individual interchanges can be modified when put through a backcross system. That is, the behavior of a given interchange depends on the genetic background against which it operates. Moreover, such alterations include not only a change from random behavior to preferential behavior but

Table 23
DISJUNCTION PATTERN ALTERATIONS FOR INDIVIDUAL INTERCHANGES OF *BLATELLA GERMANICA* WHEN PLACED AGAINST DIFFERENT GENETIC BACKGROUNDS[33]

	Orientation behavior		
Interchange	Disjunctional	Nondisjunctional	Disjunction (%)
T(3,12)	437	206	68
T(3,12), sib of dbl	465	452	50.7
T(3,12), from dbl	316	308	50.7
T(7,12)	450	435	50.9
T(7,12), in dbl	296	185	61.5
T(7,2), from dbl	303	185	62.1

Note: In the experiments summarized in this table T(3,12) has been put through a backcross system designed to produce a double translocation stock (dbl). Individuals emerging as sibs of the double are consequently designed as T(3,12) sib of dbl. On the other hand, individuals emerging from the double stock, when crossed to a marked strain, are designed as T(3,12) from dbl.

also the reverse (Table 23). This suggests that in natural populations, too, one cannot necessarily assume that a chromosome rearrangement at the present time has exactly the same properties or necessarily behaves at meiosis in the same manner as when it arose.

2. Fusions

There is a second category of translocation which, like interchange, increases linkage beyond the physical limits of a single chromosome. This is fusion. Fusions are of two major types — centric and tandem. In centric fusions interstitial segments are either very short or else nonexistent. Consequently successful centric fusion is not prohibited in acro- or telocentrics as are interchanges. On the contrary, the better-known instances of centric fusion all involve acro- or telocentric chromosomes. In tandem fusions, on the other hand, there is inevitably a long interstitial segment since this kind of rearrangement involves a distal break in one chromosome and a proximal break in a nonhomolog. Not surprisingly, therefore, no polymorphism for a tandem fusion has ever been described in any orthopteroid or, for that matter, in any other organism. Centric fusions may either involve two nonhomologous autosomes or else one autosome and a sex chromosome. Only the first of these two categories is known in a polymorphic state in orthopteroids. In most of these cases we have no more than the most basic information available. This applies, for example, to the field cricket, *Anaxipha pallidula* (Gryllodea). Here Ueshima[36] found a variable diploid number with 19, 18, and 17 chromosomes within the same population. The 19-chromosome form includes two metacentric pairs and is found in 64.3% of the population. The 17-chromosome form with three metacentrics is present in 25% of the population, while the 18-chromosome form, which includes two metacentrics and a trivalent consisting of one metacentric and two rods, occurs in the remaining 10.7% of the 28 individuals examined.

The most complete analysis available to date of a fusion polymorphism is that of Hewitt and Schroeter[37] in the acridid *Oedaleonotus enigma.* This species extends from

Table 24
POPULATION STRUCTURE OF 98 INDIVIDUALS OF *OEDALEONOTUS ENIGMA* FROM COALINGA, CALIF.[37]

Polymorphism	Karyomorph frequencies		
	Basic homozygote	Heterozygote	Structural homozygote
Pericentric inversion in chromosome 8	63	27	8
Supernumerary segment in chromosome 10	90	7	1
4͡5 Fusion	12	61	25

California to Oregon, Washington, and Idaho but with its distribution centered on California. It is characterized by the regular possession of a sex chromosome-autosome fusion in the form of a neo XY♂ neo XX♀ sex chromosome mechanism and, in this respect, is distinguishable from its close relatives, *O. borkii, O. phryneicus,* and *O. orientis,* all of which have $2n$ = 23XO♂, 24XX♀. Individuals of *O. enigma* from several localities southeast of Coalinga, Fresno County, Calif. proved to be polymorphic for an autosomal fusion, at least two pericentric inversions, a supernumerary chromosome segment on the shortest autosome, and for a supernumerary chromosome (Table 24). The fusion polymorphism which involves chromosomes 4 and 5 (Figure 8) includes a statistically significant excess of fusion heterozygotes in comparison with the expectations of a Hardy-Weinberg distribution.

The predominance of fusion heterozygotes is accompanied by a marked predominance of disjunctional orientation (97%) of the fusion trivalent at first metaphase. In agreement with this a majority of second-division cells have a normal composition (Table 25). This behavior in an established polymorphism contrasts strikingly with the behavior of a spontaneous fusion heterozygote in *Acrida lata,* where only 39.5% of the 200 first metaphase cells analyzed showed disjunctional orientation.[38]

C. Heterochromatin Variation

Particular segments of chromosomes, or in some cases entire chromosomes, may show a pronounced differential condensation at interphase of the cell cycle producing a visible heteropycnosis of the region concerned. Because they are differentially condensed, such heteropycnotic regions also stain more deeply at interphase and prophase and so are also termed heterochromatic. Where such differential behavior is characteristic of all cell types within an organism, the heterochromatic region in question is described as constitutive. Such constitutively heterochromatic regions stain intensely with Giemsa even in fixed metaphase chromosomes when they are treated with warm saline followed by barium hydroxide. Giemsa staining of this type is known as C-banding.

One of the most frequent categories of chromosome change to have occurred in the phylogeny of orthopteroids has involved changes in the amount of constitutive heterochromatin. This is especially evident in terms of the variation in C-band material which can be demonstrated in orthopteroid chromosomes. This variation takes two quite distinct forms: first, the addition of a special class of chromosomes in addition to those represented in the standard chromosome complement — these are known as supernumerary or B chromosomes; second, the presence of supernumerary chromosome segments on otherwise regular members of the chromosome complement. In both

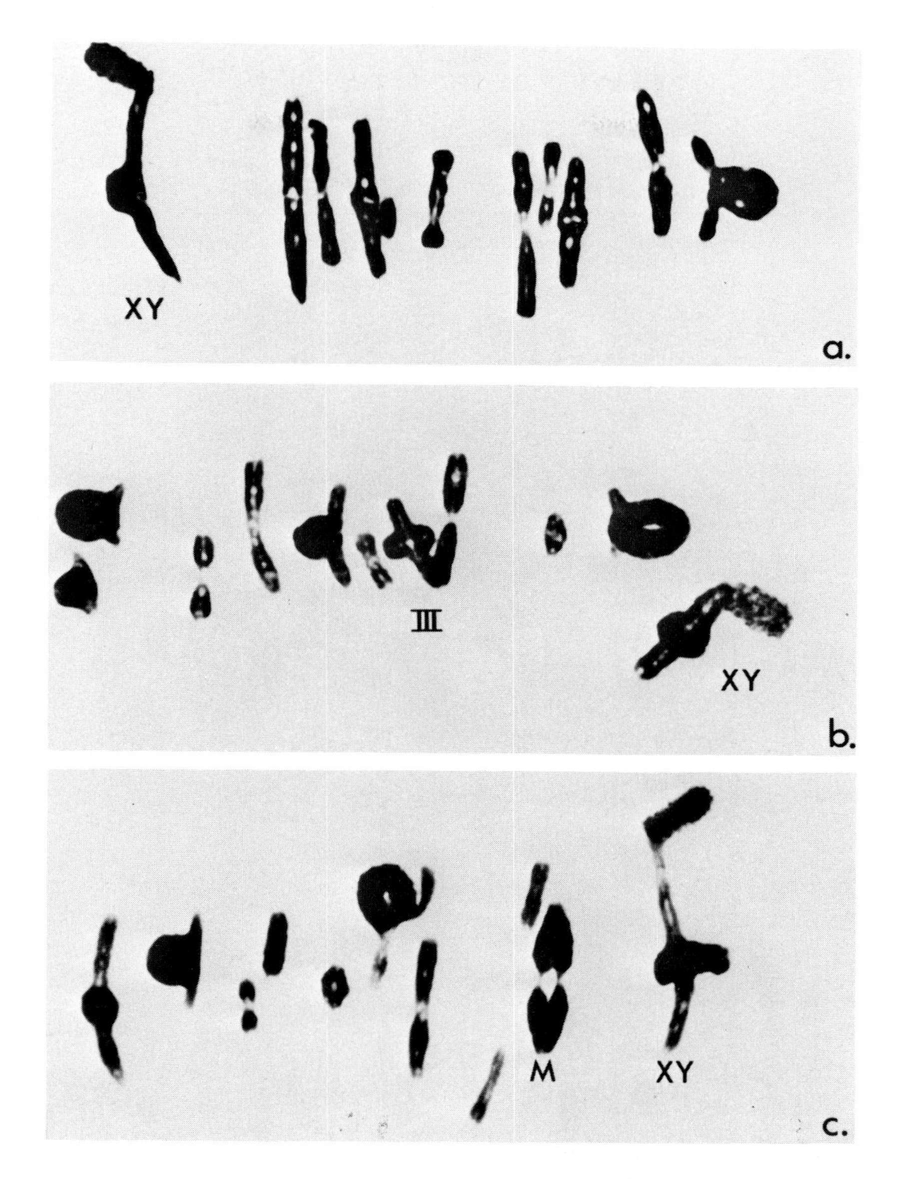

FIGURE 8. Autosomal fusion polymorphism in the grasshopper *Oedaleonotus enigma* ($2n$ = 23♂,XO). Basic homozygotes carry ten rod bivalents and a neo XY sex bivalent (a). Fusion homozygotes have 9 II + XY including one metacentric (M) bivalent (c), while fusion heterozygotes have 8 II + 1·III + XY (b). In the fusion homozygote the bivalent to the left of the metacentric is much attenuated.

cases the term supernumerary refers to the fact that the extra material is clearly dispensable to the organism itself in the sense that it can develop perfectly normally in its absence. White[8] believes that there have been extensive transfers between conventional autosomes and supernumerary chromosomes and that it is this that has led to the production of supernumerary segments. There is, however, no known evidence to support such an assumption and for this reason these two categories of chromosome variants are best considered separately.

1. Supernumerary Chromosomes

Fusion, whether centric or tandem in kind, is capable of producing a reduction in

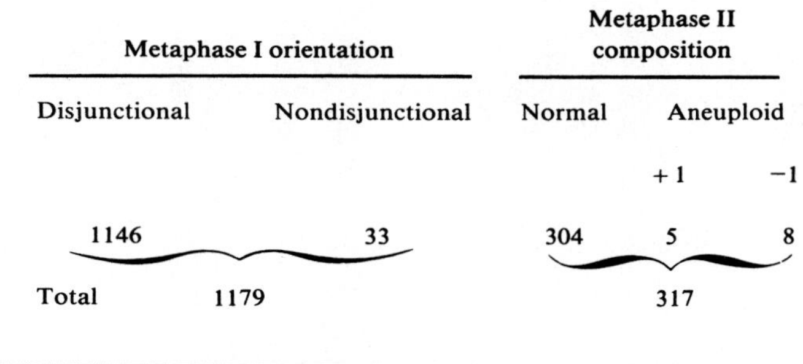

Table 25
BEHAVIOR OF THE $4\frown5$ FUSION IN
***OEDALEONOTUS ENIGMA*[37]**

Metaphase I orientation		Metaphase II composition		
Disjunctional	Nondisjunctional	Normal	Aneuploid	
			+1	−1
1146	33	304	5	8
Total 1179		317		

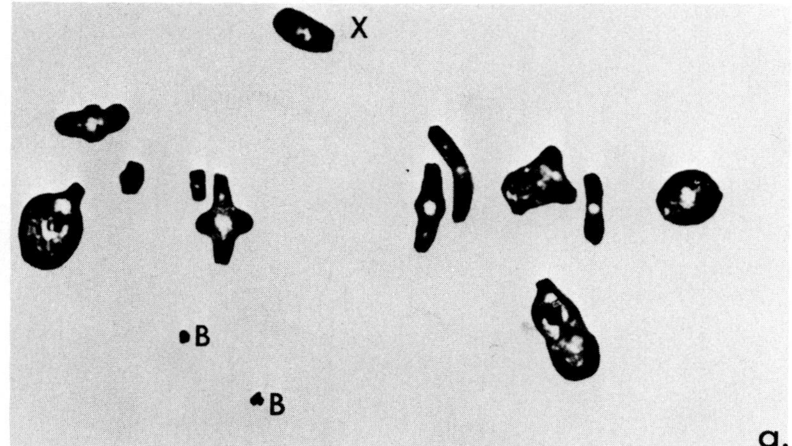

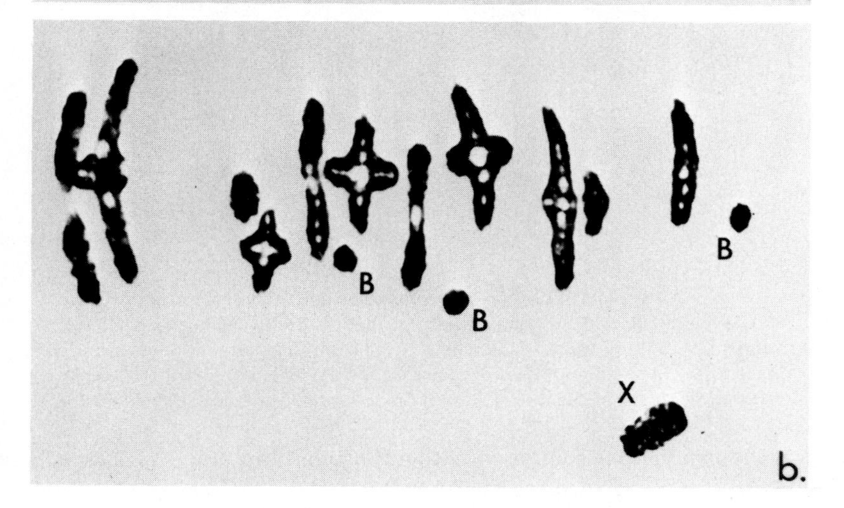

FIGURE 9. Small supernumerary (B) chromosomes at metaphase-I of male meiosis in the grasshoppers. (a) *Buforania* species 6 (11 II + 2B I + X) and (b) *Barytettix psolus* (11 II + 3B I + X).

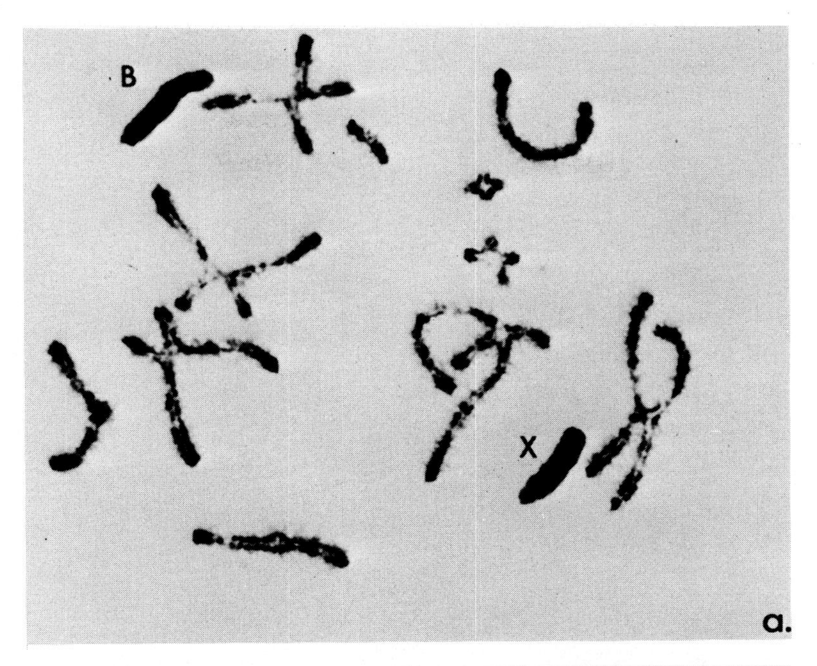

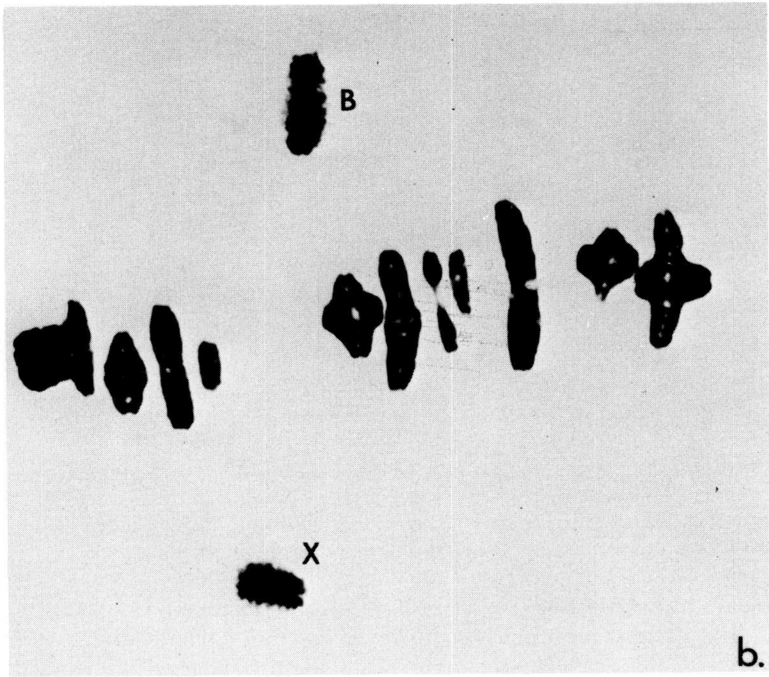

FIGURE 10. Large supernumerary (B) chromosomes in the grasshopper *Phaulacridium vittatum* (11 II + 1 B + X) at (a) diplotene and (b) first metaphase of male meiosis. In (b) the X and the B both lie off the spindle equator, a consequence of their univalency.

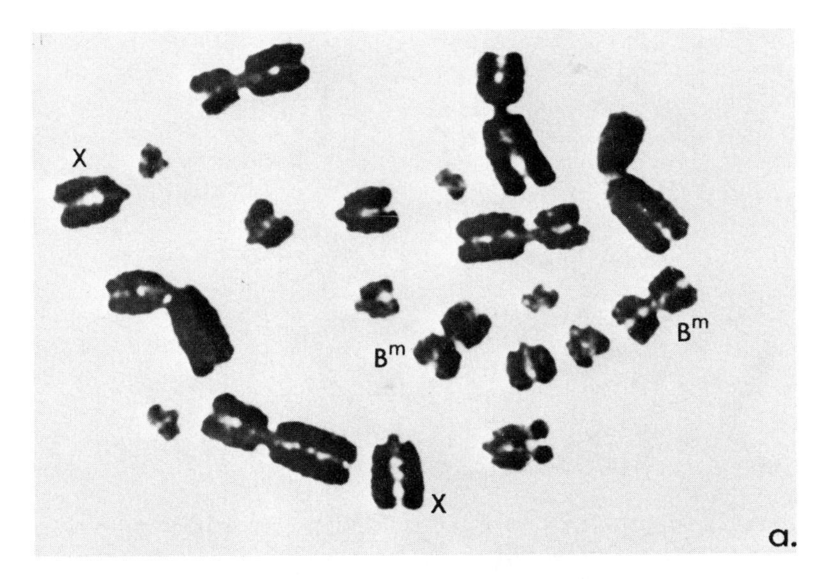

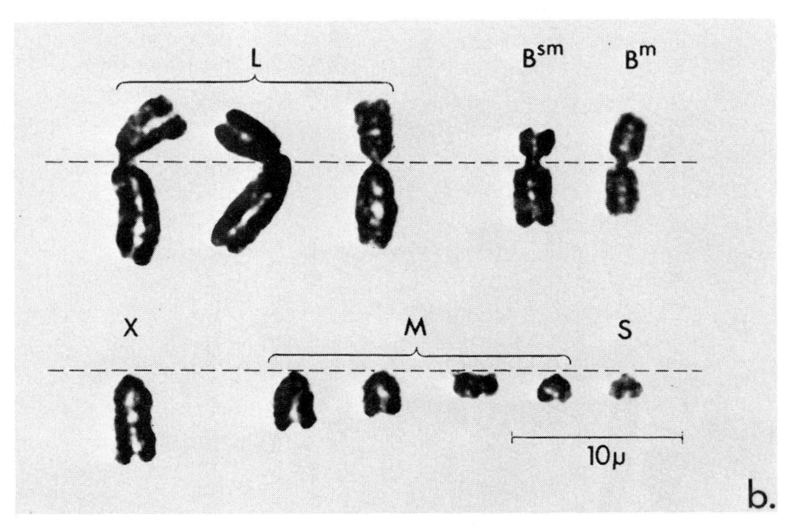

FIGURE 11. Supernumerary (B) chromosomes in the grasshopper *Myrmeleotettix maculatus* ($2n$ = 17♂XO,♀18 XX). (a) A diploid female gut ceca cell with two metacentric supernumeraries ($2B^m$); (b) a haploid karyotype with one metacentric (B^m) and one submetacentric (B^{sm}) supernumerary.

chromosome number. Methods must also exist for increasing chromosome number (see Section III). These mechanisms, however, do not include direct aneuploid increase. This is not to say that polysomy cannot be tolerated within the germ line of orthopteroids; on the contrary, germ-line polysomy has been documented in both pyrgomorphs[39-41] and acridids,[42] though the condition is certainly not common. Such polysomics, however, produce either unbalanced, normal-sized gametes, microgametes, or macrogametes, and so do not successfully transmit the extra chromosomes they carry to any of their offspring.

In striking contrast to the lack of aneuploids in nature is the presence in many different kinds of orthopteroids of polymorphisms for supernumerary or B chromosomes. This is a special category of chromosome quite different from the ordinary indispensable members of the normal complement with their strict balance and regular segregation. They do not pair at meiosis with the conventional members of the complement from which, however, they must in some sense have originated. Evidently, therefore, they are no longer homologous with their progenitor A chromosomes. Their origin is enigmatic and their precise structure is very variable (Figures 9 to 11). They range in size from the smallest to the largest elements in the complement and include telo-, acro-, and metacentric types.

Since individuals both with and without B chromosomes can coexist within the same population, it is evident that supernumerary chromosomes are not required for normal development. They are not necessarily without phenotypic effect. Thus, they sometimes[43] produce quantitative effects on morphology though even these cannot always be distinguished.[44,45] In at least one case[46] they have been associated with an alteration in gene expression.

Relatively few supernumerary polymorphisms have been analyzed in any detail. In grasshoppers, some 15% carry such polymorphisms. Some, though not all, B chromosomes in grasshoppers are mitotically unstable in the male germ line (Table 26). Others undergo preferential transmission by the female as a result of polarized movement in the egg meiosis. Both mitotic instability in the male and preferential transmission by the female may serve to increase the B-chromosome frequency from generation to generation.

In a study on a population of *Calliptamus palaestinensis* from Jerusalem, Nur[47] found the number of Bs per testis follicle to range from zero to four, though within a follicle the number was generally constant. The greater majority of follicles had 1 or 2 Bs and, although the mean number of Bs per male ranged from 0.92 to 2.58, in only three males was it lower than 1 or higher than 2. Since only 12% of the males in the Jerusalem population carried B chromosomes (Table 27), it was unexpected to find that in the majority of males the mean B number was higher than 1.5. Moreover, males with means higher than 1.5 had a higher proportion of follicles without Bs as well as follicles with 3 or 4 Bs.

In theory at least three mechanisms can explain intraindividual variation in B frequency within the male germ line:

1. Endomitotic replication of Bs
2. A higher rate of division in cells with Bs
3. Preferential nondisjunction at mitosis in the germ line

Nur[47] favored the latter explanation because it was the only mechanism capable of causing both the variation in the number of Bs and their accumulation. Thus, endomitotic replication fails to explain the presence of follicles without Bs which occurred in 11 of the 31 males analyzed by Nur. Moreover, an increased rate of division of certain cell types requires a preexisting variation in B number before such a mechanism can operate. Since in the 31 males with Bs only 3.7% (11) had cells with no Bs, whereas 53.0% (25) had cells with 2 Bs, it follows that any explanation which invokes nondisjunction as the causative mechanism for interfollicular variation must also be able to explain the large differences in the frequency of the two complementary products of nondisjunction. Indeed, to produce the observed frequencies of 0-B and 2-B classes not only must nondisjunction be preferential but it must have an efficiency of over 90%.

An equivalent pattern of behavior was subsequently reported by Nur[43] in *Camnulla pellucida*. Here, again, in 74 males with Bs which were analyzed in detail from a single population at Olema, Calif., the mean number of Bs per follicle was usually between one and two and in only two males was it below one. Again, as in *Calliptamus palaestinensis,* the frequency of Bs in the population was only 11.1% (Table 27). Four additional points were evident:

Table 26
CYTOLOGICAL CHARACTERISTICS OF B CHROMOSOMES IN ACRIDID GRASSHOPPERS

Behavior in ♂			Morphology of B			Behavior in ♀	
Germ-line mitosis	Somatic mitosis	Species				Germ-line meiosis	Somatic mitosis
Stable		*Chloealtis conspersa*				Accumulates	
		Euthystira brachyptera					
	Stable	*Melanoplus femur-rubrum*				Accumulates	
	Stable	*Mrymeleotettix maculatus*	Isochromosome or isoderivative			Accumulates	Stable
		Omocestus bolivari					
		Phaulacridium marginale					
		Trimerotropis sparsa					
			Acrocentric or Telocentric				
			L	M	S		
		Acrida lata			+		
		Buforania crassa and *B.* sp 6			+		
		Chimerocephala pacifica		+			
	Stable	*Chortoicetes terminifera*	+			Accumulates	
	Stable	*Gonista bicolor (B1)*	+				Stable
		Phaulacridium marginale	+	+			
	Stable	*Phaulacridium vittatum*	+				
		Podisma pedestris	+				
Unstable		*Atractomorpha bedeli*		+			
		Calliptamus palaestinensis		+			
		Camnulla pelucida			+		
	Unstable	*Cryptobothrus chrysophorus*			+		
	Unstable	*Gonista bicolor (B2)*			+		Unstable
	Unstable	*Locusta migratoria*			+	Accumulates	
		Melanoplus fermur-rubrum			+		
	Unstable	*Patanga japonica*		+			
Variable		*Miramella alpina*			+		
— stable/unstable		*Podisma pedestris*			+		

Table 27
SHORT-TERM STABILITY IN THE FREQUENCY OF B CHROMOSOMES PRESENT IN MALE INDIVIDUALS FROM TWO POPULATIONS OF GRASSHOPPERS[43,47]

Calliptamus palaestinensis			*Camnulla pellucida*		
Year	No. individuals analyzed	Individuals with Bs (%)	Year	No. individuals analyzed	Individuals with Bs (%)
1958	124	8.9	1963	115	14.8
1959	35	14.3	1964	140	8.6
1961	93	16.1	1965	284	11.2
			1966	36	13.9
			1967	204	13.2
	Total = 252	Mean = 11.1		Total = 780	Mean = 11.9

Note: In neither case are the individual year by year differences significant.

Table 28
ESTIMATED RATES OF ACCUMULATION OF B CHROMOSOMES IN THE MALE GERM LINE OF GRASSHOPPERS[43,49]

Species	Rate of accumulation (%)
Calliptamus palaestinensis	44
Camnulla pellucida	36
Locusta migratoria	
Hakozaki	32
Misima	27
Melanoplus femur-rubrum	21

1. The frequency of follicles without Bs was not correlated with the mean number of Bs per male.
2. Follicles with 3 or 4 Bs were more common in males with a high frequency of follicles with 2 Bs.
3. The mean number of follicles per testis in males with Bs was lower than that of males lacking Bs. In this respect normal males would be expected to have an advantage over males with Bs in respect of fecundity.
4. A higher mean number of Bs were found in male meiocytes than in cells of the gastric ceca suggesting that Bs do accumulate in the germ line. Since the two cell types were studied in different individuals of *Camnulla*, Kayano[48] subsequently carried out an equivalent comparison using the same two cell types from the same individuals in *Locusta migratoria*. He found that in six males with 1 B in the ceca cells the mean B frequency per spermatocyte was 1.01, 1.18, 1.18, 1.22, 1.25, and 1.50, respectively, while in four males with 2 Bs in the ceca cells the equivalent numbers per spermatocyte were 2.10, 2.12, 2.70, and 2.86, respectively. This leaves little doubt but that Bs do indeed accumulate in the germ line in those instances where they show interfollicular variation within the individual testis.

The fact that in *Camnulla* follicles with two or more Bs are almost ten times more

Table 29
B-CHROMOSOME FREQUENCIES IN ADULT MALES AND FEMALES OF *MELANOPLUS FEMUR-RUBRUM* IN THE WHIPPLE PARK POPULATION, ROCHESTER, N.Y.[49]

Year	♂	♀	Both sexes
1971	0.133	0.083	0.107
1972	0.157	0.119	0.140
1973	0.127	0.145	0.136
1974	0.152	0.071	0.134
Mean frequency	0.139	0.116	0.128
1977	0.007	—	—

common than follicles without Bs is again best explained on the assumption that non-disjunction is preferential and occurs in mitoses whose daughter cells have an unequal chance of giving rise to germ cells. The calculated rate of accumulation of Bs in *Calliptamus* is, however, higher than that in *Camnulla* (Table 28). In the latter species unpaired Bs, present in follicles with 1 B or 3 Bs, often lag on the spindle during anaphase I. In common with other categories of lagging univalents, these are often either lost to mini-cells, which ultimately give rise to microspermatids, or else they inhibit cytokinesis and produce macrospermatids. No such lagging occurs in *Calliptamus;* hence, here the rate of accumulation is greater.

Despite accumulation rates of over 35% in the males of both *Calliptamus* and *Camnulla*, the frequency of B chromosomes remained constant in the populations studied over the period of sampling (Table 27). This requires either an equal rate of elimination in females, a strong selection against individuals carrying Bs, or else a combination of both. The precise situation in the female has not been studied in either of the above species but it has been examined in a population of *Melanoplus femur-rubrum* from Rochester, N.Y.[49] Here the B frequency is again fairly low (Table 29) but the tendency to accumulate moderately high (Table 28). The B chromosome studied by Nur was a medium-sized metacentric. Being mitotically stable in the male, it is transmitted at the expected frequency of 0.5. During anaphase I of oogenesis, however, it moves preferentially into the secondary oocyte to be transmitted by the female at a frequency of 0.8.[50]

Since the mean B frequency was 0.128 among adults of both sexes (Table 29) but 0.155 among zygotes (Table 30), the increase in B frequency from adults to zygotes of the next generation was 0.03 (21%). The frequency of 2-B embryos in all laying females was 0.007. Consequently, selection against 2-B individuals could not have removed more than about $0.007 \times 2 = 0.014$, or only about 50% of the increase in the frequency of the B due to the preferential transmission of the B by the 1-B females. Thus, the rest of the elimination must result from a reduction in fitness of at least some of the 1-B individuals. From the fact that the B frequency of embryos produced by 0 B♀s was 0.061 (Table 30), it can be calculated that 6.1% of the sperm carried a B chromosome. Since this value is approximately that expected from the observed frequency of 1-B males in the population (Table 29) and the known rate of transmission by the male is 0.5, it follows that the B apparently has little effect on male fertility or fecundity.

On the basis of the frequency of 1-B individuals among the females (0.125) and embryos (0.141) it is evident that the survival of the 1-B female is only about 0.86 of the 0-B female. Since the frequency of 1-B males was about the same as that of 1-B females in both embryos and adults it follows that the relative survival of 1-B males

Table 30
AN ANALYSIS OF 80 LAYING FEMALES AND THEIR EMBRYOS FROM THE WHIPPLE PARK POPULATION OF *MELANOPLUS FEMUR-RUBRUM* [49]

Category	No. of females	Mean B frequency
Embryos of 0-B females	70	0.061
Embryos of 1-B females	10	0.808
Embryos of all females	80	0.155
Laying females	80	0.125

Note: The average rate of B transmission by the female is 0.808 − 0.061 = 0.747, which is not significantly different from the rate of 0.817 ± 0.052 obtained from controlled crosses.[50]

must also be 0.86. As there is no evidence that the B chromosome increased either the fertility or the fecundity of either the 1-B male or the 1-B female relative to that of 0-B individuals, Nur concluded that the B is essentially parasitic and is maintained in the population simply by virtue of the meiotic drive mechanism operative in the female. Mitotic accumulation in the male germ line can be expected to have a comparable effect.[49]

Nur arrives at the following four conclusions concerning the nature of B-chromosome polymorphisms in orthopteroids:[49]

1. A B chromosome with an average rate of transmission of less than 0.5 would have to confer an advantage on at least some of the individuals carrying it or else it would be rapidly eliminated. Kimura[51] had earlier argued that to be maintained in a population any B would have to be endowed with an accumulation mechanism so that it would be protected from loss either by selection or drift. However, he was assuming that a B chromosome, since this derived initially from a member of the standard set, must be deleterious because of the trisomic condition it creates at its inception. This assumption is open to argument. If the B at its origin is little more than a small centric fragment, as it now is in *Buforania* (Figure 9a), it would be most unlikely to have any deleterious effect.
2. When the average rate of B transmission is exactly 0.5 such a B might have no effect on fitness. It is more likely, however, that for the B to be maintained over many generations some frequency-dependent selection would have to be involved in which the B would confer a selective advantage on some individuals and a disadvantage on others.
3. Only when the average rate of transmission in both sexes is >0.5 might it be possible for the B to reduce the fitness of all individuals carrying it and, hence, be totally parasitic and still be maintained within the population. Within orthopteroids only one case, *Locusta migratoria,* has yet been described where preferential nondisjunction occurs in both males and females of the same species (Table 26). The possibility that B chromosomes are able to persist even when their effects are deleterious is aesthetically undesirable from a narrow neo-Darwinian point of view. Such a view would imply that if B chromosomes are indeed leading a virtually parasitic existence on the A complement then that complement should readjust in such a way as to lead to greater elimination of Bs.[8] Nur's evidence in *Melanoplus* is, however, compelling and indicates that the A complement may well be powerless to eliminate a B chromosome with a reasonably strong accumulation mechanism.[49]

4. A B chromosome with an accumulation mechanism might also be maintained if it increases the fitness of some of the individuals carrying it relative to that of 0-B individuals so long as selection against the remaining individuals with Bs is sufficiently strong to offset the increase in the frequency of the B through both the accumulation mechanism and the increased fitness of some of the B-containing individuals.

Apart from those cases involving preferential nondisjunction in the mitotic lineage of the male germ line there are two claims for segregation distortion in B-chromosome systems at male meiosis. In the first, the grasshopper *Phaulacridium vittatum*, the X and the B were held to move preferentially to opposite poles at first anaphase of meiosis following a persistent terminal association of the two chromosomes described as simulating a terminalized chiasma.[52] There is no doubt that the X and the B regularly engage in nonhomologous associations in this species. The precise frequency with which they do so varies somewhat from individual to individual, but a value of 50 to 60% is not uncommon. These associations are, however, of at least four categories, namely, end to end, side to side, side to end, and with partial overlap. By first metaphase of meiosis John and Freeman[53,54] found that all these classes of association had invariably lapsed. This includes the end-to-end associations[52] assumed to lead to preferential segregation of the X and the B. Moreover, analyses of second division cells failed to confirm any tendency of the X and the B to move preferentially relative to one another. Rowe and Westerman[55] likewise found the X and the B to separate at random during first anaphase. There seems little doubt, therefore, that there is no justification for Jackson and Cheung's[52] claim.

The second case involves the pygmy grasshopper *Tettigidea lateralis* (Orthoptera: Tetrigidae). Here some 33% of the males in a population from Quebec, Canada carried a B chromosome. Fontana and Vickery[56,57] analysed B-chromosome behavior in 20 males and found that in two of them there was a statistically significant excess of meta-anaphase I cells in which the X and the B moved together to the same pole. In 16 of the remaining males there was a small, but nonsignificant, excess of X and B types while in the remaining two individuals there was a slight excess of cases where the X and B moved to opposite poles. When the chi-squared values for all 20 individuals were added together the total chi-squared gave a borderline significance, but the pooled chi-squared value was highly significant with 319/542 = 58.5% of the cells with the X and the B at the same pole. Moreover, second division counts, again using pooled data from 12 individuals, confirmed a significant departure from randomness of X/B segregation.

Such a disturbance in the segregation of the X and the B should, of course, result in the B being more frequent in the X-carrying sperm as opposed to nullo-sperm. This, in turn, predicts that Bs would be more frequent in females than males provided, of course, that all sperm are equally capable of fertilization. Thus, from Fontana and Vickery's data the ratio of +X as opposed to −X sperm in B-carrying gametes should be 3/2. Despite this, however, the B frequency of males and females in the Quebec population did not differ significantly. Added to this, in samples from populations of this species in southern Ontario and southern Michigan, which carry much lower B-chromosome frequencies of 9%, there was no detectable preferential segregation of the X and the B. The actual situation, thus, needs reexamination before this claim can be accepted.

A point of some significance which requires explanation is the fact that in at least some species significant differences can be demonstrated between different populations. These are maintained over successive generations at least on a short-term basis. For example, in *Acrida lata* Kayano et al.[58] report that no significant differences could

Table 31
MEAN B FREQUENCIES IN 100 MALES PER YEAR IN EACH OF 11 JAPANESE POPULATIONS OF *ACRIDA LATA* [58]

	Population										
Year	Harozaki machi (Hk)	Ooita Univ. (Ot)	Nagasaki-shi (Ng)	Ikushi (Ik)	Kumamoto-jo (Km)	Mutsusawa (Mt)	Kurokami-cho (Kr)	Goshi-mura (Gs)	Omuta-shi (Om)	Tofuroato (Tf)	Kashii (Ks)
1957	0.520										
1958	0.410										
1959	0.350	0.330		0.420	0.310	0.210	0.290	0.150		0.070	0.070
1960	0.460	0.400		0.340	0.260	0.300	0.230	0.150		0.160	0.070
1961	0.585	0.490	0.230	0.240	0.290	0.210	0.260	0.220	0.290	0.100	0.110
1962	0.500	0.490	0.390	0.260	0.230	0.190	0.170	0.200	0.090	0.125	0.070
1963	0.400	0.440	0.410	0.220	0.270	0.420	0.180	0.230	0.190	0.040	
1964	0.400										
1964	0.360										
Mean	0.443	0.430	0.343	0.296	0.272	0.266	0.226	0.190	0.190	0.099	0.080

Note: In only two populations, Om and Mt, do the frequencies differ significantly over the period of sampling.

Table 32
A COMPARISON OF THE MEAN NUMBER OF B_2-TYPE SUPERNUMERARY CHROMOSOMES IN DIFFERENT TISSUES OF THE GRASSHOPPER *GONISTA BICOLOR*[59]

Sex	No. individuals examined	Tissue	Mean B_2 frequency per tissue
Male	13		
		Primary spermatocyte	2.86
		Gastric ceca	2.09
Female	9		
Ovariole		Ovariole wall	1.97
		Gastric ceca	1.63

be demonstrated in B-chromosome frequency in each of nine Japanese populations over periods ranging from 3 to 8 years, but the mean B frequency of these populations ranged from 0.080 to 0.443 (Table 31). On a purely parasitic hypothesis the differences imply either that different populations have different rates of accumulation or that different populations show differential tolerance to the harmful effects of the B chromosome. The Bs are mitotically stable in the male of *Acrida lata* but their behavior in the female has not been examined. An alternative explanation is that the fitness of the B chromosome is different in different populations.[58]

In *Gonista bicolor*[59] two kinds of telocentric B chromosomes are present. One, referred to as B_1, is of medium size and is mitotically stable in both male germ line and the soma of both male and female. The other, termed B_2, is very small and was mitotically unstable in the germ line of 62 of the 98 males examined. It was also unstable in the soma of both sexes (Table 32). There was a clear tendency for the B_2 to be eliminated from somatic cells since in both the gastric ceca and ovariole wall tissue a proportion of cells commonly lacked B_2 chromosomes. The existence of two behaviorally distinct types of B chromosomes, one stable and the other unstable, within the same germ line indicates quite clearly that the capacity for nondisjunction is an inherent property of the B chromosome itself. Even so this capacity can also be affected by genotype since in 36 of the 98 individuals the B_2 was stable. Sannomiya[59] suggests that the mitotic instability of the B_2 within the male germ line of *Gonista* is not likely to be associated with an accumulation mechanism, but the evidence for this claim is not convincing.

The situation in the female of *Gonista* is again not known but the B chromosomes of at least five other species of grasshopper have been shown to accumulate during egg meiosis (Table 26). The most thoroughly studied of these cases is that of *Myrmeleotettix maculatus.* Here two types of B chromosome are present in natural populations (Figure 11). One of these is an iso-chromosome (B^m) and the other an inverted derivative (B^{sm}). Accumulation of the B^m via female meiosis has been shown to be due to nonrandom distribution of B^m chromosomes on the first meiotic spindle of the oocyte.[60] This spindle is asymmetrical so that it is easy to distinguish the two poles and B chromosomes lie preferentially towards the inner spindle pole. This applies both to 1-B and 2-B females since, in the latter, B^m bivalents form in only approximately one sixth of the primary oocytes. This stems from the fact that, as the B^m is an isochromosome, two iso-rings of I form preferentially in $2B^m$ individuals.

From appropriate single pair crosses Hewitt[61] has also shown that in *Myrmeleotettix* the transmission rate of B chromosomes is not the same in all individuals (Table 33). Indeed when the B^m is introduced through the male parent there may even be an appreciable loss of Bs. Nur first advocated the use of microspermatids as a means of meas-

Table 33
MEAN TRANSMISSION RATES OF B^m-TYPE SUPERNUMERARY CHROMOSOMES IN SINGLE PAIR CROSSES OF THE GRASSHOPPER *MYRMELEOTETTIX MACULATUS*[61]

B-chromosome content of parent	Source of B parent	B introduced via	
		Male parent	Female parent
1 B	Foxhole Warren	0.381	0.582
	Tal-y-Bont	0.486	0.887
2 B	Foxhole Warren	0.264	0.645
	Tal-y-Bont	0.413	No data

Note: The Foxhole Warren population from East Anglia carries 46% of B^m chromosomes, whereas in the Tal-y-Bont population from West Wales the B^m frequency is 70%.

uring B-chromosome elimination at meiosis. Bs may also lag without elimination and in consequence give rise to macrospermatids following an inhibition of cytokinesis. Presumably no such behavior is possible in females since there is no cytokinesis at female meiosis. In *Myrmeleotettix*, although the Bs lag at first and second anaphase of male meiosis, microspermatid frequency is very low (0.2% for 1-B and 0.6% for 2-B individuals). Macrospermatid frequency is only slightly higher (4% for 2-B individuals). Neither of these values is high enough to account for the loss of Bs during male transmission. Although the female transmission data are limited to two populations, it indicates that the higher rate of transmission is correlated with a high population B-chromosome frequency while the lower rate of transmission accords with an intermediate B frequency.

An alternative in natural populations to the direct measurement of fitness or fitness components is to show that selection may well be operating, even though it cannot be measured, by correlating the frequency of different chromosome morphs with temporal or spatial differences in the environment. *Myrmeleotettix* has a wide distribution in Europe extending from the Mediterranean region to northern Scandinavia. Within Britain its distribution is correlated with latitude and temperature and is well delimited by the 16°C isotherm for mean July temperature.[62] Consequently only populations in the south of the country have B chromosomes. Added to this the derivative B^{sm} type is restricted to a more southwest distribution.

The only known case where B chromosomes have been found in populations of *Myrmeleotettix* outside Britain is in southern Sweden on the Isle of Öland in the Baltic[63] where the climate is exceptionally dry and sunny. Here the B is restricted to the southern half of the island and the frequency is very low (2% overall). In Britain and Europe generally *Myrmeleotettix* occurs in a patchwise distribution, and in relatively small populations, on sandy, broken environments such as heaths, sand dunes, and spoil heaps from heavy and precious metal mines. The steppe area of Öland, however, carries a large and fairly continuous population of *Myrmeleotettix* which is one of the dominant insect species found there. Morphologically the B in Öland is identical to the B^m type found in Britain, but whether this represents a case of independent evolution is not clear. The only alternative is that the situation in Öland and Britain is a remnant of a system that was once more widespread.

In Britain it is possible to get quite dramatic clinal changes in B frequency between populations separated by relatively short distances and these differences have been found to be maintained over a 5-year period. In some of these cases, for example, in

Table 34
EFFECT OF B CHROMOSOMES ON MEAN CELL CHIASMA FREQUENCY IN FOUR SPECIES OF GRASSHOPPERS

Species	Population		Mean cell Xa frequency −B	Mean cell Xa frequency +B
Myrmeleotettix maculatus[44,65]	Goginan Uchaf	1965	15.0	16.2
		1966	15.3	16.8
		1967	15.3	16.8
	Ynyslas	1965	15.0	16.5
		1966	15.1	15.9
		1967	15.1	16.7
	Bedd Taliesin	1965	14.8	16.8
		1966	14.5	16.8
		1967	15.4	17.2
	Llancynfelin	1965	14.2	15.9
		1966	15.1	16.4
		1967	14.9	16.1
	Yfan	1965	14.9	16.1
		1966	15.0	16.1
		1967	15.3	16.1
Eyprepocnemis plorans[45]	Alunecar		13.9	15.8
	El-Saucillo 1		14.9	16.4
	Jete		14.9	16.2
	Otivar		15.1	16.5
	El-Saucillo 2		15.2	16.4
Euthystira brachyptera[66]	Site 1	1974	13.3	15.3
	Site 2H	1978	13.4	14.2
	Site 2L	1978	13.7	14.2
	Site 1L	1978	13.8	16.5
	Site 1H	1978	13.9	14.2
Acrotylus humbertianus[67]	Manasa-Gangothri		15.4	1B 17.5
				2B 18.2
	Ranganathittu Island		15.1	17.7

West Wales, the clines appear to be related directly to the ecology of the area since B frequency is negatively correlated with rainfall.[62] In other cases such as in the Thetford area of East Anglia,[64] no obvious environmental basis for the change is apparent despite the fact that B frequency changes from 50 to 0% over a 2- to 5-mi radius.

One final feature is the fact that in four cases there is a correlation between the presence of B chromosomes within an individual and an increased chiasma frequency (Table 34), though this correlation is clearly absent in other cases.[54] That this correlation is indeed a genuine effect of the B chromosome on the system of chiasma regulation is supported by two facts:

1. In *Myrmeleotettix* one occasionally finds mosaic individuals in which, within the same testis, there are cells both with and without Bs. Since both cell types share an identical genotype in respect of their A chromosomes, any effect of a genetic component other than that provided by the B chromosome itself can be ruled out. In two such mosaics it was shown that the presence of the B again leads to an increased mean cell chiasma frequency.[62] Therefore, the relationship is causal and not casual.
2. Comparable correlations are now known in a wide range of animals and plants.

Table 35
THE EFFECT OF B^m CHROMOSOMES ON THE CHIASMA CHARACTERISTICS OF INDIVIDUALS FROM THE LAKENHEATH WARREN POPULATION OF *MYRMELEOTETTIX MACULATUS* FROM EAST ANGLIA[60]

	Mean cell chiasma frequency		Between cell variance	
Karyomorph	Male	Female	Male	Female
0 B	14.37	14.74 (13.24)	0.946	0.719
1 B	16.23	16.72 (15.10)	2.005	2.321
2 B	16.17	19.00 (17.18)	1.382	2.800

Note: The values in parenthesis for the female mean cell chiasma frequencies are adjusted means obtained by subtracting the X-bivalent contribution from the total.

Whether an effect can become a function is, of course, a different issue. On the other hand, there can be no function without an effect. Consequently, the possibility certainly exists that the effect which some B chromosomes evidently have on recombination might be of potential adaptive value, not, of course, to the individual carrying the B chromosome but to its progeny. No direct evidence exists to show that this effect is adaptive or that it might be sufficient to counteract any selection which operates against the Bs. There is a relationship between chiasma frequency and variability, so that a response to selection can be expected and, indeed, is known to obtain in plants.[68] There is no reason to doubt that an equivalent set of relationships exists in orthopteroids, too.

In *Myrmeleotettix*, the one case tested, the chiasma effect is present in both males and females (Table 35) though there is a dosage effect in the female which is not present in the male. In both cases, too, the between-cell variance is affected though the effect is more marked in females than in males. Thus, there is a sex effect as well as a B effect.

2. Supernumerary Segments

In orthopteroids, and especially in acridoids, polymorphisms for supernumerary heterochromatic segments are probably even more common than those for supernumerary chromosomes. They probably all arise from the amplification of particular chromosome segments as has been seen by Shaw[69] in *Schistocerca gregaria.* While such an amplification can, in theory, occur at almost any location in a chromosome, a majority of supernumerary segments are either procentric or terminal. The former are expected to segregate reductionally at first anaphase in bivalents heterozygous for them, while in the latter, segregation should be equational (Figure 12). A third type of unequal bivalent has been described in some acridids which shows either reductional or equational behavior. Such an unequal bivalent has been reported[70] in one of the two smallest pairs in *Calliptamus palaestinensis.* Here two types of bivalent were present. In the majority (90%) of cells a single chiasma formed between the centromere and the distally located supernumerary segment and bivalents of this type gave equational separation at first anaphase. In 7% of the bivalents, however, the two homologs were associated terminally and these bivalents gave reductional separation. To account for

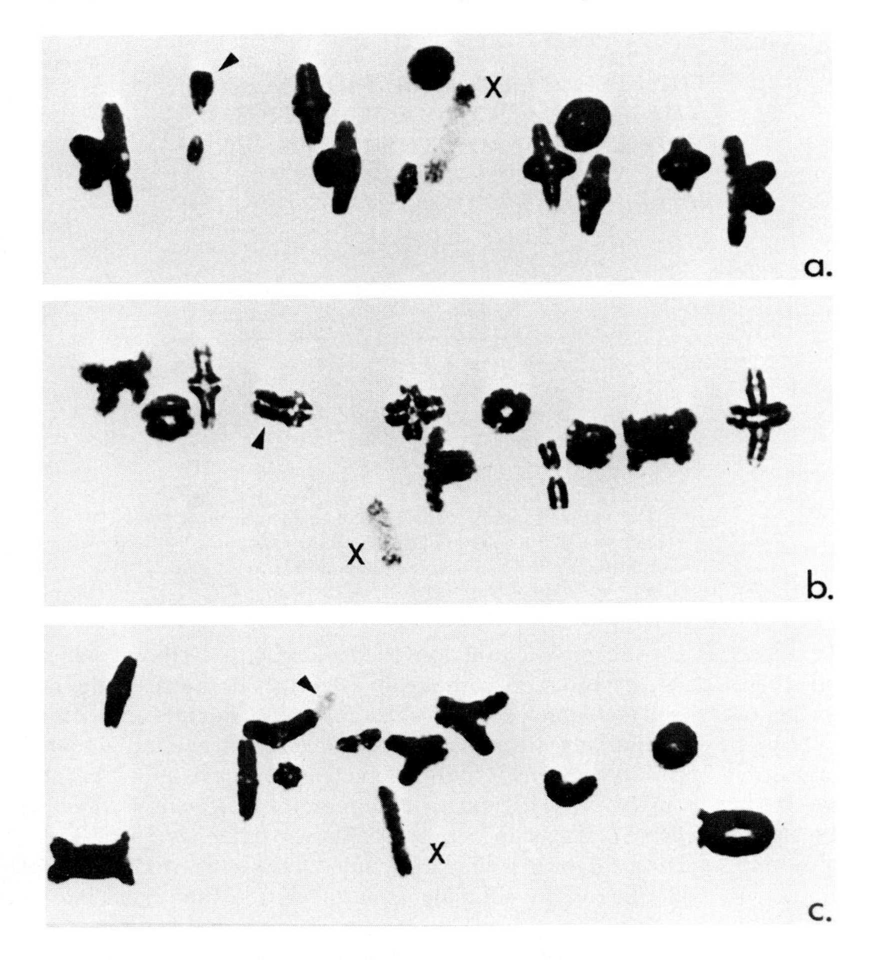

FIGURE 12. Supernumerary chromosome segments (►) in a small chromosome of *Cryptobothrus chrysophorus* (a and b) and a medium chromosome of genus nov. 66 species 2 (c). In (a) the segment is proximal in location and reductional in behavior. In (b) and (c) the segment is distal in location and equational in behavior. Both species are grasshoppers with $2n$ = 23♂,XO.

the latter class Nur[70] assumed that the supernumerary segment was not strictly terminal so that a short euchromatic segment was present distal to it which was homologous to the terminal euchromatic portion of the normal unsegmented homolog. Consequently, he believed that a chiasma sometimes occurred distal to the supernumerary segment. However, no direct evidence was offered for the actual presence of such a short distal euchromatic segment.

In *Chorthippus jucundus* the longest metacentric member of the complement is polymorphic for a distally located and heterochromatic supernumerary segment. Two types of bivalent can be identified at pachytene in heterozygous individuals. In the one the supernumerary segment simply extends beyond the euchromatic region of its homolog. In the other the segment folds back so that its tip associates with the terminal euchromatic portion of its partner homolog (Figure 13). In this latter event the association is clearly nonchiasmate. Despite this, it is persistent and can be identified at first metaphase.[71] A comparable pattern of behavior is found in *Calliptamus*, too.[70] Thus, foldback pairing of the heterochromatic block was seen by Nur in 63% of the pachytene cells. This pairing is precisely the same as that in *Chorthippus jucundus*. The one

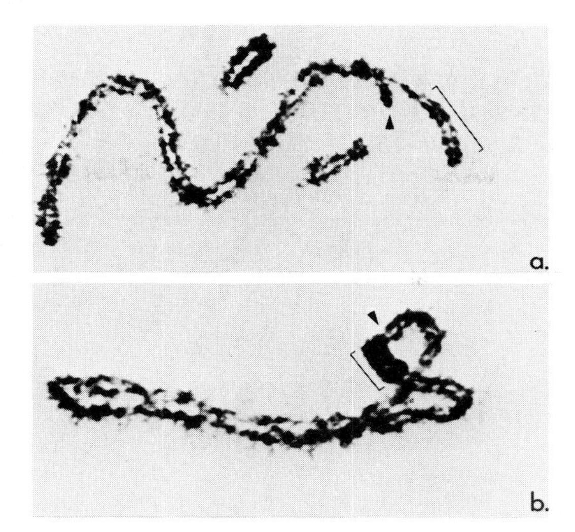

FIGURE 13. Pachytene pairing in a bivalent of the grasshopper *Chorthippus jucundus* heterozygous for a large terminal supernumerary segment (|__|) in the short arm of the largest pair of metacentric autosomes. In (a) the segment extends beyond the euchromatic tip (▲) of its pairing partner, whereas in (b) the euchromatic tip is associated side-on with the end of the segment in a nonchiasmate manner.

difference between the two cases is that in *Calliptamus* only a small number of these foldback associations persist to metaphase I, whereas in *Chorthippus* most of them do. One needs to exercise some caution, therefore, in using segregation behavior to infer the location of a supernumerary segment.

Relatively few of the segment polymorphisms that have been identified have been subjected to analysis in population terms. All 14 populations of the meadow grasshopper, *Chorthippus parallelus*, sampled in Britain are characterized by a polymorphism for terminal supernumerary segments on the two smallest autosomes (M7 and S8) which, therefore, exist in a complex polymorphic system involving nine possible combinations of karyomorphs.[72,73] All nine karyomorph classes have been observed though different populations have different frequencies (Table 36). There is also a striking difference in the observed frequency of S8 and M7 karyomorphs within populations. Thus, the S8 frequencies conform to a Hardy-Weinberg distribution, whereas those of the M7 do not; rather there is an excess of homokaryomorphs and a deficit of heterokaryomorphs.

Seven French populations[74] were also all polymorphic for supernumerary segments on the M7 and S8 chromosomes, but here the frequency of segments and the distribution of karyomorphs did not differ in different populations. Moreover, a higher proportion of S8 chromosomes were structurally normal in the French populations. In both series of populations, however, the presence of supernumerary segments leads to an increase in mean cell chiasma frequency though not in-between cell variance. In the British populations there is no dosage effect. The situation in the French populations is not entirely clear. Using pooled data from different populations Westerman[74] suggested there might be a dosage response. Since in the British populations the increase in chiasma frequency attributable to the presence of segments is not constant in different populations, Westerman's findings need to be confirmed with larger samples to avoid the necessity of pooling data.

Table 36
A COMPARISON OF KARYOMORPH AND SEGMENT FREQUENCIES IN FIVE POPULATIONS OF *CHORTHIPPUS PARALLELUS* FROM BRITAIN[73] AND SIX FRENCH POPULATIONS[74]

Populations	Karyomorph frequencies: Basic homozygotes	Segment heterozygotes	Segment homozygotes	Segment frequencies	Total individuals
British					
Ashurst					
M7	69	11	21	0.262	101
S8	56	33	12	0.282	
Ystradgynlais-1					
M7	42	1	6	0.133	49
S8	28	17	4	0.255	
Ystradgynlais-2					
M7	24	3	4	0.177	31
S8	10	13	8	0.468	
Shave Green					
M7	19	17	22	0.517	58
S8	28	24	6	0.314	
Greenham Common					
M7	29	8	23	0.450	60
S8	51	9	0	0.075	
French					
Boulogne-sur-Mer					
M7	26	3	1	0.084	30
S8	25	5	0	0.083	
Les Andelys					
M7	28	2	1	0.065	31
S8	21	10	0	0.161	
Aixe-sur-Vienne					
M7	23	1	0	0.021	24
S8	13	11	0	0.229	
Besse-en-Chandesse					
M7	18	5	4	0.241	27
S8	17	10	0	0.185	
Col de Pourtalet					
M7	18	2	0	0.050	20
S8	9	10	1	0.300	
Foret d'Ecouves					
M7	31	2	5	0.180	38
S8	26	11	1	0.192	
Mont Dol					
M7	16	0	1	0.059	17
S8	12	5	0	0.147	

Note: While both the karyomorph and segment frequencies differ significantly in the British populations, they are not different in the French.

Supernumerary segments are also present in a polymorphic state in four of the nine species of *Stethophyma* (= *Mecosthethus*) examined to date. This includes *S. grossum*, *S. lineatum*, *S. gracile*,[69,75] and *M. magister*.[76] In all four species segments have been identified in one or both of the two smallest chromosome pairs (S10 and S11). Two points of interest are noted in the analyses made in these species:

1. In the three Japanese populations of *M. magister* there are striking differences

Table 37
EFFECTS OF SUPERNUMERARY SEGMENTS ON MEAN CELL CHIASMA FREQUENCY AND CELL VARIANCE IN TWO SPECIES OF *STETHOPHYMA*[69]

Species	Mean cell Xa frequency	Mean cell variance
Stethophyma grossum		
Spain		
Basic	11.24	0.17
Segmented	11.64	0.46
Austria		
Basic	11.34	0.24
Segmented	11.95	0.75
Stethophyma gracile		
New Brunswick		
Basic	11.40	0.25
Segmented	11.92	0.39

in the frequency of S11 segments which range from a low of 0.308 in Nokonoshima to a high of 0.838 in Tufuroate.

2. While there is no effect of the S11 segment on chiasma frequency in *M. magister*, there are small but significant effects of supernumerary segments on both mean cell chiasma frequency and its variance in *S. grossum* and *S. gracile* (Table 37). This is due to an interchromosomal effect leading to the production of a larger number of ring bivalents in species in which, for the most part, there is strict proximal localization (see Section VI).

The most extreme form of segment polymorphism so far encountered in orthopteroid populations is that found in Australian populations of the pyrgomorphid grasshopper *Atractomorpha similis.* Using the C-band technique to define heterochromatic sites (Figure 14a), it is possible to demonstrate that all ten members of the haploid set ($2n$ = 19XO♂, 20XX♀) are represented in a variety of morphs in every population. The total variation is, thus, quite staggering. There is also geographical variation for both the kind and the size of the segments present though the total amount of euchromatin appears to remain constant. The most common variant, and the most obvious one, involves the addition of large terminal C-blocks, most clearly seen in three particular pairs of chromosomes — 4, 5, and 6. All three include variants which either have or else lack large distal supernumerary segments. Thus, any given individual may be homozygous for the absence of these segments (−−), heterozygous for their presence in one homolog and their absence in the other (+ −), or else homozygous (+ +) for blocks of the same or of different size. By analyzing chiasma distribution in C-banded diplotene cells (Figure 14b) from individuals with different block combinations one can demonstrate that when either one (+ −) or two (+ +) heterochromatic blocks are present terminally the chiasma positions are radically readjusted within the bivalents though meiotic pairing is in no sense disturbed (Figure 15). Specifically, in the presence of the distal blocks chiasmata are redistributed to more proximal positions.[77,78]

III. POLYTYPIC POPULATIONS

Apart from the occurrence of chromosome polymorphisms within natural orthop-

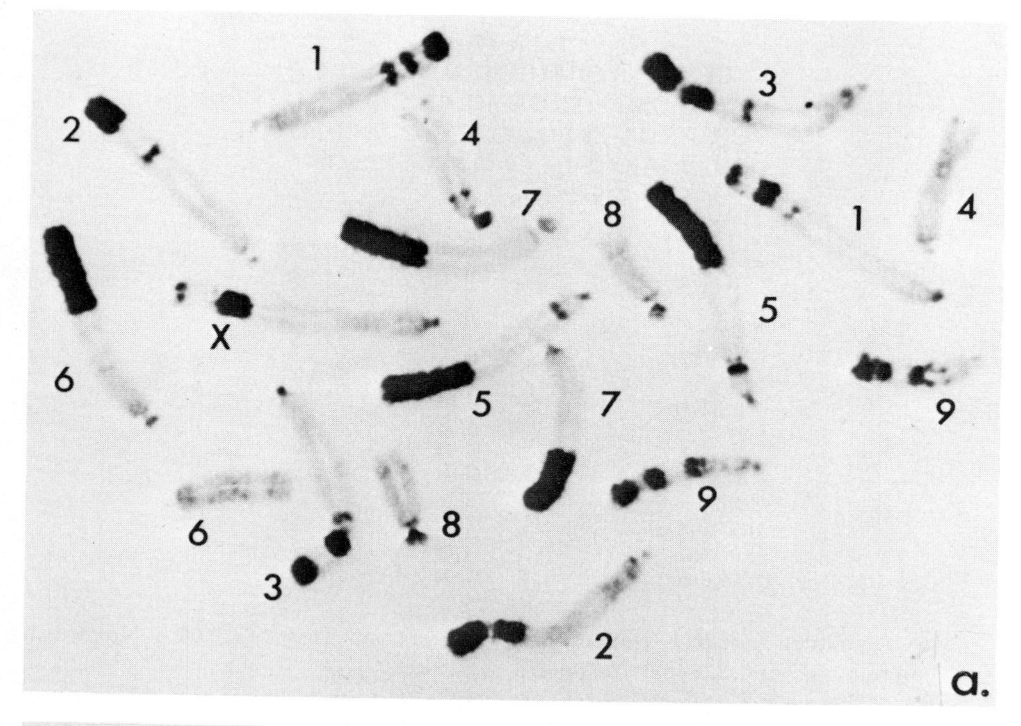

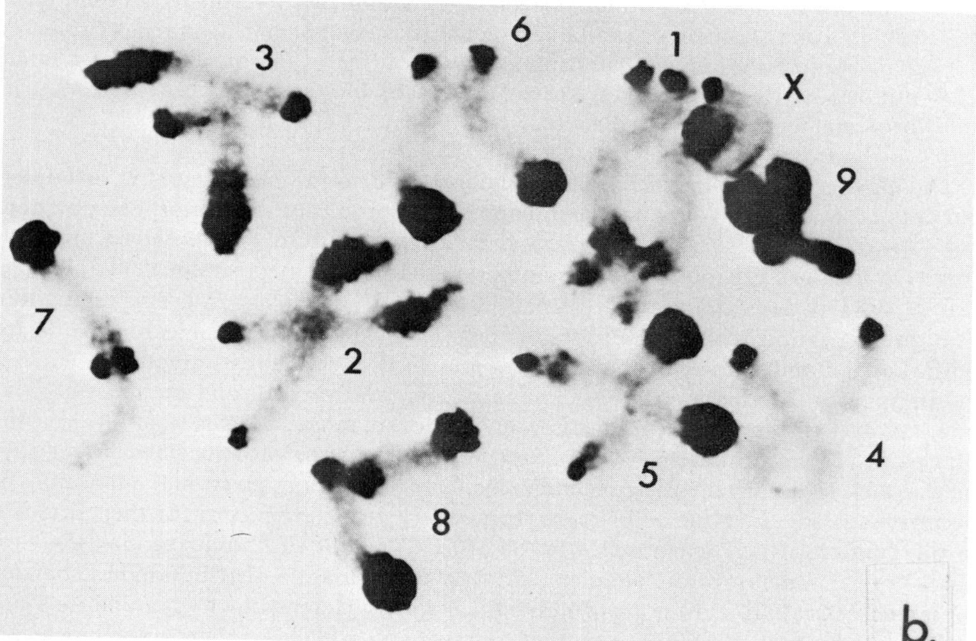

FIGURE 14. Polymorphism for heterochromatic segments in the pyrgomorph grasshopper *Atractomorpha similis* ($2n$ = 19♂,XO). C-band posture (a) A male neuroblast mitosis, autosomes 1, 2, 3, 4, 6, and 7, are all heterozygous in respect of the segments they carry; (b) a diplotene male meiocyte ($2n$ = 9 II + X), bivalents 6, 7, and 8, are similarly heterozygous.

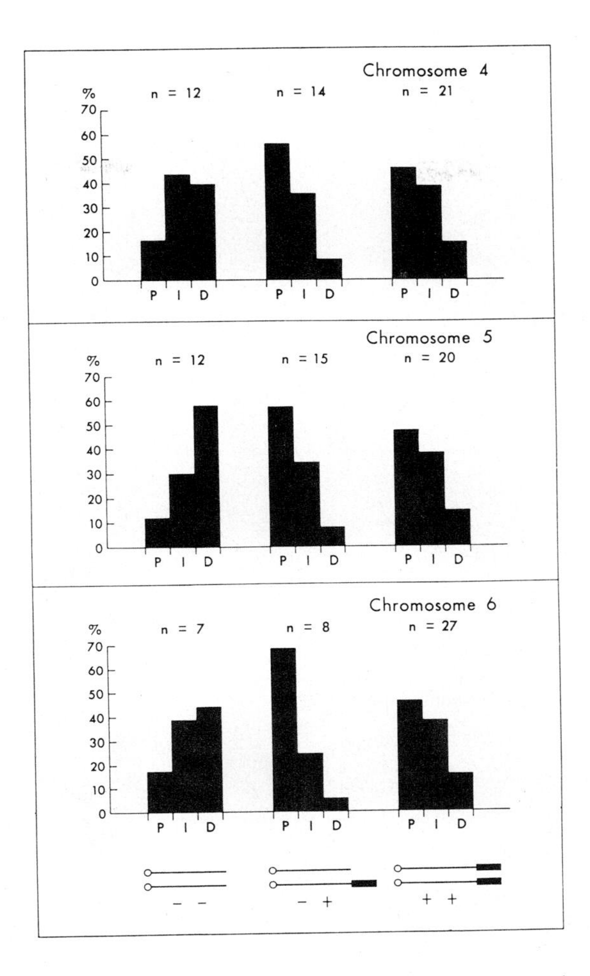

FIGURE 15. The influence of terminal heterochromatic blocks on chiasma distribution in three bivalents of *Atractomorpha similis.* For the purpose of scoring chiasmata each bivalent has been divided into three segments of equal length, namely, proximal (P), interstitial (I), and distal (D), and the numbers of chiasmata in each of these segments pooled for all monochiasmate bivalents, which in this species predominate. The total number of individuals scored in each case is given by the value (n) with ten male diplotene cells being scored per individual.

teroid populations, there are numerous cases where fixed chromosome differences are found between populations. Where these obtain within species they constitute a polytypism. Since the same kind of chromosome difference found within populations as polymorphisms may also characterize polytypic systems, it has been argued[79-81] that a polymorphism of long standing must necessarily precede the establishment of the new monomorphic state recognized as a polytypism. However, in all forms of genetic polymorphism, whether genic or chromosomal, the common denominator is the mutual dependence of the various morphs either in terms of active cooperation, as in the case of the sexes, or else in terms of reduced competition. A polymorphism will persist, and can only persist, if the several morphs make a greater contribution to posterity when they occur together than when they exist separately or alone. A return to monomorphism must, therefore, often mean the disappearance of such mutual dependence

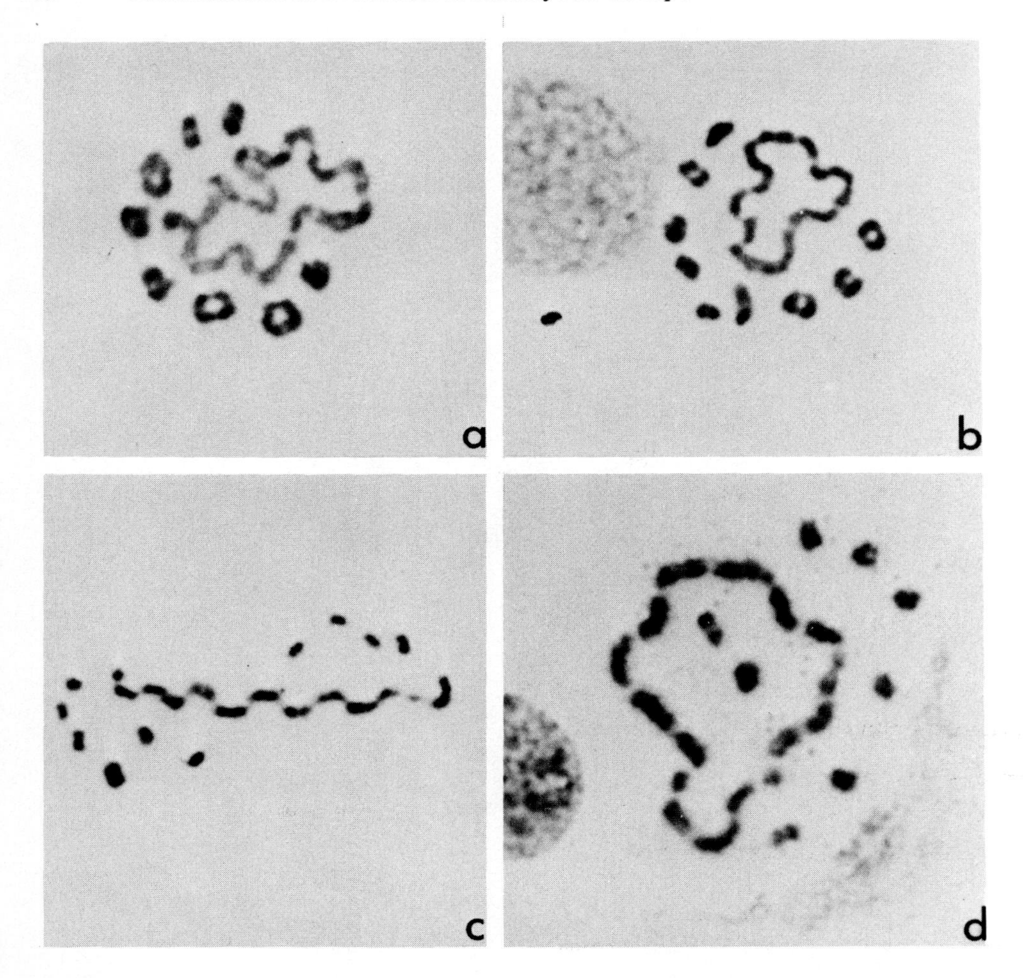

FIGURE 16. Male meiocytes of two termite species illustrating polytypism for sex-limited interchange heterozygosity. *Incisitermes schwarzi* from Dade County Nursery, Miami, has 8 II + R·XVI (a), whereas the same species from the corner of S.W. 162nd Street and Old Cutler Road, Miami has 9 II + R·XIV (b). *Kalotermes approximatus* from Kendrick has 10 II + CXIII (c) while at Yulee, Fla. it has 8 II + R·XVI (d). Photographs kindly supplied by Dr. Peter Lukyx, University of Miami, Miami, Florida.

and the substitution of self-sufficiency. The replacement of sexuality by female parthenogenesis (Section V) is not impossible, but the longer the morphs stay together the more mutually dependent will they become. Indeed, in the cases to be considered there is no evidence that any of them have ever existed in a balanced polymorphic state, though presumably they have all passed through a transient polymorphic state.

A. Interchange Systems

Multiple chromosome associations at meiosis, indicative of the presence of reciprocal translocations (Figure 16), are now known in three different families of termites, namely, the Kalotermiditidae,[82-84] the Termiditae,[85] and the Rhinotermitidae[86] (Table 38). That the multiples in question are indeed the result of a series of interchanges rather than some other form of association is confirmed by two facts.[83] First, whole chromosome arms are regularly and completely paired at pachytene. Second, typical chiasmata can occasionally be seen between the arms of adjacent chromosomes in the multiple configurations. In all these cases, the interchanges in question are fixed as permanent heterozygotes in the male and absent in the female.[82,83,85]

Table 38
THE OCCURRENCE OF INTERCHANGE MULTIPLES IN MALE TERMITES[82-86]

Species	2*n*	Meiotic configuration
Kalotermitidae		
Incisitermes schwarzi	32	9.II + R.XIV
		8.II + R.XVI
Kalotermes approximatus	32	9.II + C.XVI
		8.II + R.XVI
	34♀, 33♂	11.II + C.XI
		10.II + C.XIII
		9.II + C.XV
		9.II + C.XIV
		8.II + C.XVI
		7.II + C.XIX
Crypotermes brevis	36♂, 37♀	17.II + C.III
Neotermes castaneus	38	16.II + R.VI
Rhinotermitidae		
Reticulitermes lucifugus	42	19.II + C.IV
		18.II + C.VI
		17.II + C.VIII/R.VIII
Termitidae		
Apicotermitinae		
Acidnotermes praus	42	19.II + 1.IV
Microcerotermes fuscotibialis, pravulus, and sp. 1		
Microcerotermes sp. 2	44	20.II + 1.IV
Termitinae		
Noditermes lamanianus	38	17.II + 1.IV
Procubitermes sp. 1		
Crenetermes albotarsalis		
Cubitermes exiguus, sankurensis, sp. 1, and *weissi*		
Ophiotermes mandibularis	42	19.II + 1.IV
Pericapritermes sp. 1		
Phoracotermes macrothorax		
Unguitermes bouilloni		
Macrotermitinae		
Macrotermes bellicosus	42	19.II + 1.IV
Odontotermes redemanni, snyderi, and sp. 1		
Protermes minimus		
Pseudocanthotermes militaris		
Nasutitermitinae		
Nasutitermes arboreus	42	19.II + 1.IV

There is, moreover, a direct relationship[83] between the number of metacentrics in a complement and the size of multiple configurations present. Where, as in *Incisitermes milleri*, there are no metacentrics, no translocation complex is present at male meiosis. In *Calcaritermes nearcticus* and *Cryptotermes brevis* there is a single metacentric and this forms the central member of a chain of three multiple at male meiosis. In *Neotermes castaneus* and *I. schwarzi* there are nine metacentrics and here rings of 14 or 16 are formed.

The occurrence of interchange multiples in at least three families of termites indicates that interchanges have been extensively exploited in termite evolution. Their restriction to the male suggests that the interchanges have been grafted onto a morphologically indistinguishable pair of chromosomes responsible for sex determination. The chromosomes carrying the translocated segments segregate as a unit at male meiosis

Table 39
GEOGRAPHIC VARIATION FOR SEX-LINKED TRANSLOCATION HETEROZYGOTES IN MALE TERMITES[83,84,86]

Species	Multiple type	Distribution
Incisitermes schwarzi		
2n = 32♂, ♀	8.II + R.XVI	N. Miami Beach, Coral Gables, S. Miami, New Providence Is. (Bahamas)
	9.II + R.XIV	Perrine, Florida Keys, Dania, Marco Island, Fla.
	10.II + C.XI	Yucatan Peninsula, Mexico
Kalotermes approximatus		
2n = 34♀, 33♂	11.II + C.XI	Sarasota, Fla.
	10.II + C.XIII[a]	Brooksville, Melbourne, Kendrick, Gainesville, Palatka, Fla.
	10.II + C.XIII[b]	Gainesville, Fla.
	10.II + C.XIII[c]	Valdosta, Fla., Tifton, Ga., Gardens Corner, Myrtle Beach, S.C., Wilmington, New Bern, N.C., Cape Henry, Va.
	10.II + C.XIII[d]	Hosford, Pensacola, Fla.
2n = 32♀ 32♂	9.II + C.XIV	Savannah, Ga., Charleston, S.C.
33♂	9.II + C.XV	Daytona Beach, Fla.
2n = 32♀, 32♂	8.II + C.XVI	Jacksonville, Fla.
	8.II + R.XVI	Yulee, Fla.
33♂	8.II + C.XVII	St. Augustine, Fla.
	7.II + C.XIX	St. Augustine, Fla.
Reticulitermes lucifugus		
2n = 42♂, ♀	19. II + C.IV	Udine (Friuli), Tombolo (Tuscany), Barberino (Tuscany), Polizzi (Sicily)
	21.II	Squinzano (Apulia)
	19.II + C.IV	Squinzano (Apulia)
	18.II + C.VI	Squinzano (Apulia)
	17.II + C.VIII	Sennori (Sardinia)
	17.II + R.VIII	Sennori (Sardinia)

and pass only to the male determining sperm. Normally the heterozygous state represents the essence of impermanence but when associated, as it is here, with the differentiation of the sexes it becomes obligatory since, in effect, it creates a kind of partially balanced lethal system in which one class of homozygotes, equivalent to the YY state, does not form. As a consequence of this, structural heterozygotes, nevertheless, breed true, a complete contradiction of the conventional rules of inheritance. Such a state of permanent heterozygosity is, of course, expected to have strong genetic effects in the male, including the close linkage of genes that assort independently in the female.

In at least three species, *Reticulitermes lucifugus,*[86] *I. schwarzi,*[82,83] and *Kalotermes approximatus,*[83,84] there is geographic variation for the number of translocations present within colonies (Table 39). In *I. schwarzi,* for example, there are at least three kinds of colony. One found in North Miami Beach, Miami, Coral Gables, South Miami, and on New Providence Island in the Bahamas has a ring of 16 chromosomes. A second, south of the Miami area in Perrine and the Florida Keys, has a ring of 14 chromosomes (Figure 16a and b). A third arrangement, involving a chain of 11 chromosomes, has been found in a colony near Tulum in the Yucatan peninsula of Mexico. Males and females from this colony have been successfully mated with those from colonies in south Florida and have produced female offspring confirming that the variation is indeed polytypic. The situation has been most thoroughly analyzed in *Kalotermes approximatus,*[84] a species with a limited geographic range along the coastal plains of southeast U.S.

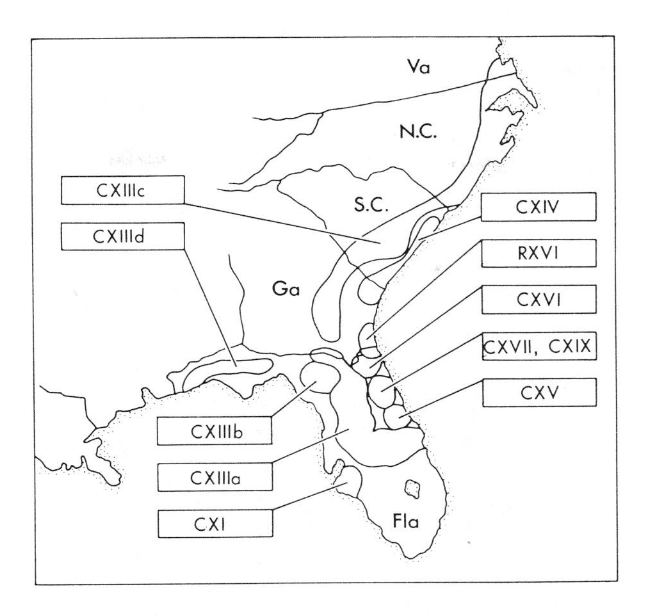

FIGURE 17. Geographic distribution of male interchange multiples in the termite *Kalotermes approximatus.* (After Syren, R. M. and Luykx, P., *Chromosoma*, 82, 65, 1981. With permission.)

Table 40
POLYTYPISM FOR FUSION DIFFERENCES INVOLVING SEX CHROMOSOME SYSTEMS[87-89]

Species	Male constitution	
	Race 1	Race 2
Leiotettix politus	14, XY	13, X_1X_2Y
Leiotettix sanguineus	23, XO	22, XY
Karruacris brownii	20, XY	19, X_1X_2Y
Podisma pedestris	23, XO	22, XY

A majority of colonies of *K. approximatus* have $2n = 33♂,34♀$, though some from northern Florida, Georgia, and South Carolina have $2n = 32$ in both sexes. In 37 colonies from different geographical localities covering most of the range of the species, all males are fixed heterozygotes for an extensive series of interchanges involving up to half of the entire haploid genome and producing ring and chain multiples of from 11 to 19 members (Figures 16c and d). No direct observations have been made on female meiosis in this species, but, since males homozygous for the interchanges do not occur, it is clear that females do not transmit interchange complexes to their offspring. Apart from the interchanges some centric fusions have also occurred which are responsible for the presence of colonies with a diploid number of 32.

There is considerable geographical variation in the translocations which occur as a series of polytypic races (Figure 17). Moreover, the distribution of translocation complexes is evidently nonrandom. Adjacent populations are closely related in karyotype suggesting a sequential series of additions to the multiples. Thus, each of the higher order complexes tends to occur near the next lower one from which it can be derived, though what maintains the integrity of the colonies within a given geographical area is not known.

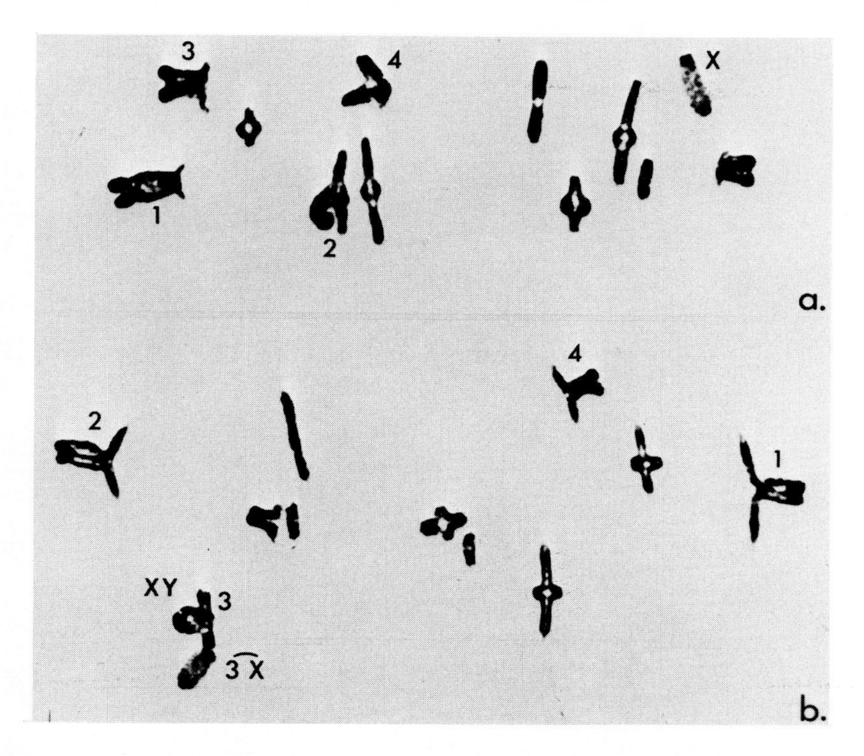

FIGURE 18. Polytypic chromosome variation in the young mare grasshopper *Podisma pedestris* at male meiosis in the $2n = 11\ II + X,XO$ race (a) and the $2n = 10\ II + XY$, neo XY race (b). The autosome involved in the fusion is the third longest member of the complement so that the fusion can be represented as $\widehat{X3}$ = neo X and 3 = neo Y.

B. Fusion Systems

Two rather different kinds of fusion are present as polytypisms in orthopteroid populations. First are autosome/autosome ($\widehat{AA}$) fusions of the type met as polymorphisms, and second autosome/sex ($\widehat{AX}$, $\widehat{AY}$) chromosome fusions. These differ from the $\widehat{AA}$ type in that heterozygotes for the rearrangement are confined to the heterogametic sex in the same way that interchanges in termites are confined. Where such a fusion originates in a basic XO♂, XX♀ system it gives rise to a neo XY♂, neo XX♀. This, in turn, may be converted by a further fusion into an X_1X_2Y♂, $X_1X_1X_2X_2$♀ system. Both such systems are found in polytypic states (Table 40), but only one has been studied in any detail. This is the polytypism present in the young mare grasshopper *Podisma pedestris*, a wingless species of acridid widely distributed across Russia and in Siberia and Mongolia. It is also found in western Europe as essentially an alpine form, seldom present below 1000 m, and occurring in isolated populations, often quite small in size, throughout the northern Appenines of Italy and in the French and Swiss alps. These European populations are of two kinds (Figure 18). A majority are 23(XO)♂, 24(XX)♀, the chromosome condition that characterizes the species over most of its range. However, a region of exclusively $2n = 22$ populations exists in a restricted area of the Alpes Maritimes at the extreme southwest limit of the distribution of the species.[89,90] The XO and XY forms are usually well separated but the two races meet in a long narrow hybrid zone running east-west for about 100 km at the northern boundary of the distribution of the neo XY form along a high mountain ridge. This zone appears to consist of intrusions of the neo XY form into essentially XO territory and is characterized by the presence of females with 23 chromosomes and a constitution of the type X/$\widehat{XA}$/A (Figure 19). Such females occur at varying frequencies within particular

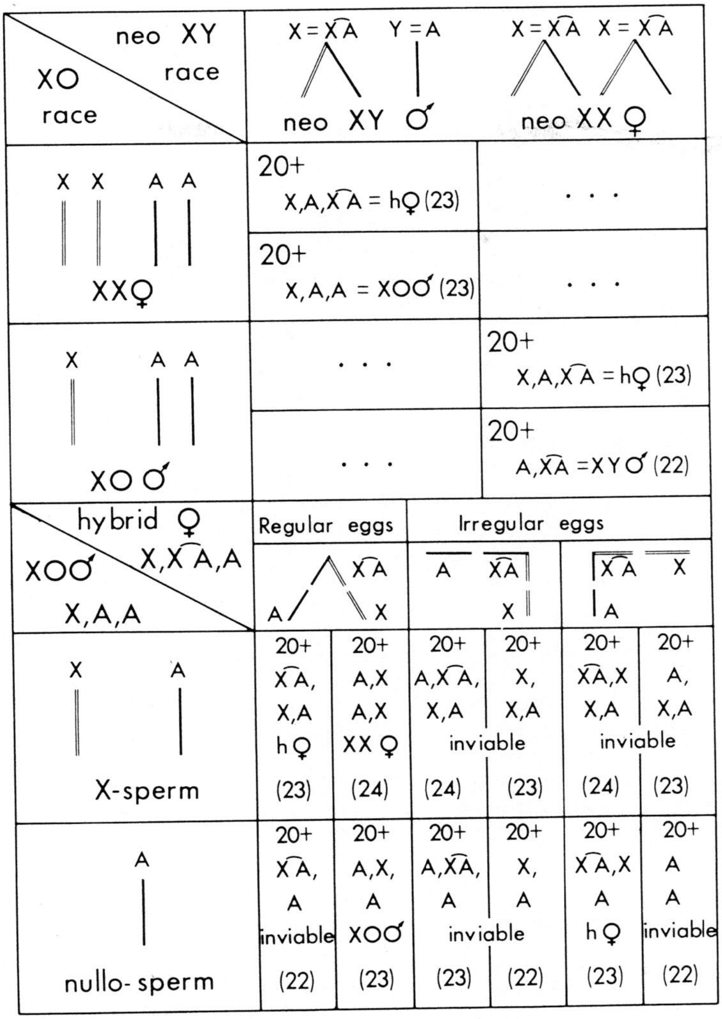

XO race \ neo XY race	X = X͡A, Y = A → neo XY ♂	X = X͡A, X = X͡A → neo XX ♀
X X A A — XX♀	20+ X,A,X͡A = h♀ (23)	. . .
	20+ X,A,A = XO♂ (23)	. . .
X A A — XO ♂	. . .	20+ X,A,X͡A = h♀ (23)
	. . .	20+ A,X͡A = XY♂ (22)

XO♂ X,A,A \ hybrid ♀ X,X͡A,A	Regular eggs: X͡A	Regular eggs: A, X	Irregular eggs: A X͡A	Irregular eggs: X	Irregular eggs: X͡A X	Irregular eggs: A
X A — X-sperm	20+ X͡A, X,A h♀ (23)	20+ A,X A,X XX ♀ (24)	20+ A,X͡A, X,A inviable (24)	20+ X, X,A inviable (23)	20+ X͡A,X X,A inviable (24)	20+ A, X,A inviable (23)
A — nullo-sperm	20+ X͡A, A inviable (22)	20+ A,X, A XO♂ (23)	20+ A,X͡A, A inviable (23)	20+ X, A inviable (22)	20+ X͡A,X A h♀ (23)	20+ A A inviable (22)

FIGURE 19. The consequences of hybridization in *Podisma pedestris.* The upper portion of the figure shows the F_1 products of crossing the two races while the lower represents the backcross products of mating a hybrid female with an XO male.

hybrid populations.[91] The two races also interbreed successfully in laboratory crosses to produce equivalent F_1 hybrids.

At the end of the natural hybrid zone, near Col de Tende, there is a cline, about 800 m wide,[92] which runs from the lower limit of the species distribution at 1500 m to a height of some 2800 m (Figure 20). The behavior of laboratory-reared hybrids has been analyzed[93] in a series of controlled crosses between individuals from chromosomally pure populations spanning this cline, as well as between individuals from a mixed population within it. The first of these crosses can be expected to demonstrate the cumulative effect of all the genetic differences which exist between the two races, while the second shows the effect of the chromosome difference which exists between them. The term fitness as applied to orthopteroid insects implies a summation of all the life cycle components that are capable of producing a differential contribution of heredity to the progeny generation. It, thus, involves embryonic development, hatching

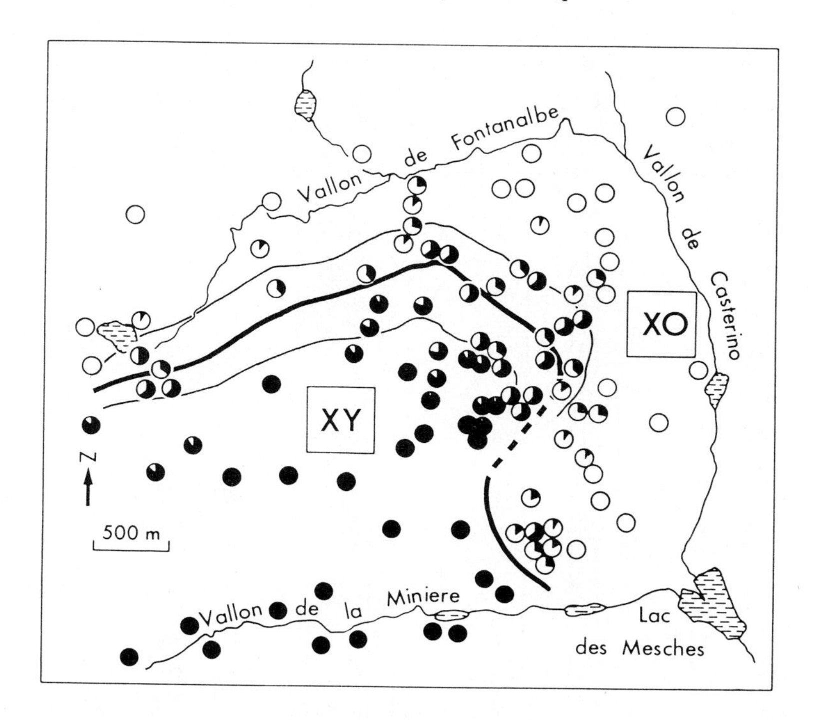

FIGURE 20. The structure of the natural hybrid zone of *Podisma pedestris* in the Col de Tende region of the Alpes Maritimes. XO populations are demarcated by open circles and neo XY populations by solid circles. The center of the zone is defined by a heavy line and the 23 and 77% levels by thinner lines. (After Hewitt, G. M. and Barton, N. H., *Insect Cytogenetics,* Symp. No. 10, Royal Entomological Society of London, 1980, 149. With permission.)

of the embryo, nymphal and adult viabilities, mating success, fecundity, and fertility. Fitness measured under laboratory conditions need not accurately reflect fitness in nature.

Crosses between pure populations — In these crosses the number of eggs per pod did not differ significantly between hybrid and nonhybrid crosses, but there was a large difference in hatching success associated with the chromosome difference. Hybrid crosses also showed an excess of embryos which ceased development very early and of embryos which died as they were hatching (Table 41). There were no consistent differences in survival, rate of development, fertility, or rate of egg laying between the various classes after hatching though the numbers involved are small. It is clear that the fraction of apparently normal embryos varies greatly between crosses but there is no consistent pattern to this variation (Table 42). A hybrid female is expected to produce a sex trivalent of the type X,$\widehat{XA}$,A at meiosis, and if this were to undergo nondisjunction it would produce aneuploid embryos. The data indicates that aneuploidy from nondisjunction of the sex trivalent (X,$\widehat{XA}$,A) in hybrid crosses is unimportant as a contributor to embryonic death by comparison with other causes. Thus, the trivalent seems to segregate efficiently and shows only a low frequency of nondisjunction. While laboratory crosses show heavy selection against F_1 hybrids it is clear that this selection is not related to the XO/XY difference as such.

Crosses between individuals from within the cline — There is no evidence for any marked nondisjunction of the sex trivalent in terms of the frequency of aneuploid embryos. Pooling data from both sets of crosses gives an upper limit of 1.6% for the nondisjunction rate expressed as a fraction of the surviving embryos.

Table 41

THE FATE OF EGGS PRODUCED BY SINGLE-PAIR MATINGS USING NYMPHS FROM PURE POPULATIONS OF *PODISMA PEDESTRIS* AND FROM COMPARABLE MATINGS BETWEEN INDIVIDUALS FROM A MIXED POPULATION WITHIN THAT CLINE[93]

	Total no. eggs	Successful hatch	Died at hatch	Failed to hatch	Ceased development at or before diapause	Little or no embryo material
Between pure populations						
Nonhybrid crosses	651	0.53	0.15	0.08	0.07	0.17
Hybrid crosses	496	0.34	0.20	0.09	0.06	0.31
Between individuals from a mixed population within the cline						
Nonhybrid crosses	233	0.25	0.30	0.03	0.25	0.17
Hybrid crosses	111	0.12	0.16	0.07	0.37	0.28
BX hybrid	244	0.25	0.18	0.01	0.43	0.13

Note: Pure populations span the cline of XO and neo XY forms in the Col de Tende region of the Alpes Maritimes in southeast France.

While the two chromosome races of *Podisma pedestris* cannot be distinguished on morphological grounds, they are distinguishable in respect of their recombination systems. The fusion responsible for the development of the neo XY system affects recombination in two quite distinct ways:

1. It reduces the number of linkage groups and, hence, the number of randomly assorting units.
2. It changes both the pattern and the frequency of the chiasmata that form in the autosomally derived limb of the neo XY (Table 43). Thus, there are more distal and fewer proximal chiasmata in the autosomal limb of the fused X compared to the equivalent free autosome in the XO form. There is also an overall reduction in the mean chiasma frequency of the autosome involved in the fusion from 2.2 per bivalent in the XO state to 1.8 per bivalent in the XY state.

In terms of pure autosomal fusions ($\widehat{AA}$) there are several cases of polytypism on record (Tables 44 and 45). In the two phasmatids concerned, *Crenomorpha chronus* and *Didymuria violescens,* narrow zones of overlap between adjacent cytological forms have been found. Hybrids occur in some of these zones which are both viable and at least partially fertile. Only one has, however, been analyzed in detail, namely that of *Didymuria*. This is a species endemic to the coastal and montane eucalypt forests of southeast Australia. Populations are typically sparse, although they can reach plague proportions occasionally in particular areas. *Didymuria* is a sedentary insect; the adult female has very small wings and cannot fly at all; the adult male is capable of only limited flight.

Within its distribution range Craddock[96,97] has shown that the species is differentiated into ten chromosome races which differ widely in chromosome number (26 to 40 in females), morphology, and sex chromosome mechanism (Table 45). The distribution of these races is parapatric (Figure 21) and the pattern of distribution has been found to be constant over successive seasons. The status of these ten taxa is equivocal since they cannot be distinguished morphologically. The available evidence indicates that they are polytypic variants within a single species.

Table 42

A CYTOGENETIC ANALYSIS OF EMBRYOS OBTAINED FROM SINGLE-PAIR CROSSES IN *PODISMA PEDESTRIS*. (PATERNAL KARYOTYPE IS GIVEN FIRST)[93]

Cross type		No. crosses	Total embryos analyzed	Abnormal embryos: Haploid	Polyploid	Aneuploid	Dead	Total abnormal	Normal embryos: XO	XY	XX	X/XA	XA/XA	Total normal
Between individuals from distant populations														
Pure	XY × XX	4	40	3	0	0	26	29	—	8	—	—	3	11
	XO × XX	6	171	4	0	6	47	57	42	—	72	—	—	114
F_1	XO × XY	2	104	10	0	1	88	99	—	2	—	3	—	5
	XY × XO	2	34	0	0	0	17	17	11	—	—	6	—	17
BX	XO × F_1	3	154	1	2	0	114	117	12	11	5	9	—	37
	XY × F_1	7	384	1	1	7	169	178	40	51	—	59	56	206
Totals		29	887	19	3	14	461	497	105	72	77	77	59	390
Between individuals from a mixed population														
Pure	XY × XX	12	297	1	3	6	118	128	—	95	—	—	74	169
	XO × XX	2	28	4	0	0	7	11	6	—	11	—	—	
F_1	XO × XY	4	92	0	0	5	45	50	—	25	—	17	—	42
	XY × XO	3	115	3	2	4	58	67	29	—	—	19	—	48
BX	XO × F_1	8	206	10	0	8	113	131	19	21	16	19	—	75
	XY × F_1	11	255	9	3	8	94	114	35	34	—	41	31	141
Totals		40	993	27	8	31	435	501	89	175	27	96	105	492

Table 43
A COMPARISON OF CHIASMA DISTRIBUTION IN MONO-, BI- AND TRI-CHIASMATE BIVALENTS OF THE L3 EQUIVALENT IN XO AND NEO XY POPULATIONS OF *PODISMA PEDESTRIS*[90]

	Singles			Doubles					Mean cell
							Triples	Mean Xa frequency	Xa
Race	D	I	P	2D	1D	PD	PID	in L3 equivalent	frequency
XO (free L3)	1	6	5	32	55	40	60	2.23	20.37
XY ($\widehat{X3}$ Fusion)	47	1	0	57	68	14	11	1.79	19.35

Note: D = distal, I = interstitial, and P = proximal.

Table 44
KARYOTYPE VARIATION IN MALES FROM POLYTYPIC RACES OF TWO ORTHOPTEROIDS[94,95]

Race	♂ 2*n*	No. of metacentrics	Total arm no.
Ctenomorpha chronus			
Race 1 — central and	32	18	50
southern highlands	30	20	
Race 2 — Sydney coastal			
Race 3 — northern montane	30	18	48

Race	♂ 2*n*	Fusion A	Fusion B	Total metacentrics	
Dichroplus pratensis					
Race 1	15	+, hom.	+, hom.	4	19
Race 2	16	+, het.	+, hom.	3	
Race 3	17	—	+, hom.	2	
Race 4	16	+, hom.	+, het.	3	
Race 5	17	+, het.	+, het.	2	
Race 6	18	—	+, het.	1	
Race 7	17	+, hom.	—	2	
Race 8	18	+, het.	—	1	
Race 9	19	—	—	0	

Note: Hom. = homozygous; het. = heterozygous

While the morphologically undistinguishable chromosomes in the two different races need not be genetically equivalent, a comparison of the karyotypes of the ten races enables provisional inferences concerning the chromosome rearrangements that may have been involved in the differentiation of the races. In making such comparisons Craddock assumes that the lower chromosome number races are derived. In support of this assumption is the fact that the 39:40(m) system occurs in most members of the order Phasmatodea and can, therefore, be argued to be ancestral for the group (but see Section IV.B). This is also supported by the extensive distribution of this race, its ecological latitude between the extremes tolerated by the species, and the occurrence of apparently relict isolates of the race. Moreover, unpublished DNA measurements

Table 45
CYTOLOGICAL CHARACTERISTICS OF THE 10 RACES OF *DIDYMURIA VIOLESCENS*[97]

Race	Haploid no. of chromosome arms	Female karyotype			Male meiosis	
		m	sm	a	Mean cell Xa frequency	Sex mechanism
Coastal or lowland						
39:40 (m)	30	1	9	10	19.56 ± 0.06	XO
39:40 (sa)	27	—	7	13	19.56 ± 0.09	
Mountain						
37:38	28	2	7	10	19.18 ± 0.05	
35:36	28	3	7	8	19.16 ± 0.07	X_mO
31:32	26	5	5	6	16.43 ± 0.07	
32(XY)	26	4	6	6	17.13 ± 0.07	
30(XY)	25	5	5	5	16.42 ± 0.06	XY primary
28(XY)w	21	5	2	7	17.51 ± 0.08	
28(XY ring)	21	5	5	4	17.83 ± 0.10	XY secondary
26(XY ring)	21	5	5	3	17.06 ± 0.08	

referred to by Craddock indicate that however the ten races are related, there has been no major loss or addition of chromosome material. She, therefore, suggests that three kinds of chromosome change appear to have operated in *Didymuria:*

1. Centric fusions constitute the major rearrangement involved in the production of the numerically descending series of XO races (39 → 37 → 35 → 31).
2. The lack of constancy in arm number indicates that some of the autosomes involved in the fusion series display additional pericentric inversion differences.
3. Tandem fusions appear to have been involved in the formation of both the primary neo XY and the secondary XY ring systems.

Ethological barriers to mating appear to be absent both in the field, where narrow hybrid zones have been identified, and in the laboratory. Consequently Craddock[97] attempted to test the validity of her conclusions through the analysis of meiosis in F_1 interracial hybrids which were characterized by trivalent associations (Table 46). Where centric fusion differences are involved such trivalents consist of one metacentric and two rods, whereas in a tandem fusion the trivalent includes two metacentrics and one rod. On the basis of the evidence from these crosses she proposes the system of derivation summarized in Figure 22.

There were three forms of chromosome disturbance at meiosis in the male F_1 hybrids studied:

1. Failure of pairing led to the production of univalent chromosomes. This was not confined to the chromosomes involved in structural rearrangements, although pairing failure in chromosomes which normally associated as bivalents was more frequent in laboratory hybrids. Asynapsis in the trivalents ranged from 3.3 to 77.8% in laboratory hybrids but was much less marked, 0 to 15%, in field hybrids.
2. Malorientation of trivalents at first metaphase occurred in all hybrids, but Craddock[96] gives no actual frequency data.
3. Finally, occasional nonhomologous associations were present which also sometimes led to irregular segregation.

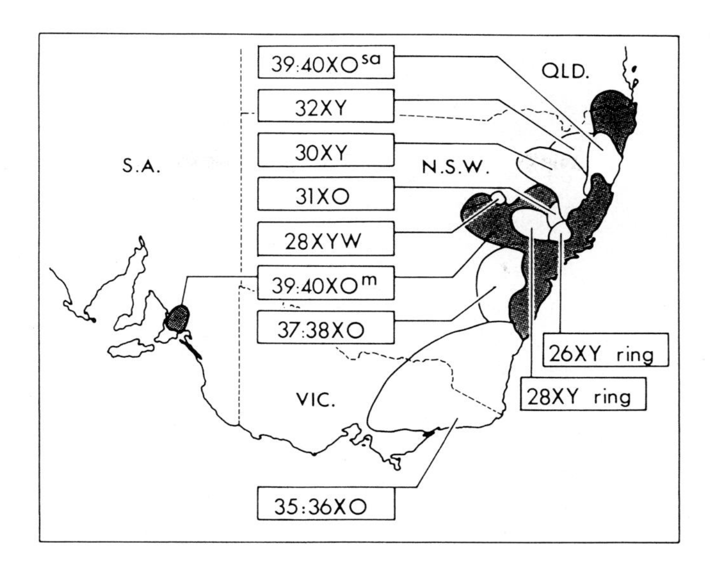

FIGURE 21. The distribution of the ten races of the stick insect *Didymuria violescens* in southeastern Australia. (After Craddock, E. M., *Genetic Mechanisms of Speciation in Insects,* White, M. J. D., Ed., Australia and New Zealand Book Co., Sydney, 1974, 24. With permission.)

Table 46
A SUMMARY OF THE PAIRING BEHAVIOR IN LABORATORY BRED F_1 INTERRACIAL HYBRID MALES OF *DIDYMURIA VIOLESCENS*[97]

Cross type	2*n* F_1 ♂	Most frequent MI configuration	Major structural difference between parental races
40(m) × 35 36 × 39 (m)	37	2.III + 1.5.II + X	2 Centric fusions
32 × 35	33	2.III + 13.II + X	2 Centric fusions
32 × 39 (m) 40(m) × 31	35	4.III + 11.II + X and 3.III + 11.II + 1 ub + 1 I + X	4 Centric fusions
28 × 26 (XY ring) 26 × 28 (XY ring)	27	1.III + 11.II + XY ring	1 Autosomal translocation
32 × 30 (XY)	31	15.II + X	1 X-autosome translocation
30 × 26 (XY ring) 26 × 30 (XY)	28	2.III + 11.II	1 Autosomal translocation, 1 X-autosome translocation, and 1 neo Y fusion
32 × 26 (XY ring)	29	2.III + 11.II + X	1 Neo Y fusion and 1 autosomal translocation
40(m) × 28 (XY ring)	33	Irregular meiotic pairing	5 Autosomal translocations, 2 X-autosome translocations, and 1 neo Y fusion

Note: In each cross the maternal parent is listed first; ub = unequal bivalent.

Despite these meiotic irregularities Craddock makes it clear that the effective reduction in the fertility of the hybrids in *Didymuria* was small and she succeeded in producing both backcross and F_2 generation progenies. This quite spectacular chromosome variation occurs within the confines of a single species and is clearly polytypic in pattern.

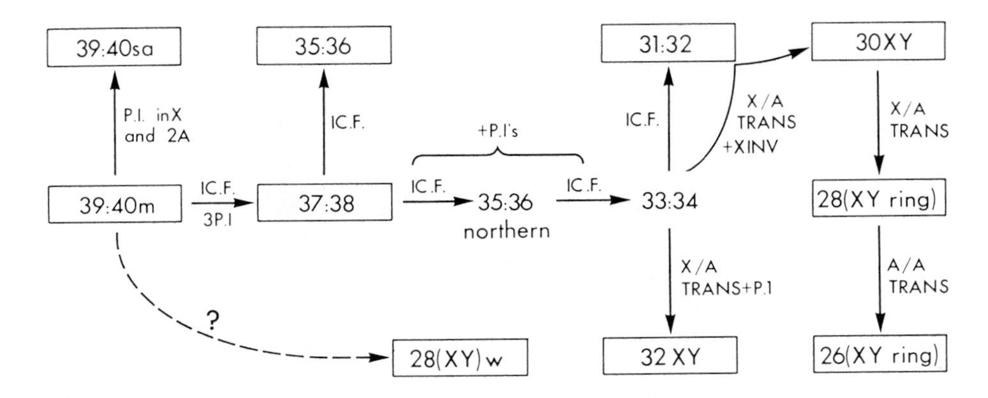

FIGURE 22. Presumed evolutionary relationships between the chromosome races of the phasmatid *Didymuria violescens*. The two karyomorphs not boxed represent extinct intermediate forms. P.I. = pericentric inversion, C.F. = centric fusion. (After Craddock, E. M., *Chromosoma*, 53, 1, 1975. With permission.)

C. Fission Systems

In attempting to interpret the structural rearrangements involved in karyotype evolution it has, in the past, been standard practice to accept Muller's[98] formulations based on his assumption that centromeres and telomeres are immutable entities. These formulations imply that:

1. All viable rearrangements, both within and between chromosomes, require at least two breaks with the subsequent reforming of the broken ends.
2. Rearranged chromosomes must have two telomeres but only one centromere to be mechanically stable.
3. While breaks can occur at any point along a chromosome they occur with a higher frequency in inert chromocentral regions than elsewhere.

While these formulations certainly apply to some forms of rearrangement it is equally clear that they do not apply to others. Specifically there is now unambiguous evidence that:

1. Dicentrics can be stable when the two centromeres are very close together, as they are expected to be in some kinds of fusions[99,100] or else where one of the two centromeres is inactivated following the rearrangement that produced the dicentric.[101]
2. Fission can result from a one-break rearrangement in cases where the centromere of a metacentric is able to produce two functional and stable telocentrics. This carries the added implication that centromeres and telomeres can be coincident structures in stable telocentrics.[102]
3. The belief that heterochromatic chromosome regions are necessarily preferred in the production of rearrangements appears to be an oversimplification. Thus, interchange breaks occur within heterochromatic regions in *Blaberus discordalis*[25] but in euchromatic regions in *Periplaneta americana*.[22] Both systems, however, give stable interchange systems.

The morabine grasshopper *Warramaba scurra* exists as two chromosome races in southeast Australia. One of these carries an AB metacentric and has a $2n = 15$, while the other has two separate A and B rod chromosomes and a $2n = 17$ (Figure 13).

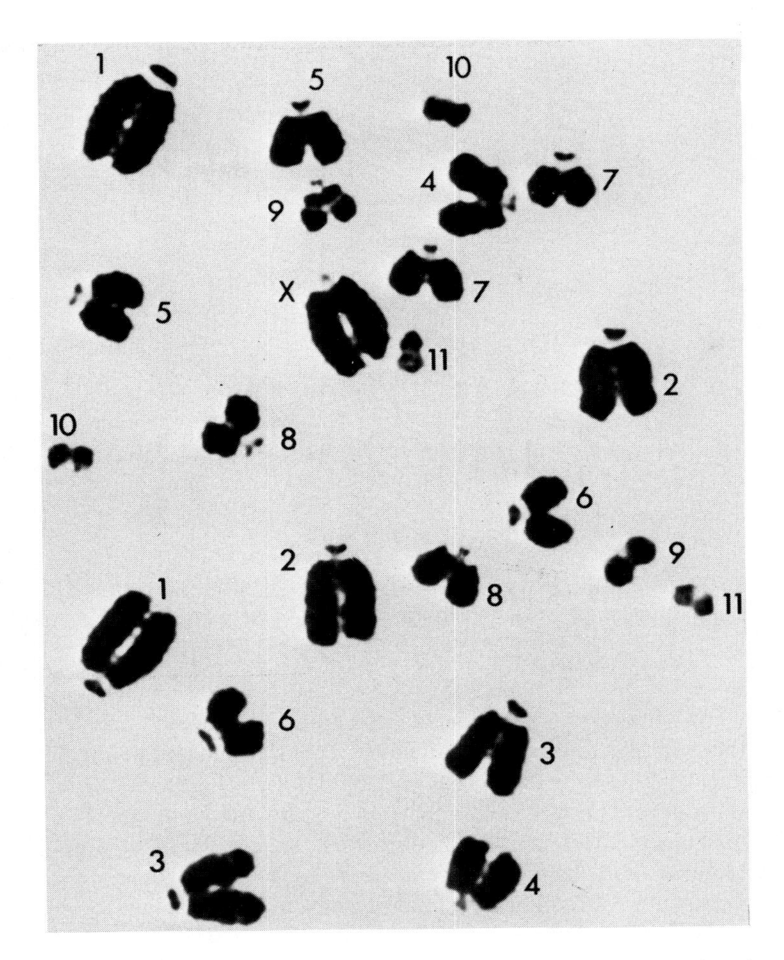

FIGURE 23. A colchicine-treated gut ceca cell of the grasshopper *Gastrimargus musicus* to illustrate variation in chromosome structure within a single complement ($2n$ = 23♂,XO). Note chromosomes 1, 2, 3, and 6 have substantial short arms. In 4, 5, and 7 the arms are smaller. They are even smaller in 6, 9, and the X and are absent in 10 and 11.

White[103] claims that the $2n = 17$ race is derived from the $2n = 15$ by dissociation in which a Blundell CD chromosome is assumed to have donated a centromere for one of the two arms of the AB metacentric. To do so would, of course, imply the loss of a CD chromosome which, unless it were initially present in a trisomic state, would lead to monosomy. Germ-line polysomy is known to occur in some grasshoppers[39-41] but there is no recorded instance of this in morabines. White[103] discounts the possibility of simple fission of the AB metacentric because, in both *Drosophila* and in man, stable telocentrics resulting from simple breaks through the centromere have never been observed. The absence of such an event in *Drosophila* led Muller to formulate his rules for karyotype change. But the fact that stable telocentrics do not exist in these two organisms is no justification for excluding their possible existence elsewhere in the entire animal and plant kingdoms. Thus, fission mutants affecting either the entire germ line or else a substantial part of it have been described in *Myrmeleotettix maculatus*[104,105] and *Chortoicetes terminifera*.[105] Added to this the microspreading technique for visualizing the synaptonemal complex (SC) also defines the position of the centromere, which appears as a prominent dense round body associated with the lateral

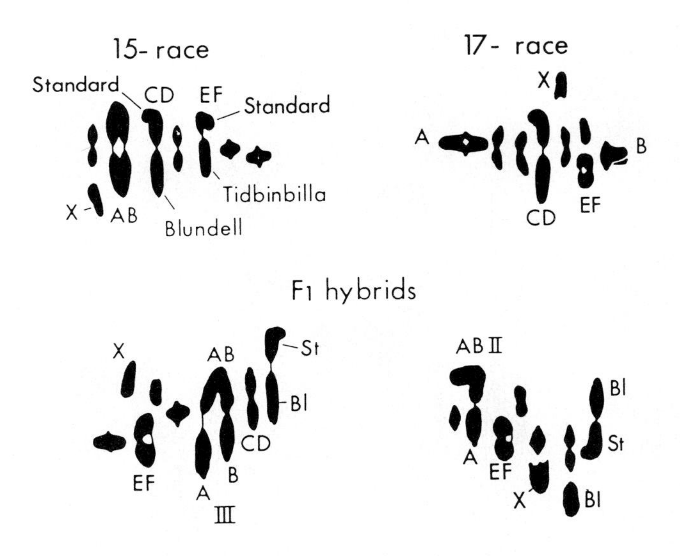

FIGURE 24. The consequences of hybridization between the 15 (AB) and the 17 (A + B) chromosome races of *Keyacris* (formerly *Moraba*) *scurra*. In the F_1 hybrid either an A/A B/B trivalent is present or else an A B/A bivalent and a B univalent.[13,103]

axis of the SC. In *Melanoplus differentialis* Solari and Counce[106] find that the centromeres of the three smallest autosome pairs are located at one end of the SC and appear strictly terminal, overlapping the actual end of the lateral element. In all eight other autosome pairs a short section of the SC lies beyond the centromere. This not only confirms the existence of telocentric chromosomes as regular, stable members of the chromosome complement, but contradicts the opinion that chromosomes with different types of centromere organization cannot coexist in the same complement[107] for which there has long been clear evidence even from conventional preparations (Figure 23). For this reason, the noncommital term "rod chromosome" has been used to describe those situations where it has not yet been possible to accurately determine whether the centromere is located terminally (telocentric) or subterminally (acrocentric). Perhaps the most significant fact of all in relation to the actual situation in *W. scurra* comes from the report of an indisputable mosaic mutant for an AB fission in this species.[2]

Whereas dissociation remains a hypothetical explanation which requires a donor chromosome that has not been adequately accounted for, fission is an event that has been directly observed. Since the direction of evolutionary change has usually to be inferred, it is often a matter of speculation whether a fission or a fusion has been involved. For this reason these two classes of rearrangement are sometimes referred to under the common and noncommital heading of Robertsonian rearrangements.

In *W. scurra* the 17 and 15 races meet along a front of at least 250 km. Unfortunately the effect of grazing has decimated the natural vegetation of this area so that there is only limited interaction between the two taxa. Even so White and Chinnick[108] did find $2n = 17$ and $2n = 15$ populations within 820 m of one another with no sharp discontinuity between the two forms at the zone of contact. Only one natural hybrid with $2n = 16$ was collected in the contact zone and in a population of otherwise $2n = 15$ forms. They proposed that the narrowness of the contact zone was determined by selection against heterozygotes. However, the evidence obtained both from the one natural hybrid and from several laboratory hybrids that have since been produced (Figure 24) clearly indicates that the heterozygotes suffer little infertility. Indeed the

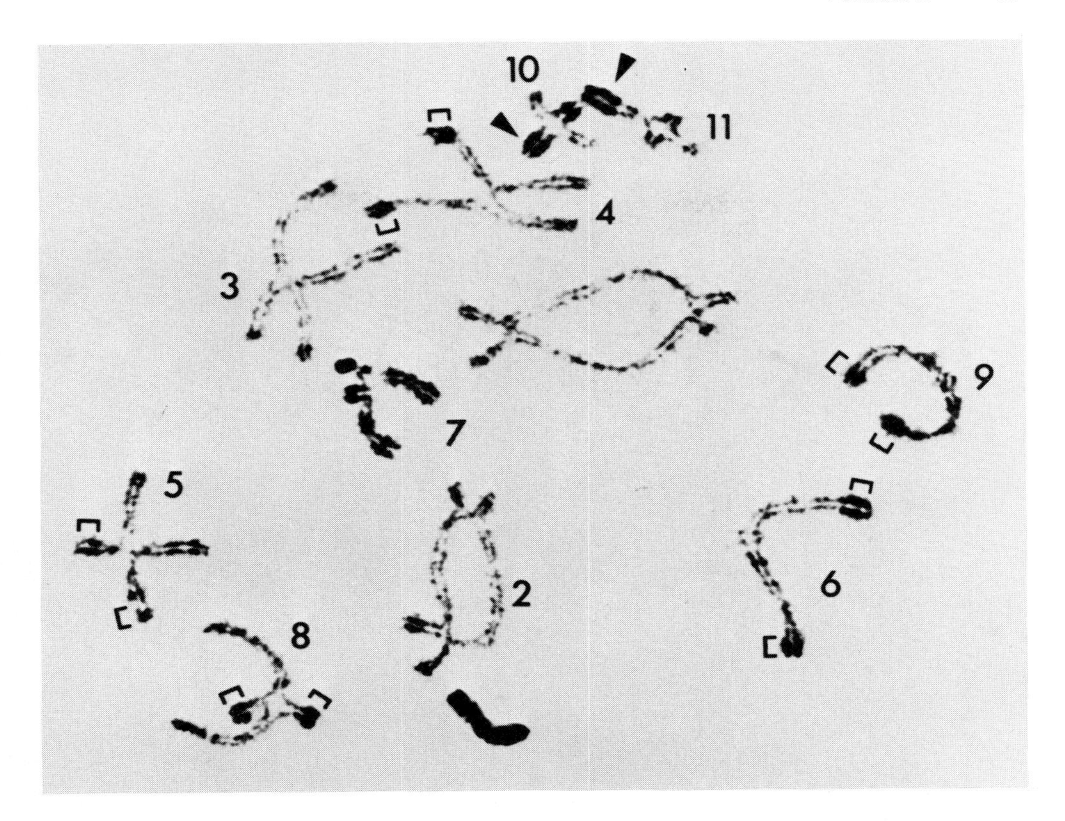

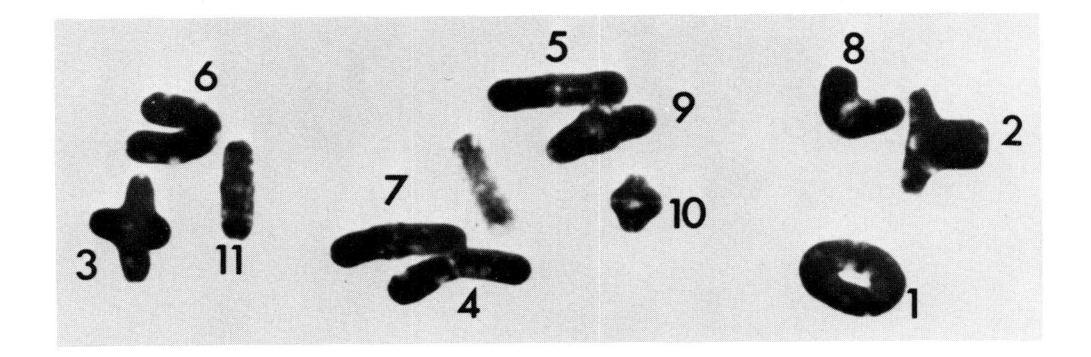

FIGURE 25. (a) Diplotene and (b) first metaphase of male meiosis in the northern (N) race of the grasshopper *Cryptobothrus chrysophorus* ($2n = 23♂$,XO). In (a) note the presence of distal heterochromatic blocks (⌊__⌋) in bivalents 4, 5, 6, 8, and 9. Note also the heterozygous supernumerary segment in bivalent 11 and homozygosity for blocks of two different sizes in bivalent 10 (▶). Additionally bivalent 7 is precociously condensed (megameric). At metaphase I (b) those bivalents with terminal heterochromatic blocks have proximal chiasmata. Bivalent 11 is homozygous for a proximal supernumerary segment and so looks larger than bivalent 10.

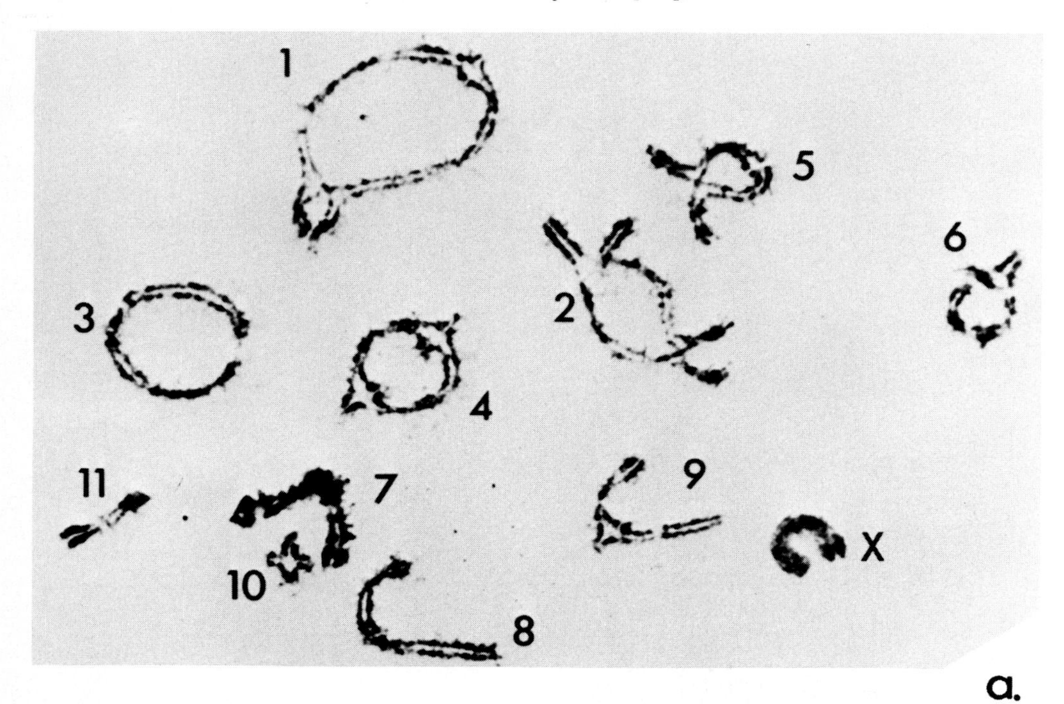

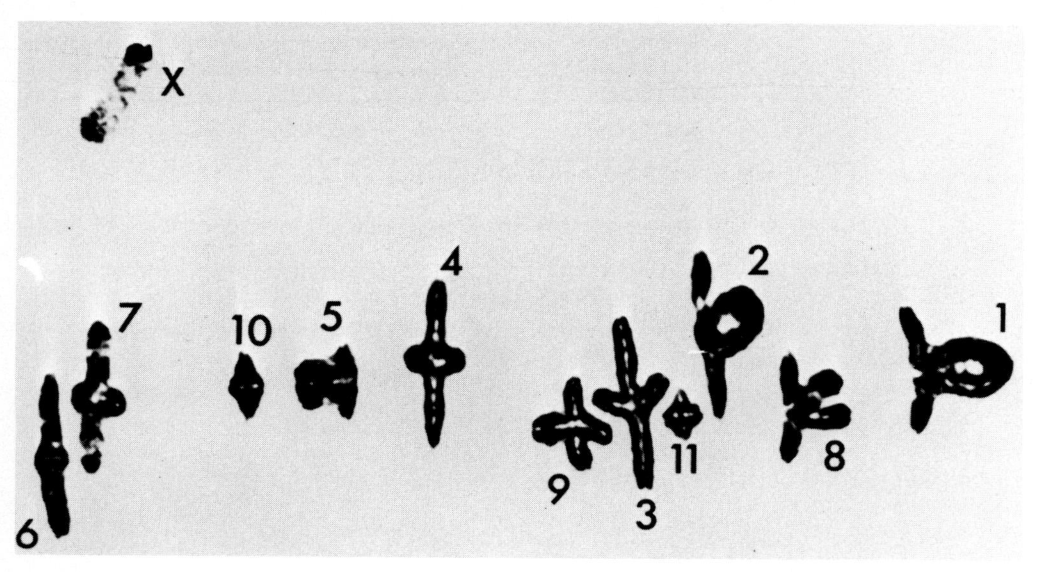

FIGURE 26. (a) Diplotene and (b) first metaphase of male meiosis in the southern (S) race of *Cryptobothrus chrysophorus* ($2n$ = 23♂,XO). In (a) there are no distal heterochromatic blocks on bivalents 4, 5, 6, 8, and 9 (compare Figure 25 [a]) although bivalent 7 is again megameric. Neither are there any supernumerary segments on bivalents 10 and 11. At metaphase I (b) chiasmata are not localized (compare Figure 25 [b]) and bivalents 10 and 11 again lack supernumerary segments.

trivalent produced in them has 98 to 99% regular orientation. Thus, in terms of the available evidence, the two chromosome forms are no more than geographical races in a polytypic pattern of relationship.

There are quite striking differences between both the frequency and the position of chiasmata in the two races. Thus, the chiasma frequency in the separate A and B bivalents of the 17-chromosome race is some 50% greater than in the fused AB metacentric of the 15-chromosome race. The AB bivalent contains relatively few proximal chiasmata but the separate A and B bivalents each show quite a high frequency of such chiasmata.

D. Heterochromatin Variation

Polytypism for fixed differences in heterochromatin content have not often been described in orthopteroids, but this may well reflect a lack of adequate population sampling over the whole distributional range of a species. There are certainly indications that fixed differences for the presence/absence of heterochromatic short arms occur in different karyotypes within the *W. viatica* complex,[109] and this may well turn out to be an example of polytypic variation. The one really convincing case is to be found in *Cryptobothrus chrysophorus*,[110,111] a species endemic to southeast Australia living in the leaf litter associated with open stands of eucalypt trees. Populations of this species fall into two quite distinct geographical clusters. One of these, the northern (N) race, is confined to northern New South Wales and southern Queensland. The other, the southern (S) race, occupies the remainder of the distribution in New South Wales, Victoria, and Tasmania. The karyotype of the N race is characterized by distal blocks of heterochromatin on chromosomes 4, 5, 6, 8, and 9 (Figure 25) which are absent in the S race (Figure 26). The two races interbreed successfully in the laboratory.[112] Superimposed on this polytypic pattern of variation is an extensive polymorphism for the presence and character of additional blocks of heterochromatin on the two smallest members of the complement (10 and 11). These polymorphisms are present in both the N and S races but they are more common and more extensive in the N.

The presence of the fixed terminal heterochromatic blocks in the N race, like the distal polymorphic blocks of *Atractomorpha similis* referred to earlier, leads to pronounced modifications in the chiasma distribution pattern. In the presence of the blocks chiasmata are restrained from occurring in the euchromatic regions adjacent to the blocks and in consequence occupy more proximal positions (Figures 25 to 27).

Synthetic F_1 hybrids have been obtained from selected N/S interpopulation crosses in the laboratory.[112] Meiosis in these F_1 males is characterized by the presence of univalency in from 22 to 32% of the meiocytes. This is genotypic rather than structural in causation and may involve from one to three chromosome pairs in any one cell. As a result, from 17 to 36% of the sperm produced are diploid or tetraploid. Despite this F_1 males produce F_2s when allowed to intersib mate, although many of the embryos produced fail to develop either as a result of genotypic imbalance following recombination at the F_1 meiosis or else because they are aneuploid in constitution. In the F_2s that do survive, meiotic pairing is much improved and F_3s were successfully obtained.

IV. INTERSPECIES DIFFERENCES

At the cytological level the most sensitive indication of similarity or difference is that provided by meiotic pairing.[113] An analysis of meiotic pairing in hybrids allows for a quantification of hybrid fertility in terms of segregation behavior and offers some scope for deciding the relative contributions which genic vs. structural differences make to pairing and segregation. Meiosis depends for its proper functioning on the genetic similarity between the sets of homologous chromosomes present in a given

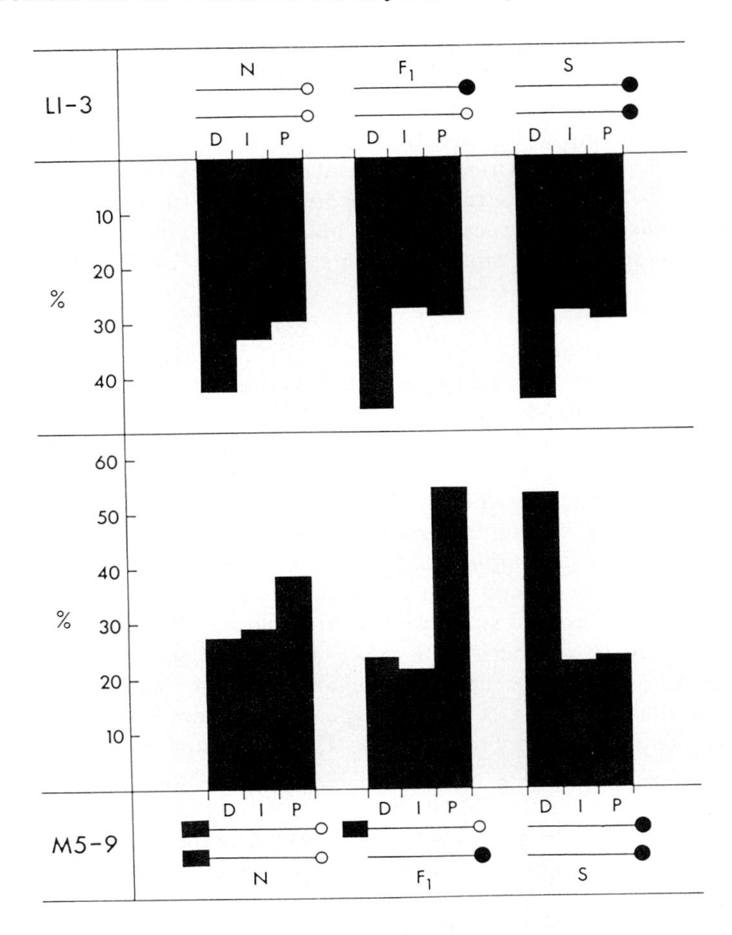

FIGURE 27. The influence of terminal heterochromatic blocks on chiasma distribution in male diplotene meiocytes of the southern (S) and northern (N) races of *Cryptobothrus chrysophorus* and in their F_1 hybrids. In each of the two races ten cells were scored from each of 20 individuals. Sixteen F_1 hybrids were scored giving a total of 160 cells. For the purpose of defining chiasma location the euchromatic portion of each bivalent was divided into three segments of equal length, proximal (P), interstitial (I), and distal (D), and the numbers of chiasmata in each of these three categories pooled over all chiasma classes (from one to three per bivalent).

complement. If the chromosomes do not pair or else do not segregate regularly for other reasons, then either aneuploid or polyploid gametes will result. In orthopteroids, polyploid gametes do not even fertilize. Aneuploid gametes lead to aneuploid zygotes and, subsequently, to embryonic death. This, in turn, leads to infertility or, in extreme cases, to sterility. Evidently this is not the case in polymorphic or polytypic states where the chromosome differences which are present must be able to pass through meiosis without producing severe aneuploid problems. Let us then examine what happens when species are crossed either in nature or in the laboratory.

A. The Evidence from Interspecies Hybrids

Where no premeiotic barriers to mating exist, and sometimes even when they do, hybridization between members of normally distinct breeding groups offers a means of discovering what happened at the cytological level during the divergence of the taxa in question.

Table 47
THE CONSEQUENCES OF CROSSING DIAPAUSING (d) AND NONDIAPAUSING (nd) RACES OF *T. COMMODUS* WITH EACH OTHER AND WITH *T. OCEANICUS*[114]

Cross	Fecundity	Egg viability (%)	Character of F_1 hybrid
T. commodus × *T. commodus*			
(d)♂ × (nd)♀	High	75	Usually sterile
(d)♀ × (nd)♂	High	70	Sterile
T. commodus × *T. oceanicus*			
(d)♂ × *T. oceanicus* ♀	Low	65	Sterile
(d)♀ × *T. oceanicus* ♂	Low	85	Sterile
(nd)♂ × *T. oceanicus* ♀	High	80	Fertile
(nd)♀ × *T. oceanicus* ♂	High	50	Fertile

1. Inversion Differences

Teleogryllis commodus, the common field cricket of Australia, has a wide geographic distribution, occurring in very different climates. Two races are now distinguished on the basis of morphometrics, physiology, mating behavior, and acoustic behavior. One of these, (nd), includes populations from localities within the tropical region of Australia and does not have an embryonic diapause. The other, (d), includes populations from the temperate zone of Australia where embryos have the capacity to enter diapause. Both races have $2n = 26 + XO$♂, $26 + XX$♀, as, too, does the related species *T. oceanicus* which is widely distributed over a large number of Pacific islands and is nondiapausing. In all three taxa the 13 pairs of autosomes are conveniently grouped into three large (L1 to 3), five medium (M4 to 8), and five small (S9 to 13) types. The two races of *T. commodus* differ significantly in mean cell chiasma frequency which in the (d) race has a value of 14.4 ± 0.06 but is lower, 13.55 ± 0.55, in the (nd) race.[114] Since diapause is a physiological device that has evolved as an adaptation enabling survival in southern latitudes of Australia, it follows that the (d) race must have been derived from the (nd).

As Fontana and Hogan[114] point out, the karyotype of the northern (nd) race differs in many respects from that of the southern (d) race so that the two are readily distinguishable. These differences involve mainly the centromere position and the size of most of the chromosomes. In the (nd) race only L1, the largest autosome, and the X are regularly metacentric, though the L1 and one of the medium autosomes, probably M7 or M8, may carry a pericentric inversion in a polymorphic state. These two inversions may or may not occur together in the same individual. In the (d) race, five of the autosomes (L2, M6, M7, and S9) are meta- or submetacentric while four others (L3, M5, M8, and S10) are subacrocentric which implies that the two races are differentiated in respect of inversion differences; only M4, S11, S12, and S13 are acrocentric. Additionally pericentric inversion polymorphisms appear to be common in the medium bivalents and there are also polymorphisms for unequal bivalents. *T. oceanicus*, which is also nondiapausing, has a karyotype which is very close to that of the (nd) race of *T. commodus*, differing from it in only two respects: the L3 is a metacentric in *T. commodus*, while the M5 is a subacrocentric. *T. oceanicus* is also polymorphic for a pericentric rearrangement in one of the medium-sized bivalents.

Fontana and Hogan[114] studied a series of hybrids involving the three forms (Table 47, Figure 28). Only one of these crosses, involving the (nd) race of *T. commodus* and *T. oceanicus*, gave fertile F_1 hybrids. Very few eggs were obtained from the *T. com-*

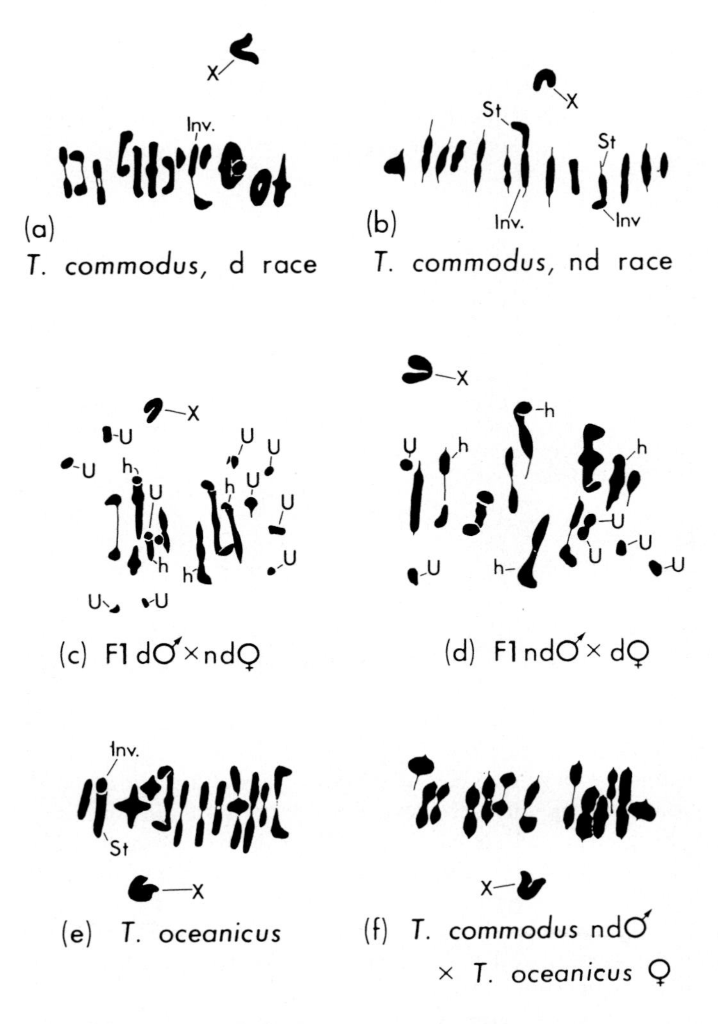

FIGURE 28. Metaphase-I meiocytes from parental (a, b, and e) and F_1 hybrid males (c, d, and f) of the cricket *Teleogryllus*. In parentals, bivalents heterozygous for a pericentric inversion are represented as Inv. and their standard homologues as St. In hybrid individuals, (h) denotes a heterozygous bivalent and (U) a univalent. (After Fontana, P. G. and Hogan, T. W., *Aust. J. Zool.*, 17, 13, 1969. With permission.)

modus (d) race and *T. oceanicus* cross and their viability was low. The small number of F_1 hybrids produced were sterile. The F_1 hybrids from the (d) × (nd) races of *T. commodus* were also sterile. This sterility stemmed from two main causes:

1. Irregular synapsis involved both incomplete pairing and occasional illegitimate pairing between nonhomologs. Thus, some 85% of the first metaphase cells in (d)♂ × (nd)♀ crosses had from 2 to 14 univalents. Cells with 2, 4, 6, or 8 univalents were most common; less commonly 10 to 12 univalents were present and on one occasion 14 were seen.
2. Not only was there, as expected, an extensive production of unbalanced spermatids, but, additionally, there was also an unexplained and massive degeneration of cells at a more or less advanced stage of spermiogenesis.

It would appear, therefore, that the (d) and (nd) geographic races of *T. commodus*

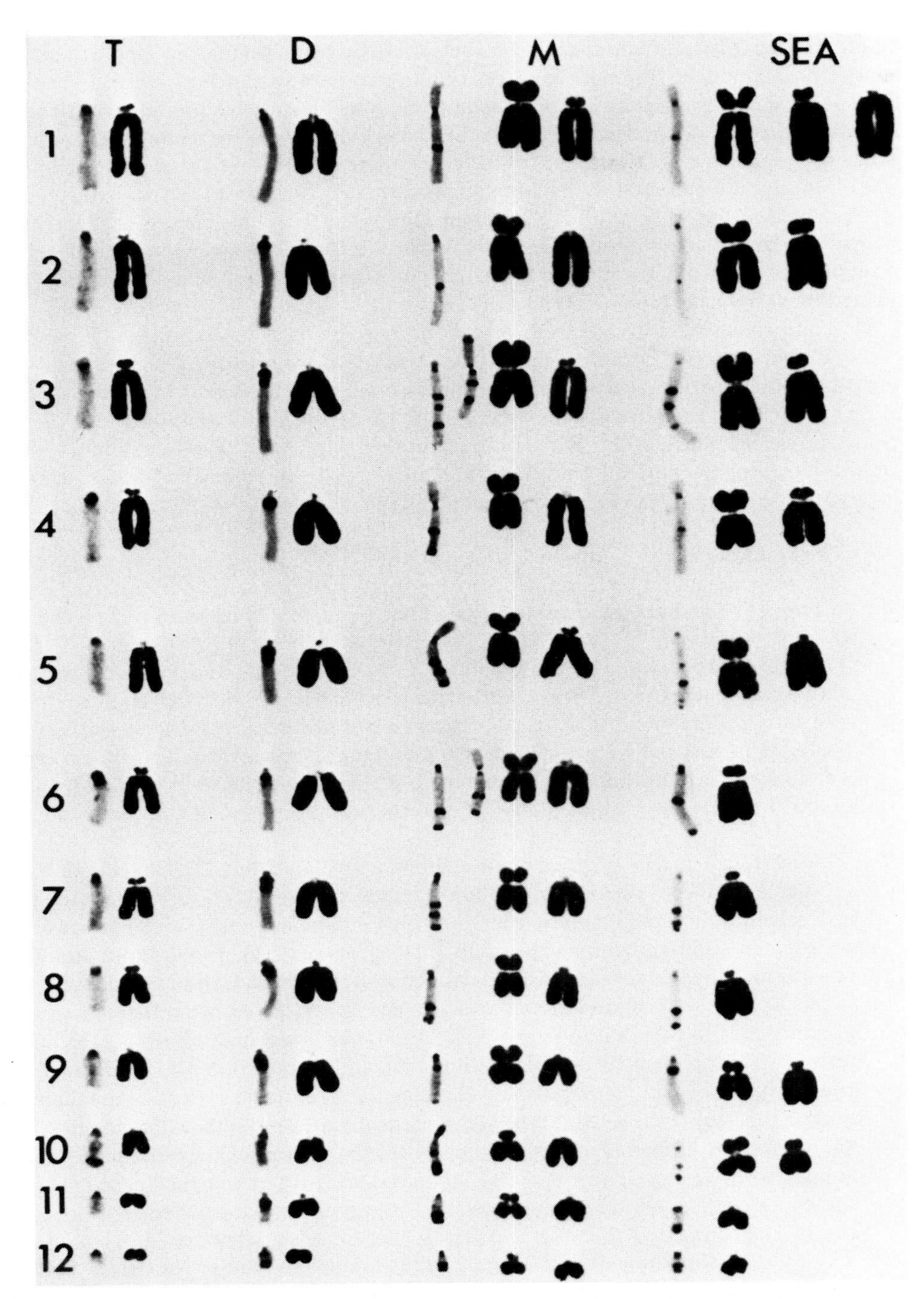

FIGURE 29. Karyotype differences between the four chromosome races of the grasshopper *Caledia captiva* ($2n$ = 23♂,XO). The chromosomes are numbered in decreasing size order from 1 to 12 with the X at position 3. In each case the C-band pattern is shown to the left and the equivalent conventionally stained chromosome to the right. *Note:* T = Torresian, D = Daintree, M = Moreton, and SEA = Southeast Australian. (Photograph kindly supplied by Dr. David Shaw, Australian National University, Australia.)

should be regarded as full species characterized by both premating (ecological, seasonal, and acoustic differences) and postmating (failure of pairing, aneuploid sperm, and sperm degeneration) isolating mechanisms. On the other hand, the (nd) race of *T. commodus* and *T. oceanicus* from Tahiti have evidently no reproductive incompatibility or premating isolating mechanisms and these are best regarded as geographical races. Finally, the meiotic breakdown seen at male meiosis in both the (d) × (nd) crosses of *T. commodus* and in the (d) *T. commodus* × *T. oceanicus* crosses evidently owe their origin largely to genotypic and not structural differences, since there was a general reduction of chiasma frequency in all chromosomes so that where bivalents were formed these invariably had only a single chiasma.

A second case where inversion differences are involved in interspecies differences is in the *Caledia captiva* complex within Australia. It includes four geographically distinct taxa which differ in respect of centromere location[115] as well as in the distribution and amount of constitutive heterochromatin as defined by C-banding.[116] All four forms, however, share a common diploid number with $2n$ = 23,XO♂ and 24,XX♀. All the chromosomes in all four taxa are individually distinguishable and they are differentiated in the following terms (Figure 29):

1. *The Daintree taxon* — all the chromosomes are telocentric and have large blocks of heterochromatin next to the centromere. This form occupies a restricted region in tropical northern Queensland extending from the Daintree River to the McIlwraith Ranges.
2. *The Torresian taxon* — chromosomes 1, 2, X, 4, 6, 7, and 8 are acrocentrics with short euchromatic arms. This form, which surrounds the Daintree taxon and is sympatric with it at one location in the McIlwraith Ranges, but with no evidence of natural hybridization, extends from Arnhem Land in the Northern Territory as far south as the Brisbane River valley in eastern Queensland. In this taxon there are only small blocks of heterochromatin around the centromere regions.
3. *The Moreton taxon* — present in a limited area of some 225 km in southeast Queensland and to the east of the Mary and Brisbane rivers, there are interstitial and terminal heterochromatic segments on all the chromosomes. This taxon can be further subdivided into three distinct cytotypes: (1) populations along the coastal margin and offshore islands are consistently homozygous for metacentric forms of 1, 2, 4, 5, 6, and 10 and have an acrocentric X carrying three interstitial C-bands; (2) in the southwestern part of its distribution populations are characterized by a metacentric X but are otherwise indistinguishable from those of the coastal margin; and (3) populations located in the northeast part of the distribution of this taxon show an extensive polymorphism for all those autosomes which are homozygous metacentrics in the other two regions. These polymorphisms appear to have arisen through the introgression of acrocentric Torresian autosomes across the contact zone between these two taxa and where hybrid zones can still be found.[117-119]
4. *The Southeast taxon* — has a wide distribution in New South Wales and Victoria though the precise limits remain to be defined. It is characterized by an independent series of small pericentric inversions in chromosomes 4, 6, and 9 and larger inversions in 1 to 5 and 10 which appear identical to those in the Moreton taxon, at least in terms of arm ratios. It also shares with the Moreton taxon the presence of interstitial C-bands. All the inversions in the southeast Australian taxon, both large and small, occur as polymorphisms in all populations.

Crossing experiments in the laboratory indicate that the Daintree taxon is effectively

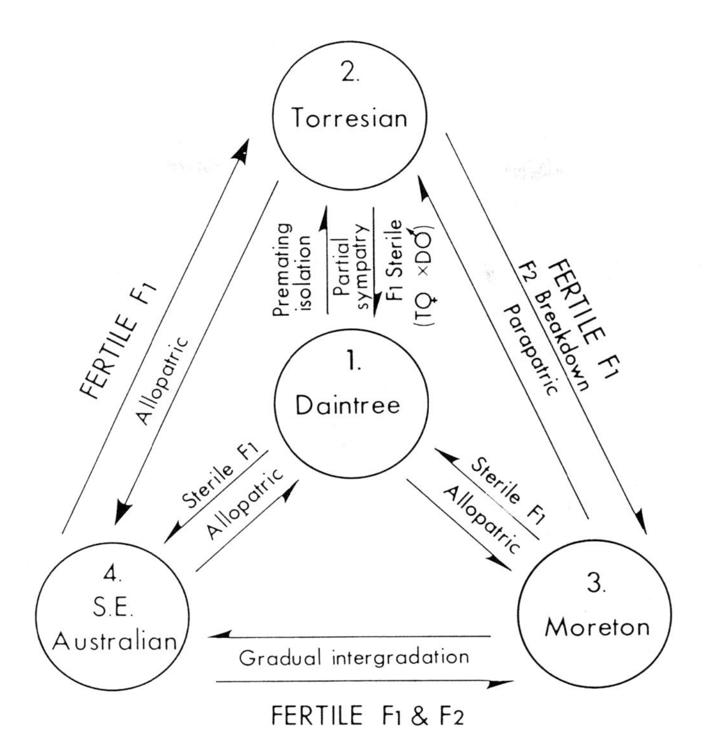

FIGURE 30. The breeding relationships of the four chromosome races of *Caledia captiva.* (After Shaw, D. D., Moran, C., and Wilkinson, P., *Insect Cytogenetics,* Symp. No. 10, Royal Entomological Society London, 1980, 171. With permission.)

isolated from all three other taxa within the complex (Figure 30). This may be either by a prezygotic mechanism as in the Torresian taxon or by postzygotic mechanism as in the Moreton and southeast Australian taxa. The Daintree taxon is, thus, best regarded as a distinct sibling species. The results of the intertaxon hybridization experiments[120-122] may be briefly summarized as follows:

1. In the Daintree × Torresian crosses mating occurs only when a Torresian female is used. Both the male and the female F_1 products of the Torresian ♀ × Daintree ♂ cross were sterile. This was true whether the individuals used in the cross came from close sites (50 km apart) or distant sites (1000 km apart). Female F_1s showed no ovariole development and males either had no testes, or if they were present they lacked meiocytes. Premeiotic spermatogonia were observed but only as polyploid cells.
2. In the Moreton ♀ × Daintree ♂ crosses gonadal development was nonexistent in the F_1s, while in the reciprocal crosses male meiosis was grossly abnormal and some 70% of the meiocytes were tetraploid. This doubling appeared to result from endoreduplication at the last premeiotic mitosis; $8x$ and $16x$ cells were also seen but far less commonly. During meiosis the $4x$ cells invariably gave rise to 22 autosomal bivalents, presumably because sister products of endoreduplication were juxtaposed at leptotene and so preferred in pairing. The two X chromosomes also sometimes formed a bivalent presumably as a consequence of the diphasic behavior in the $4x$ cells. Such X bivalents were present in about half the $4x$ meiocytes. In the remainder two X univalents were present. In the diploid

meiocytes there was no regular bivalent pairing and cells contained mixtures of univalents, bivalents, and multiple configurations of various kinds. Because the Daintree chromosomes carried prominent blocks of procentric heterochromatin, they could be identified at diplotene-diakinesis. Using this distinctive property it was possible to show that in many meiocytes two or more nonhomologous members of the Daintree complement were associated in pseudobivalent or pseudo-multiple configurations. Moreover, the mode of association was either by chiasmate-like X-type nonsister chromatid exchanges or by unconventional U-type nonsister chromatid exchange. The X chromosome was sometimes involved in both pseudo-bivalent and pseudo-multiple configurations. The exchange events between nonhomologous members of the Daintree complement were predominantly adjacent to the blocks of procentric heterochromatin. Similar exchanges were also seen between nonhomologous Moreton chromosomes but here were more interstitial in location. From this sequence of behavior it was clear that the normal events of meiosis had been completely disrupted in these hybrids.

3. In southeast Australian ♀ × Daintree ♂ hybrids diploid male meiocytes were again extremely irregular, behaving in the same manner as the Daintree ♀ × Moreton ♂ cross.
4. The Torresian and southeast Australian taxa hybridize freely and produce fully fertile F_1s. Both F_2 and backcross progeny are viable although there is in excess of 25% mortality during their embryonic development.
5. The Moreton and Torresian taxa, which differ by a series of pericentric inversions involving a minimum of 7 of the 12 pairs of chromosomes, also interbreed freely in the laboratory, but up to 35% of the first metaphase cells in male F_1 hybrids are characterized by two to six univalents per cell. This represents an approximately 20-fold increase over both parental controls and intrataxon crosses. Additionally, multiple associations were present in 4 to 10% of the meiocytes in the intertaxon hybrids. A majority of these (53/95) again involved intragenomic pairing, this time of nonhomologous Torresian chromosomes. A further nine cases involved intragenomic pairing of Moreton chromosomes and only two were definitely identified as Moreton/Torresian associations. The remaining 31 cells could not be classified. The intragenomic pairing in the Torresian genome again involved associations of the centromere regions, where the heterochromatic blocks are present. Those between the Moreton chromosomes, on the other hand, involved the terminal regions of the short arms and these regions do not C-band.

Apart from the laboratory hybrids ten presumptive field F_1 male hybrids were identified mitotically in one of the narrow zones of hybridization known to occur between the Moreton and Torresian taxa. Their meiosis was subsequently analyzed, along with a control sample of nonhybrid individuals from the same zone. The most common meiotic anomaly was the occurrence of univalents present in 25 to 37% of all cells examined. The mean frequency of univalent pairs per cell in control and F_1 hybrid males was highly variable. Half of the field-collected hybrids fell within the range of the controls. Given that all the F_1 hybrids were heterozygous for the same chromosome differences this extreme variation implies that pairing failure arises from genotypic differences between the parental forms responsible for production of the hybrids. In keeping with this conclusion there was no evident relationship in the laboratory hybrids between the incidence of pairing failure and either the size of the chromosomes involved or the size of the inversion difference involved. Indeed, as in the case of the pericentric inversion polymorphisms described in Section II.A.2, pairing between relatively inverted regions in these *Caledia* hybrids was straight and only three possible cases of reverse loop pairing were seen in over 2700 meiocytes analyzed. Thus, crossing

Table 48
PRODUCTIVITY OF F_1, F_2, AND BACKCROSS (BX) FEMALES IN HYBRIDS BETWEEN THE MORETON (M) AND TORRESIAN (T) TAXA OF *CALEDIA CAPTIVA* OVER A 28-DAY LAYING PERIOD

Cross	Parents ♀		Parents ♂	No. egg pods per ♀	No. eggs per pod	No. eggs per ♀	No. hatchlings per ♀
Controls	MAX	×	MAX	1.70	14.10	24.49	24.00
	MMX	×	MMX	1.52	15.86	24.10	23.86
	TT	×	TT	1.26	17.03	21.46	21.46
F_1	MAX	×	TT	2.80	7.89	22.09	21.65
	TT	×	MAX	0.70	12.27	8.59	8.25
	MMX	×	TT	1.50	5.80	8.70	8.70
	TT	×	MMX	0.45	13.49	6.07	5.95
F_2	(MAX × TT)	×	(MAX × TT)	2.30	11.04	25.39	0
	(MMX × TT)	×	(MMX × TT)	0.65	9.07	5.90	0
BX	(MAX × TT)	×	MAX	0.85	9.75	8.29	4.70
	MAX	×	(MAX × TT)	2.10	9.58	20.12	0
	(MAX × TT)	×	TT	0.49	7.00	3.43	2.60
	TT	×	(MAX × TT)	0.70	9.53	6.67	0
	MMX	×	(TT × MMX)	0.20	4.40	0.88	0.50
	(TT × MMX)	×	TT	0.20	12.20	2.44	1.00

Note: MAX refers to Moreton individuals carrying an acrocentric X chromosome and MMX to Moreton individuals with a metacentric X.[121]

over within structurally inverted regions can be dismissed as a cause of sterility in these hybrids as also in any grasshopper species distinguished by fixed differences of an inverted kind which behave in a like manner. As expected the univalents present in the Moreton × Torresian hybrids gave rise to macrospermatids following lagging. They also led to irregular segregation less frequently.

As far as the laboratory-reared F_1 hybrids were concerned (Table 48), when females carrying a Moreton acrocentric X were used as parents (MAX♀ × T♂) they produced more egg pods than control (MAX♀ × MAX♂) crosses but with only half the number of eggs per pod. When females carrying a Moreton metacentric X were used as parents (MMX♀ × T♂) they produced the same number of pods as their (MMX♀ × MMX♂) controls but again with fewer eggs per pod. Both F_1 crosses utilizing Torresian females produced very low numbers of egg pods per female but with only slightly lower numbers of eggs per pod. Hatchability of all F_1 eggs was greater than 95%. By contrast all eggs produced by the F_2 crosses were completely inviable and failed to hatch. Likewise those backcrosses (BX) involving an F_1 (MAX♀ × T♂) as male parent were also totally inviable. The major factor responsible for the high level of embryonic breakdown in both F_2 and BX generations appears to be the recombination and segregation of the chromosomes of the Moreton and the Torresian genomes.

A third case, where inversions have been held to play a role in differentiating taxa, is that of hybrids between the European grasshoppers *Chorthippus biguttulus* and *C. brunneus (C. bicolor)*. These two species are sympatric in distribution and mixed populations of them are common. Although not separated by any ethological barrier in nature they, nevertheless, remain distinct and hybridize extremely rarely although they share an equivalent pattern of mating behavior. Indeed interspecific copulation is difficult to obtain even under laboratory conditions. It can, however, be facilitated if males of the same species as the female used in a given cross are placed within hearing distance of, but separated from, a mixed pair.[123]

Klingstedt[124,125] studied a number of hybrid individuals between these two species.

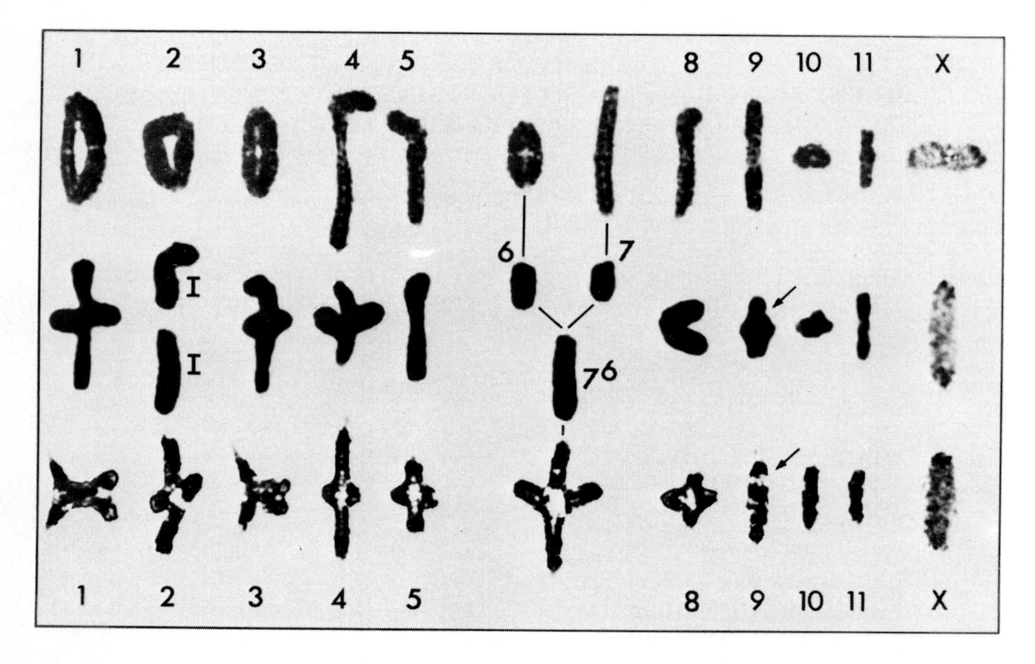

FIGURE 31. A comparison of male meiotic karyotypes in *Trimerotropis thalassica* ($2n$ = 23♂XO, upper row), *T. occidentalis* ($2n$ = 23 XO, lower row), and a natural F_1 hybrid between them ($2n$ = 22♂XO, center row). In the zone of sympatry from which the hybrid was obtained the population of *T. thalassica* included individuals homozygous for two or more of autosomes 1, 2, 3, 6, 7, and 8 and heterozygous for one or more of autosomes 3, 4, 5, 6, 7, and 8. Thus, the individual shown is homozygous for 1, 2, 3, and 6 and heterozygous for 4, 5, and 8. The population of *T. occidentalis*, on the other hand, either lacked inversions completely, as in the individual shown here, or else included homozygotes or heterozygotes for 1 or, in the latter case, double heterozygotes for 1 and 5. In the hybrid, autosome 2 is represented by two univalents (I), one of which carries an inversion. Autosomes 6 and 7, together with their tandem fusion product 7^6, are also represented as univalents. The arrows on bivalent 9 indicate the nucleolar organizing secondary constrictions. The X is metacentric in *T. thalassica* and telocentric in *T. occidentalis* and the F_1 hybrid.

One of these was wild caught, the rest lab-reared. All were identified on morphological grounds. He claimed that hybrid males exhibited a number of cytological abnormalities including difficulties in anaphase separation at the first meiotic division. These led either to the production of restitution nuclei, and, hence, to polyploid sperm or else to breakage and loss of chromosome material. Additionally, dicentric bridges and acentric fragments seen at anaphase I were assumed to result from crossing over from inversions, since Darlington[126] had earlier concluded that inversions were present in *C. brunneus.* Lewis and John,[127] however, showed that their observations were explicable in terms of either spontaneous chromosome breakage or else of the formation of side arm bridges, both of which are particularly common in these grasshoppers under normal conditions.

Klingstedt pointed out[124] that a large number of seemingly normal sperm were present and that, while some degree of sterility might be expected, "this must be mainly genic in origin, though the structural differences possibly present will also cause some loss of functioning gametes."[125] White,[8] however, states that "although Klingstedt's experiment was not carried beyond the first generation, it seems likely that the hybrids would have been entirely sterile in spite of the complete pairing of the chromosomes."

Perdeck[123] carried out a cytological examination of a number of laboratory-produced hybrid males and found the same abnormalities in F_1s, backcrosses, and parentals with but little change in frequency. No signs of hybrid sterility or breakdown were obtained and Perdeck concluded that the contribution of cytological differences to the isolation of these species is negligible. Rather, the most important factor responsible for reproductive isolation between them is the specificity of song which initiates mating interaction between male and females. Thus, ethological isolation, in this case, is due to specific song stimuli and hybrids are easily detected by virtue of the fact that their song is intermediate between that of the parental species.

2. Fusion Differences

While a majority of trimerotropine grasshoppers have $2n = 23$ chromosomes in the male[128] a few have $2n = 21$. Both types may, however, be characterized by polymorphisms for pericentric inversions. This is true, for example, of *Trimerotropis thalassica* ($2n = 23$) and *T. occidentalis* ($2n = 21$) which are sometimes found in sympatry. In one such situation at Jaspar Ridge, Calif., a natural male hybrid was discovered combining the morphological features of the two forms. The testis was relatively small and contained only a few dividing cells. In these cells from one to five univalents were present. When only one univalent occurred it was invariably accompanied by a markedly heteromorphic bivalent, but in a majority of cells all three of these chromosomes were univalant. Although no trivalent ever formed it was clear from the presence of two distinguishable kinds of heteromorphic bivalents, as well as from a comparison of parental karyotypes (Figure 31), that the two species were distinguishable by a single tandem fusion.[129] The presence of univalents in the hybrid meiocytes leads to infertility in two rather different ways: first, by the production of macrospermatids following the lagging of univalents and consequent failure of cytokinesis; second, by the production of the aneuploid spermatids as a result of the irregular segregation of univalents.

A large number of F_1 hybrids between *T. verruculatus*, $2n = 21$ (= *Circotettix verruculatus*) from Mount Desert Island, Maine, and *T. suffusa*, $2n = 23$, from Woodland Park, west of Colorado Springs, Colo., were produced experimentally.[130] Attempts to mass mate these F_1 individuals, however, proved unsuccessful. Both parental species again show polymorphisms for pericentric inversions but no record exists of the two ever overlapping naturally. In every hybrid at least two pairs of chromosomes failed to synapse. All chromosomes were affected in this way though some formed univalents more commonly than others. The number of spermatocytes per individual with normal pairing ranged from 9.8 to 94%. There were two other conditions which would be expected to have led to the production of abnormal sperm and, hence, to infertility:

1. Irregular segregation of the trivalent which forms in the $2n = 22$ hybrid and which involves a metacentric of *T. verruculatus* with two rod chromosomes of *T. suffusa*. Unfortunately Helwig[130] does not provide frequency data for the occurrence of such irregular segregation. The presence of the trivalent makes it clear that the two parental forms are distinguished structurally in respect of a single centric fusion.
2. The occasional presence of multiple chromosome associations of IV or III involving clearly nonhomologous chromosomes and which may also show irregular segregation.

Through the courtesy of David Lightfoot the author examined three natural hybrids, identified again by their intermediate morphology, between the two sympatrically occurring species *T. cyaneipennis*, $2n = 21$, and *T. suffusa*, $2n = 23$, and collected in Oregon. As in the case of Helwig's hybrids there was a fusion trivalent present (Figures 32 and 33) together with a number of bivalents heterozygous for pericentric inversions which characterize both parental species. In the three hybrid individuals the actual frequency of convergent orientation of the trivalent varied considerably with

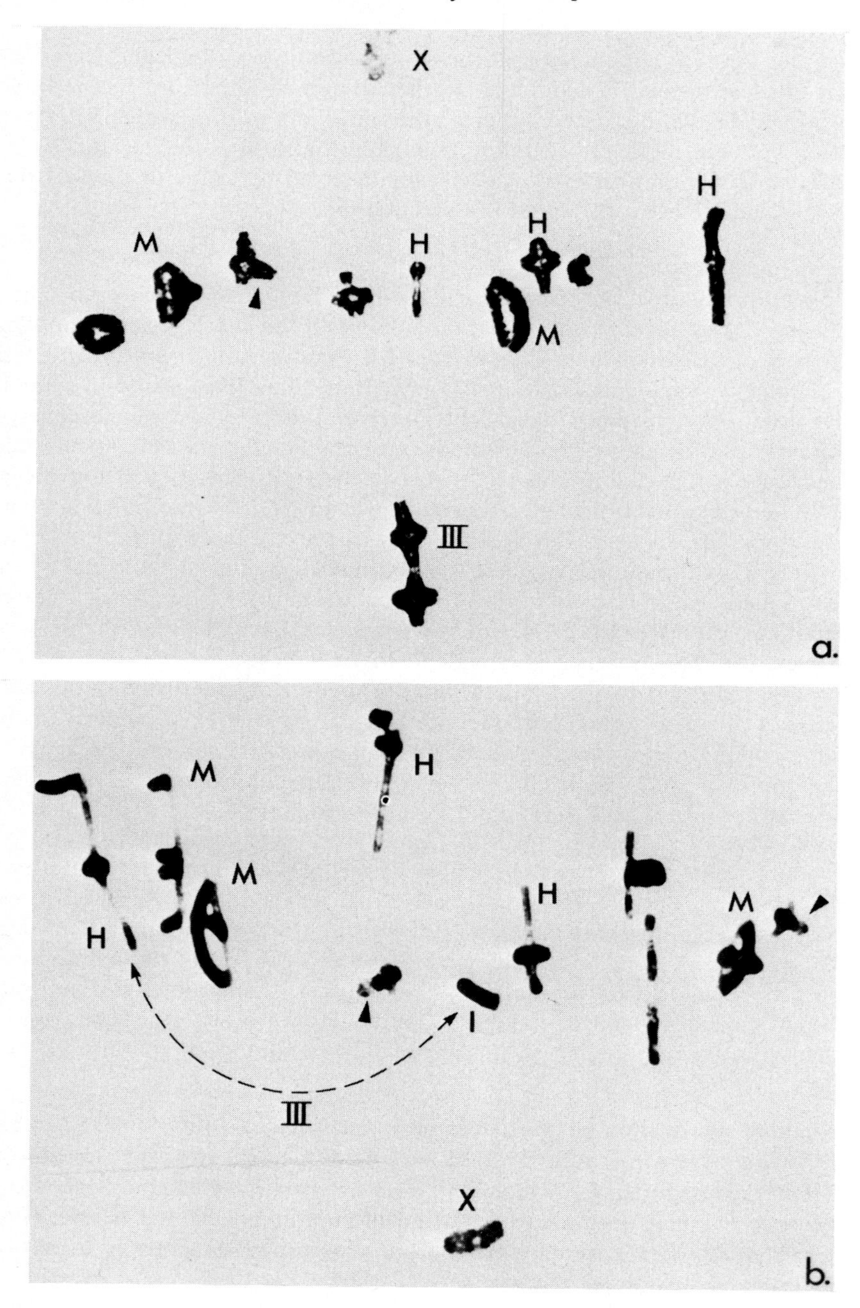

FIGURE 32. First metaphase of male meiosis in two separate natural hybrids from a sympatric population of *Trimerotropis suffusa* ($2n$ = 23♂,XO) and *T. cyaneipennis* ($2n$ = 21♂,XO). M represents a metacentric bivalent, H a bivalent heterozygous for a pericentric inversion, and I a univalent. In (a) there is a trivalent consisting of one metacentric and two telocentrics which, like the X univalent, lies off the spindle and is oriented nondisjunctionally. Additionally, one of the bivalents (▶) carries a nonsister U-type exchange in place of a chiasma. In (b) no trivalent is present but there is a heterozygous bivalent involving the fusion metacentric and one of its telocentric equivalents with the other telocentric forming a univalent. The two small bivalents are heterozygous for supernumerary segments (▶).

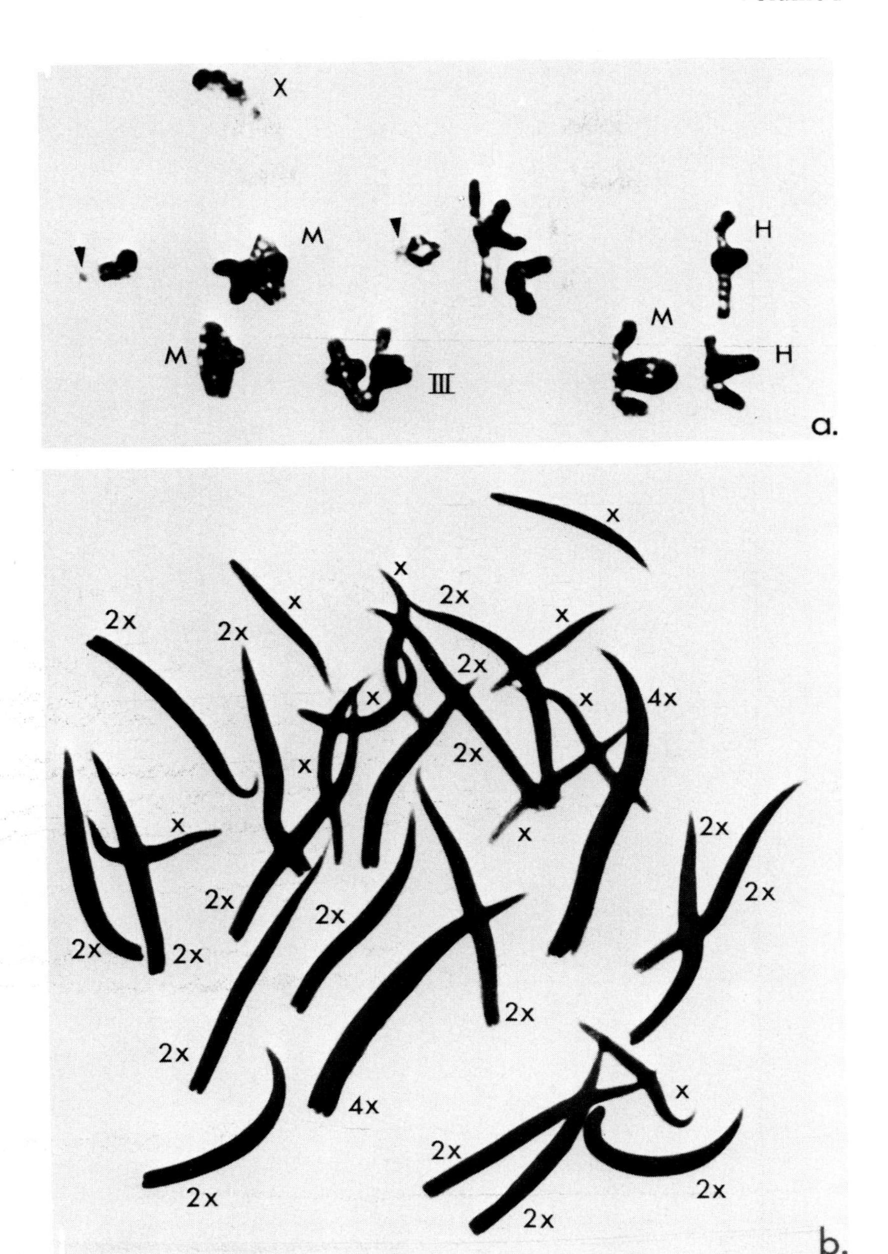

FIGURE 33. (a) First metaphase of male meiosis in the same natural hybrid between *T. suffusa* and *T. cyaneipennis* as shown in Figure 32 (b) but with the trivalent in disjunctional orientation. (b) A sample of spermatids with 9 haploid (x), 16 diploid ($2x$), and 2 tetraploid ($4x$) nuclei.

values of 80, 45, and 28%, respectively (Table 49). This in itself, however, gives no realistic indication of the effects on fertility, since where univalents were present, in addition to convergent multiples, the gametes are expected to be abnormal despite the regular segregation of the multiple. Moreover, assuming that female hybrids behave in a comparable manner, then one expects the actual frequency of balanced zygotes to be the product of the frequencies of unbalanced gametes in the two sexes. Such

Table 49
MEIOTIC BEHAVIOR OF THREE NATURAL MALE HYBRIDS TAKEN FROM A SYMPATRIC POPULATION OF *TRIMEROTROPIS CYANEIPENNIS* ($2n$ = 21♂) AND *T. SUFFUSA* ($2n$ = 23♂) IN OREGON

	Pairing behavior										Orientation behavior of trivalent		
Hybrid male no.	III + 9.II	(II + I) + 9.II	(3.I) + 9.II	III + 2.I + 8.II	(II + I) + 2.I + 8.II	III + 4.I + 7.II	(II + I) + 4.I + 7.II	(3.I) + 2.I + 8.II	(II + I) + 6.I + 6.II	Total cells	Con-vergent	Linear	Total cells
79/36	29	4	1/7	36	5	13	2	—	1	90	66	12	78
79/37	37	12	—	1	—	—	—	—	—	50	19	19	38
80/32	19	25	2	16	29	2	1	6	—	100	19	18	37

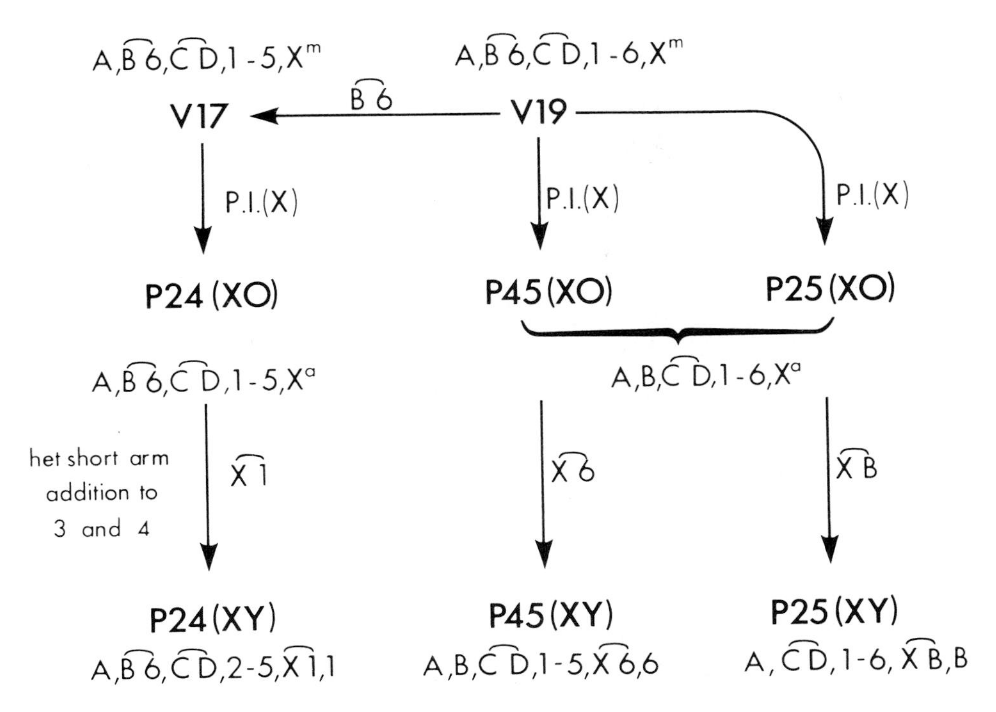

FIGURE 34. The presumed relationship of *Vandiemenella* karyotypes differentiated in respect of fusions.

calculations do not take into account any recombinational inviability that may arise from the mixing of the two parental genomes following meiosis in the F_1 hybrids.

Fusion differences are also implicated in the coastal complex of morabines belonging to the genus *Vandiemenella*, formerly *Morraba*, in Australia. It includes 11 taxa of uncertain affinity which are parapatrically distributed through the coastal region of Southeast Australia, namely, V19, V17, P24(XO), P24(XY), the P24(XY) translocation race, P45b(XO), P25(XO), P45b(XY), P25(XY), P45c, and P50. On the assumption that the karyotype of V19 is ancestral for the group, White believes these forms to be related as summarized in Figure 34.

Hybrids between some 11 different combinations of these taxa have been examined (Table 50, Figure 35) though usually only in small numbers.[131-133] Some of them, and especially those where primary hybrid zones are known to exist in nature, show relatively small impairments of viability and fertility. The most marked effects are in laboratory synthesized hybrids (compare Craddock[97]). Even here, however, one finds that forms distinguished by the same, or similar, structural differences often behave quite differently. For example, in the three hybrids between P24XY and P25XY the frequency of univalency in the B/B6/6 fusion combination ranges from 48 to 99%. This suggests that genetic differences between the individuals may have contributed significantly to the irregular behavior of the number 6 chromosome which determines this univalency.

3. *Interchange Differences*

The genus *Eyprepocnemis* has a wide distribution in Africa, the Middle East, and Spain. Within Africa a complex taxonomic situation occurs involving numerous so-called subspecies and species. John and Lewis[134] analyzed a number of laboratory-reared hybrids between *Eyprepocnemis plorans meridionalis* from Tanzania and *E. plorans ornatipes* from French Sudan. Here abnormalities were extensive enough to suggest that the two forms are best regarded as sibling species. Both parental types

Table 50
MEIOTIC BEHAVIOR IN FIELD AND LABORATORY-BRED MALE F_1 HYBRIDS BETWEEN MEMBERS OF THE COASTAL TAXA OF THE *VANDIEMENELLA* COMPLEX DIFFERENTIATED IN RESPECT OF FUSIONS[131,133,a]

Cross type	Total cells analyzed	MI cells with (%) Univalents	MI cells with (%) Maloriented trivalents
Field hybrids			
V17 × V19	86	10.4	1.1
	110	8.0	21.0
	76	26.3	2.1
Laboratory hybrids			
V17 × P24(XO)	100	8.0	
	100	1.0	
	100	1.0	
V17 × P24(XY)	1171	0.8	
P24(XY) × V17	1185	2.4	
V19 × P24(XY)	156	20.6	
	186	45.0	
P24(XO) × P25(XO)	200	69.0	
	200	92.0	
	200	67.5	
P25(XO) × P24(XO)	115	81.7	
	115	72.2	
P24(XO) × P25(XY)	100	29.0	
P24(XY) × P25(XY)	50	38.0	
P25(XY) × P24(XO)	50	82.0	29.0
P25(XY) × P24(XY)	100	99.0	
	80	65.0	
	50	48.0	

Note: In the laboratory crosses the maternal parent is given first.

[a] See Figure 34.

share $2n = 23(XO)♂, 24(XX)♀$ and have superficially very similar karyotypes in which all the members are rod chromosomes.

F_1 hybrids are fully viable but their meiosis is characterized by the production of unequal bivalents, the presence of univalents, and of multiple configurations (Figure 36, Table 51). The multiples range in size from III to IX and fall into two types:

1. Those in which successive members of the multiple are associated by interstitial chiasmata. These, in conjunction with the occurrence of unequal bivalents, make it clear that the parental forms must differ by fixed interchanges. All three pairs of small chromosomes, four of the five pairs of medium-sized chromosomes, and one of the two long chromosome pairs may form unequal bivalents giving a maximum of eight per cell. Their production and the variation in the number of them seen in any given cell can be explained by the fact that the members of an interchange system can exist as bivalents in a number of different ways. The argument that the unequal bivalents may be due to duplications and deletions of heterochromatic segments[110] can be countered by the simple fact that there are no major heterochromatic blocks in the chromosomes of *Eyprepocnemis* and

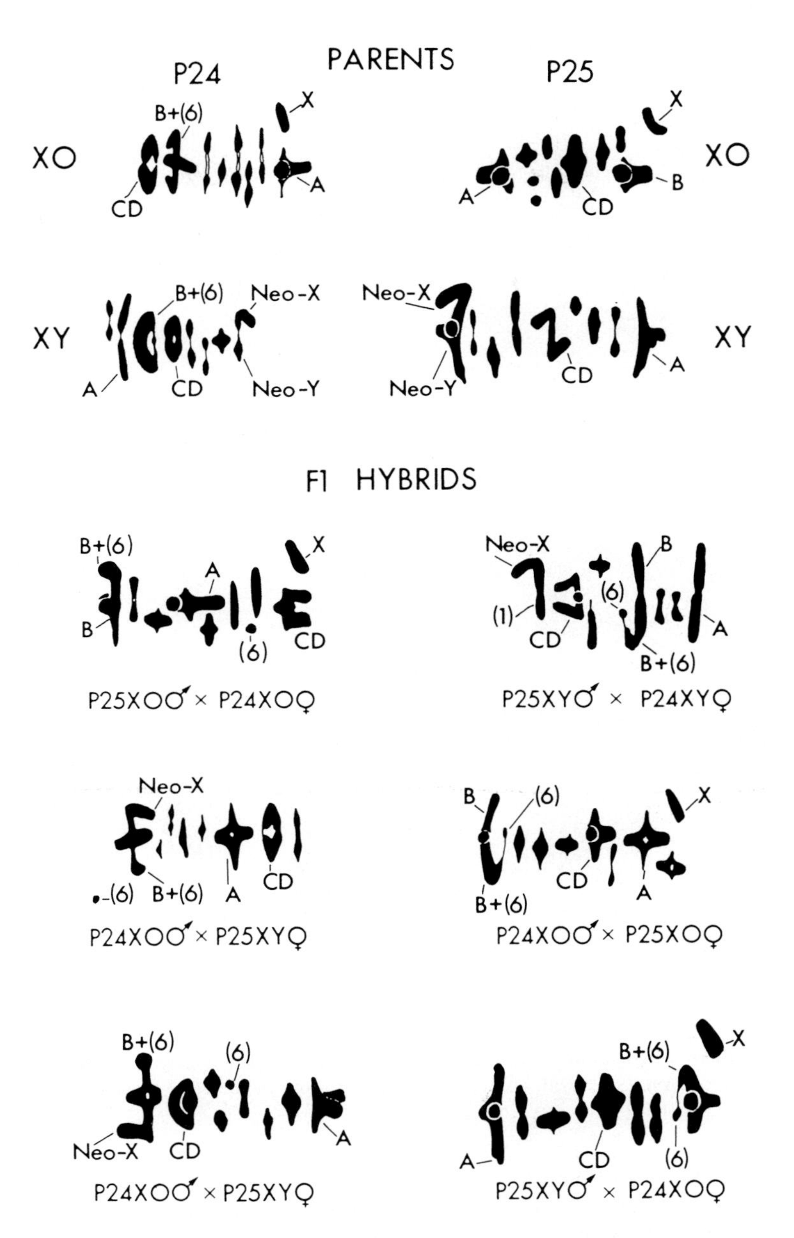

FIGURE 35. Metaphase I of male meiosis in the P24 and P25 races of the *Vandiemenella* complex and their hybrids. (After White, M. J. D., Blackith, R. E., Blackith, R. M., and Cheney, J., *Aust. J. Zool.*, 15, 263, 1967. With permission.)

that at pachytene in F_1 hybrids the differences between members of an unequal bivalent depend on differences in the length of their euchromatic segments.[129]

2. Those in which the successive members of the multiple are associated in a purely terminal manner not only at metaphase I but at diplotene, too. This most usually involves centromere-to-centromere associations but equivalent centromere-to-end arrangements also occur, as occasionally do end-to-end associations. The number of different terminal associations in which a given chromosome can be found to exist is, however, too high to be accounted for on the basis of inter-

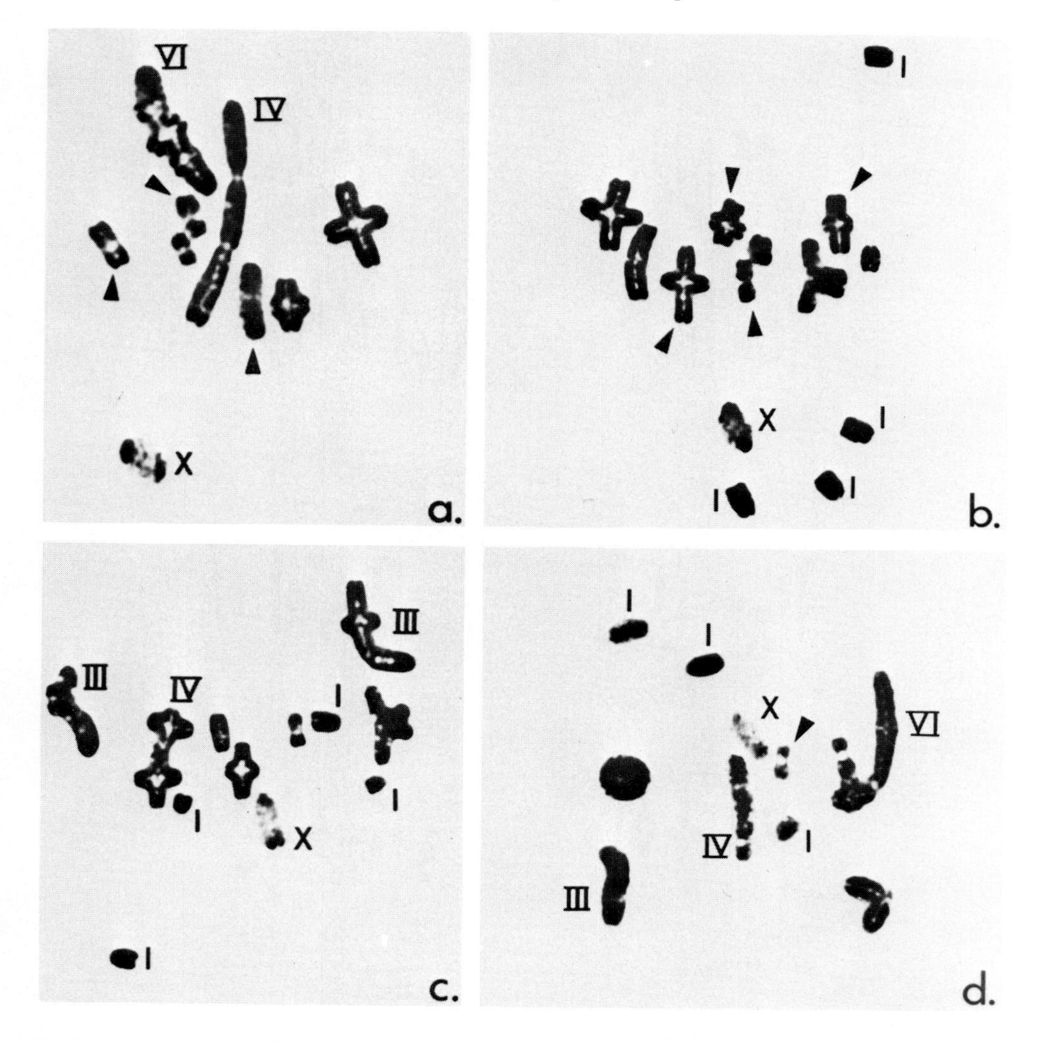

FIGURE 36. First metaphase of meiosis in male F_1 hybrids between *Eyprepocnemis plorans ornatipes* and *E. plorans meridionalis.* Univalents are designated by 1 and the most obvious unequal bivalents are arrowed (▶).

change, and John and Lewis[134] viewed them as persistent nonhomologous associations which arise as a consequence of the unusual genotypic conditions present in the hybrid state. White,[8,107] on the other hand, has claimed that these associations are chiasmate and reflect a segmental homology of chromosome ends based upon the repeated translocation of short terminal end duplications throughout the complement. They lead to extensive disturbance of normal segregation which, in turn, gives rise to unbalanced, aneuploid gametes. This is compounded by the presence of from 2 to 20 univalents which may also segregate irregularly. Alternatively, by lagging, these univalents impede cytokinesis and give rise to nonfunctional macrospermatids. This combination of events leads to extensive F_1 sterility. Consequently no F_2 is ever produced and backcross hybrids are similarly highly infertile.

B. The Evidence from Comparative Karyomorphology

Atchley[135] has emphasized that the use of chromosome number and morphology to

Table 51

AN ANALYSIS OF MALE MEIOTIC BEHAVIOR IN F_1 AND BACKCROSS HYBRIDS INVOLVING *EYPREPOCNEMIS PLORANS ORNATIPES* (EPO) AND *EYPREPOCNEMIS PLORANS MERIDIONALIS* (EPM), BOTH OF WHICH HAVE $2n = 11\ II + X$[134]

			No. cells with univalents								No. cells with single multiples								No. cells with 2 or 3 multiples of the same kind						
																			III		IV		V	VI	
	Total males	Total cells analyzed	0	2	4	6	8	10	20	Total	III	IV	V	VI	VII	VIII	IX	Total	2	3	2	3	2	2	Total
Epm♀ × Epo♂	18	1058	86	161	139	55	10	—	—	451	212	230	25	7	2	6	1	483	49	8	45	10	1	11	124
Epo♀ × Epm♂	6	196	12	33	32	23	6	4	1	112	37	36	3	3	—	—	1	80	16	2	4	3	—	—	25
F_1♀ (Epm × Epo) × Epo♂	11	572	258	146	25	—	—	—	—	429	57	75	4	3	—	—	—	139	4	—	—	—	—	—	4

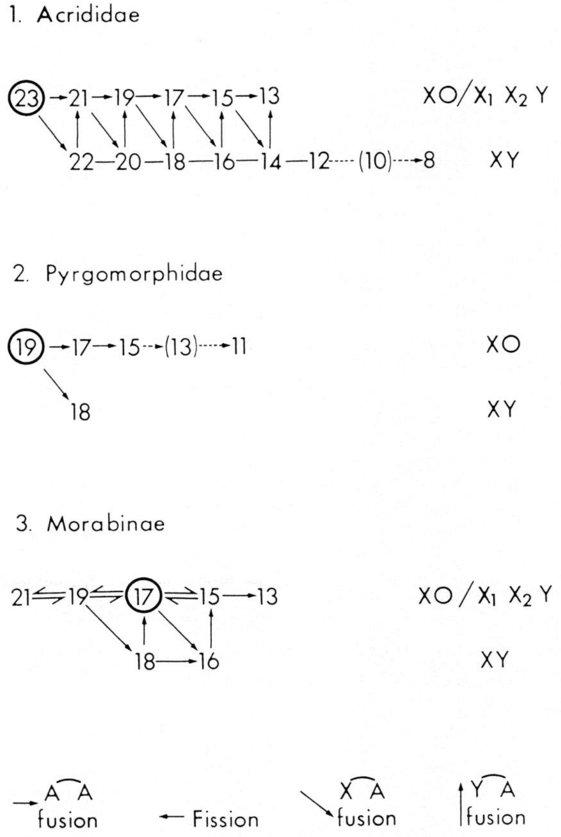

1. Acrididae

(23)→21→19→17→15→13 XO/X_1 X_2 Y

22—20—18—16—14—12····(10)···→8 XY

2. Pyrgomorphidae

(19)→17→15···→(13)····→11 XO

18 XY

3. Morabinae

21⇌19⇌(17)⇌15→13 XO / X_1 X_2 Y

18→16 XY

→ A⁀A fusion ← Fission ↘ X⁀A fusion ↑ Y⁀A fusion

FIGURE 37. Presumed chromosome phylogenies in three groups of grasshoppers.

infer lineage relationships and phylogenies, on the one hand, and to speculate on genetic divergence, on the other, is fraught with danger since it does not offer a rational means of defining chromosome homology. It is, of course, for this very reason that the karyotype is defined as the phenotypic appearance of the chromosomes in contrast to their genetic content. Karyotype comparisons also necessitate assumptions concerning the direction of evolutionary change. Unfortunately the reversibility of many types of chromosome change makes the detection of their direction difficult.

The only sensible means of defining chromosome homology by cytological means is through the analysis of meiotic pairing behavior in hybrids and even this is not without its complications. Thus, both intragenome pairing and nonhomologous pairing can lead to false conclusions concerning the kinds of change which may have occurred between related forms.

Regrettably much of the literature on orthopteran cytology deals with direct karyotype comparisons between only a small number of individuals sometimes even in quite remote genera. Consequently it is well near impossible to put this information into a sensible evolutionary context and some of the attempts to do so are both simplistic and naive (see Kumaraswamy and Rajasekaresetty, for example[136]). Thus, as a prerequisite to any consideration of karyotype evolution one needs to be able to define a primitive karyotype. Here one meets with considerable conceptual difficulty since it is unlikely that any contemporary species possesses a karyotype which has not in itself diverged to some extent from that of its ultimate progenitor. This is compounded by

Table 52
KARYOTYPE RELATIONSHIPS IN THE GENUS
DICHROPLUS[95,137]

Species	Relative DNA content	♂ $2n$	No. metacentrics	No. chromosome arms
D. conspersus	—	23(XO)	0	23
D. elongatus	129.7			
D. punctulatus	55.6	22(XY)	1	23
D. bergi	100.0	22(XO)	1	23
D. sp. nov.	—	21(X_1X_2Y)	2	23
D. dubius	—	18(XY)	5	23
D. obscurus	—	18(XO)	1	19
D. pratensis	—	8(XY)	4	13
D. silveiraguidoi	100.0			

Note: In contrast to the extensive distribution shown by the other species, *D. silveiraguidoi* is confined to a small zone of northern Uruguay. See also Table 44.

the problem of justifying a karyotype as ancestral in a phylogenetic sense as opposed to modal, that is one to which a species may have reverted to during the course of evolution. Here the conventional way out of this paradox is to assume that the most common condition is most likely to be the primitive condition. This assumption is, of course, not only untested but untestable.

For example, on this basis White[107] has argued for the chromosome phylogenies summarized in Figure 37. As they stand these schemes do not distinguish between centric and tandem fusions. They also omit the possibilities of pericentric inversion, paracentric inversion, equal interchange, the growth of heterochromatic arms, and changes in DNA content. Yet, as Table 52 indicates some of these processes must be involved, in some way, in the evolutionary changes that have gone on within the karyotype even within a single genus like *Dichroplus*[95,137] to say nothing of the many hundreds of other species of acridids. A second problem situation arises in the Podismini (Cantantopinae: Acrididae). Here the member species fall into two groups on cytological grounds:

1. Species with either $2n = 22 + XO/XX$ or else $2n = 20 +$ neo XY/neo XX and where, with the exception of the neo X chromosome, all the chromosomes are rods (Figure 18)
2. Species with $2n = 20 + XO/XX$ but again with all rod chromosomes (Figure 38a)

White assumes that this difference can be explained by a centric fusion followed by a pericentric inversion. However, since it is one of the smaller autosome pairs which appears to be "missing" as a separate entity when karyotypes are compared both within the Podismini (see References 138,139 and Figure 38a) and in broader terms (Figures 38b and c), the real explanation of the situation may well be quite different.

Equivalent problem cases exist in other orthopteroids. The data available for the North American Decticinae[140] (Table 53) indicate that simple centric fusion is not adequate to explain the observed differences in karyotype since *Atlanticus* and *Plagiostira* appear again to have lost a small pair of autosomes. In *Neduka*, too, the situation does not fit a simple pathway of centric fusion and the same is true of the Nemobiinae,[141] a group of small ground crickets (Table 54). In the Old World representatives

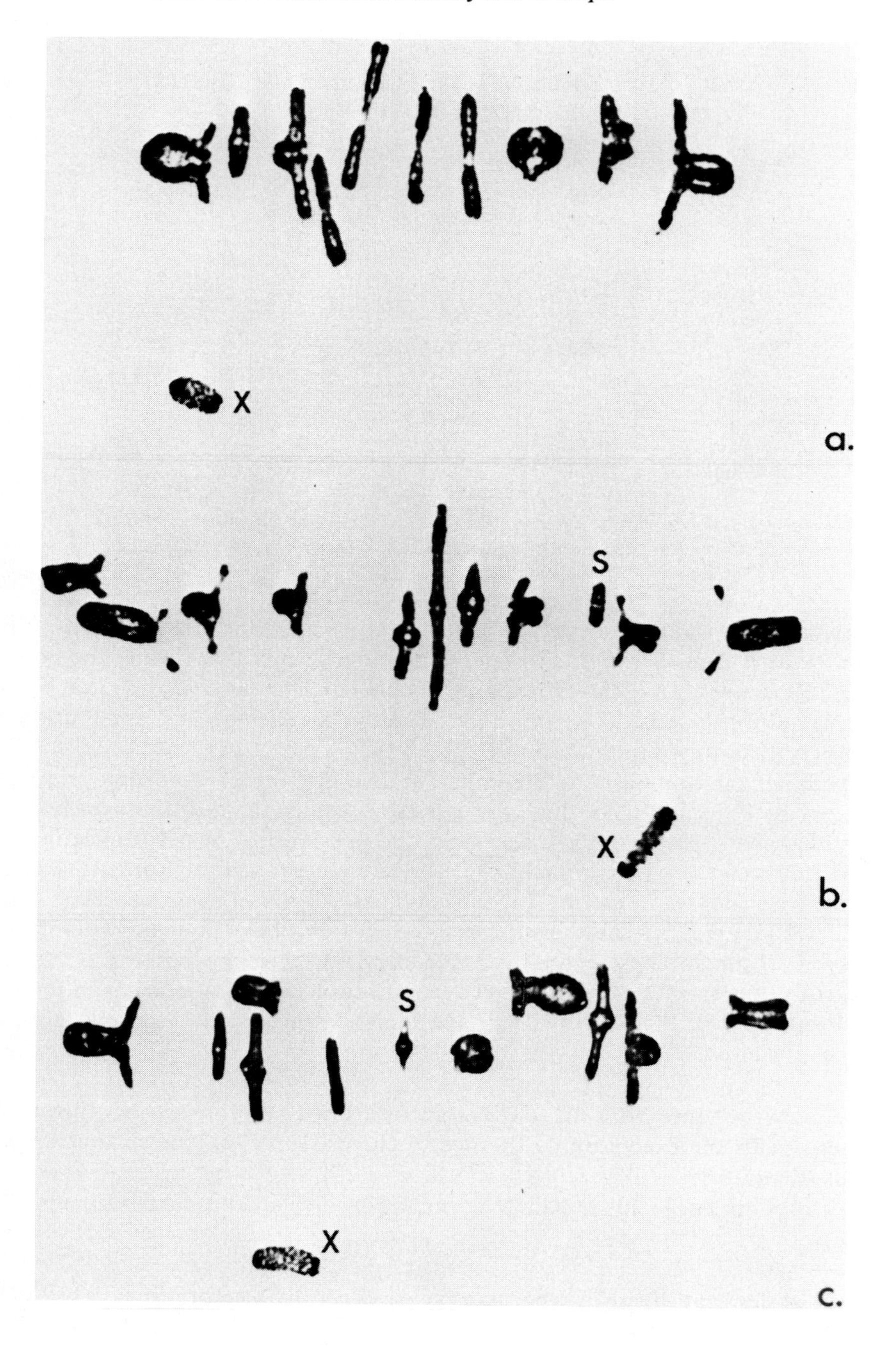

FIGURE 38. Metaphase I of male meiosis in (a) *Micropodisma* species ($2n$ = 10 II + X), (b) *Arcyptera fusca* ($2n$ = 11 II + X), and (c) *Calliptamus wattenwylianus* ($2n$ = 11 II + X). Note the absence of the small (S) bivalent in (a).

of this subfamily the diploid number ranges from 19 to 7 forming an odd-numbered series of 19, 17, 15, 11, and 7, with 15 as the most common number. If the number is accepted as primitive, then both increases and decreases in number have to be accounted for while at the same time retaining an all-rod complement of autosomes in all cases except the $2n$ = 7 form which has three metacentrics. Finally, in the tettigoniid

Table 53
KARYOTYPE VARIATION IN THE TETTIGONIINAE[140]

Genus	♂ 2n	No. of Rods	No. of Metacentrics	No. arms
Decticita, Decticus, Gampsocleis, Pholidoptera, Platycleis	31(XO)	31(X)	0	31
Metrioptera brachyptera	31(XO)	31(X)	0	31
Metrioptera saussureana	29(XO)	27(X)	2	31
Anabrus, Atelophis, Clinopleura, Steiroxys	29(XO)	27(X)	2	31
Idiostatus apollo, bechteli, gurneyi, inermis, kathleenae, magnificus	29(XO)	27(X)	2	31
Idiostatus elegans, californicus, fuscopunctatus, hermani, inermoides, rehni	29(XO)	26	3(X)	32
Idiostatus nevadensis	27(XO)	25(X)	2	29
Idionotus	27(XO)	25(X)	2	29
Atlanticus, Plagiostira	25(XO)	20	5(X)	30
Neduba diabolica, macneilli, sp. (67—38)	25(XO)	23(X)	2	27
Lanciana albidicornis	25(XO)	25(X)	0	25
Capnobotes, Zacycloptera	23(XO)	16	7(X)	30
Neduba diminuta	23(X_1X_2Y)	21(X_2)	2(X_1Y)	25
Neduba ovata armiger	23(XO)	23(X)	0	25
Neduba sp. (67—86)	22(XY)	17(Y)	5(X)	27

Table 54
KARYOTYPE VARIATION IN THREE GENERA OF NORTH AMERICAN NEMOBIINAE[141]

Genus	Species	♂ 2n (all XO)	No. of Rods	No. of Metacentrics
Neonemobius	*palustris*	19	19	1(X)
Ailonemobius	*allardi* } *fasciatus* } *griseus*	15	14	1(X)
Eunemobius	*carolinus*	7	0	7

Note: 7 is the lowest number recorded in the Orthoptera.

Odontura XO and neo XY species occur with the same diploid number, namely *O. marocana*, 26(XO) and *O. stenoxippa*, 26(XY). Here the origin of the neo XY form is not easily accounted for by centric fusion and its real relationship to the XO form is not known. Alicata et al.[142] suggest the neo Y may have originated from a B chromosome, but this is no more than a convenient solution for an inconvenient problem.

It has often been assumed in the past that the Acrididae, for example, shows a considerable degree of conservatism in karotype with a majority of the species having $2n$ = 23XO♂24XX♀. There are, however, striking and very significant differences in DNA content between species whose karyotypes are superficially similar (Table 55). There are also visible differences in heterochromatin content as judged by C-banding though it is equally clear that these differences, in themselves, do not explain the differences in DNA content.[146] Thus, apparent karyotype stability is fully compatible with a marked reorganization of the genome in terms of its DNA content.

Table 55
VARIATION IN NUCLEAR DNA CONTENT IN ACRIDID GRASSHOPPERS[2,143-145]

Species	Subfamily	♂ 2n	DNA content
2C values in absolute units (10^{-12}gm)			
Heteracris adspersus	Eyp.	23	12.68
Locusta migratoria	Ac.	23	12.70(12.18)
Chortoicetes terminifera	Ac.	23	14.44(13.90)
Austroicetes pulsilla	Ac.	23	14.60
Aiolopus thalassinus	Ac.	23	15.50
Chorthippus longicornis	Gomph.	17	17.16
Chorthippus vagans	Gomph.	17	17.36
Schizobothrus flavovittatus	Ac.	23	17.40
Schistocerca gregaria	Cyrt.	23	17.42
Gomophocerus sibiricus	Gomph.	23	17.90
Schistocerca pallens	Cyrt.	23	17.92
Humbe tenuicornis	Oed.	23	19.00
Macrotona australis	Cat.	23	19.70
Cryptobothrus chrysophorus— S race	Ac.	23	19.70
Schistocerca americana	Cyrt.	23	20.00
Chorthippus brunneus	Gomph.	17	20.30(18.92)
Gastrimargus musicus	Ac.	23	20.90
Valanga irregularis	Cyrt.	23	21.90
Schistocerca cancellata	Cyrt.	23	22.00
Chorthippus albomarginatus	Gomph.	17	22.56
Cryptobothrus chrysophorus— N race	Ac.	23	23.80
Peakesia hospita	Cat.	23	24.30
Phaulacridium vittatum	Cat.	23	24.90
Acrida conica	Ac.	23	25.10
Caledia captiva	Ac.	23	25.30
Myrmeleotettix maculatus	Gomph.	17	28.20(25.32)
Chorthippus parallelus	Gomph.	17	28.60(26.78)
Omocestus viridulus	Gomph.	17	30.6
Stauroderus scalaris	Gomph.	17	32.68
Arbitrary units			
Eirenephilus longipennis	Cat.	23	301 ± 4
Oedaleus asiaticus	Oed.	23	353 ± 4
Chorthippus apricarius	Gomph.	17	363 ± 6
Chorthippus longicornis	Gomph.	17	386 ± 6
Chorthippus biguttulus	Gomph.	17	390 ± 5
Chorthippus hammarstroemi	Gomph.	21	392 ± 5
Omocestus viridulus	Gomph.	17	402 ± 6
Chorthippus brunneus	Gomph.	17	408 ± 9
Chrysochroan dispar	Gomph.	17	408 ± 6
Stenobothrus lineatus	Gomph.	17	435 ± 6
Euthystira brachyptera	Gomph.	17	456 ± 8
Psophus stridulus	Oed.	23	521 ± 5
Stauroderus scalaris	Gomph.	17	549 ± 4
Bryodema tuberculatum	Oed.	23	682 ± 12
Angaracris barabensis	Oed.	23	763 ± 14

Note: Ac = Acridinae; Cat. = Catantopinae; Cyrt. = Cyrtacanthacridinae; Eyp. = Eyprepocnemidinae; Gomph. = Gomphocerinae; and Oed. = Oedipodinae.

Table 56
THE INCIDENCE OF THELYTOKOUS PARTHENOGENESIS IN ORTHOPTEROID INSECTS[1,8]

System	Species	Taxonomic position
Apomictic	*Pycnoscelus surinamensis*	Blattodea
	Carausius morosus	Phasmatodea
	Sipyloidea sipylus	Phasmatodea
	Saga pedo	Tettigoniidae
Automictic		
Fusion at cleavage	*Bacillus rossius*	Phasmatodea
	Clitumnus extradentatus	Phasmatodea
Premeiotic doubling	*Warramaba virgo*	Eumastacidae
Unknown	*Baculum artemis*	Phasmatodea
	Cloncopsis gallica	Phasmatodea
	Epibacillus lobipes	Phasmatodea
	Leptynia hispanica	Phasmatodea
	Phobaeticus sinetyi	Phasmatodea
	Timema genevievae	Phasmatodea
	Brunneria borealis	Mantodea
	Poecilimon intermedius	Tettigoniidae
	Xiphidiopsis lita	Tettigoniidae

Finally, the considerable variation in genome size found both within the 23- and 17-chromosome forms, as well as between them, suggests that the two are unlikely to be related by simple Robertsonian change.

V. PARTHENOGENETIC SYSTEMS

A number of orthopteroid species are capable of forced thelytoky, or female parthenogenesis, under experimental conditions when virgin females are isolated from males. In many such cases a small proportion of unfertilized eggs commence development with a haploid complement. Diploid nuclei soon make their appearance as a result of chromosome reduplication involving an extra round of DNA synthesis between successive mitoses. The success of the event depends on the measure of diploidization attained. Many embryos die or develop abnormally because of haplo/diploid mosaicism. Bergerard and Seuge,[147] for example, have shown that regions of partheno-produced embryos which are inhibited in development have a higher proportion of haploid cells than do normally developing regions.

Obligatorily parthenogenetic populations in orthopteroids are less common and the known cases are summarized in Table 56. They fall into two distinct classes:

1. *Automictic (meiotic) parthenogens* — A conventional meiosis occurs but is compensated for in one of two ways: either by a premeiotic doubling of chromosome number or else by the fusion of cleavage nuclei or by the halves of divided chromosomes remaining together in early cleavage nuclei.
2. *Apomictic (ameiotic) parthenogens* — There is no conventional meiosis which is replaced by either one *(Saga pedo)* or two *(Carausius morosus, Pycnoscelus surinamensis)* equational type divisions. The two divisions in the latter species are necessary because there is an endomeiotic DNA duplication during early first prophase of meiosis.

In only two cases, one automictic and one apomictic, detailed information is available relating to the evolutionary importance of parthenogenetic populations in orthopteroids:

(1) *Warramaba virgo*, the wingless Australian morabine grasshopper, is the only known thelytokous species within the suborder Caelifera. It was first discovered in eastern Australia in sandy areas of western New South Wales and northeast Victoria where it lives on acacia shrubs.[148,149] Its only known relatives, however, are all found in western Australia[150] where *virgo* itself was also subsequently found. It has, therefore, been argued that it arose in western Australia and then extended its range, a not uncommon occurrence in parthenogenetic forms. Of the bisexual species currently present in western Australia, the still undescribed P169 and P196 are most similar to *virgo*. These, in turn, are presumed to have been derived from a karyotype of the kind which is present in *W. picta*, an XO/XX species with $2n♀ = 18$. P169 is a neo XY/XX species with $2n = 16♀$ while P196 is an $X_1X_2Y/X_1X_1X_2X_2$ species with $2n = 14♀$. Phenotypically females of P196 resemble *virgo* closely though P169 is a rather divergent member of the genus.

Hewitt[151] first suggested that *virgo* owed its origin to hybridization between progenitors with karyotypes similar to those currently present in P169 and P196. This is now supported by two principal facts:

1. Twenty-six clones of *W. virgo* are now known which differ from one another either in gross structural features of their karyotype or else in the details of their C-banding.[152] Some of those extend over considerable areas while others are found only at single localities. In general only one clone is present at any one locality but two clones are known to coexist in some localities and in one case three clones do so. Even so it is possible to discern a standard karyotype which occurs at many localities. Here eight of the members are indistinguishable from those of P169, while the other seven cannot be distinguished from those of P196 (Figure 39). It is, thus, an X_1X_2 species with $2n = 15$.
2. P169 and P196 still coexist in a number of localities in a broad zone of sympatry in western Australia, though there is no evidence that they are hybridizing in nature at the present time. Added to this all attempts to hybridize "sympatric" individuals from the natural overlap area give zero egg hatchability which implies that there is now effective post-mating isolation between the two species where they occur together. However, by using individuals of P169 from the overlap zone and individuals of P196 from outside the zone it has been possible to produce synthetic female hybrid individuals which closely resemble pure *virgo* in phenotype as well as karyotype. These can be reared to adulthood despite the fact that they show poor viability.[153]

Normal *virgo* reproduces by a form of automictic parthenogenesis which is uncommon in insects. This involves a premeiotic doubling of the chromosome complement so that the oocyte contains $2n = 4x = 30$ chromosomes. Meiotic synapsis is then restricted to sister products of this endoduplication process. This leads to the formation of 15 chiasmate bivalents. The chiasmata are, of course, without any genetic significance and serve only to secure orderly segregation leading to the production of diploid eggs which develop into diploid females. In synthetic *virgo* individuals the ovaries are abnormal and very few oocytes undergo normal development. They do, nevertheless, have a limited capacity for parthenogenetic development and appear to possess the premeiotic doubling mechanism.

Some male hybrids also result from crosses between P169 and P196 though, which-

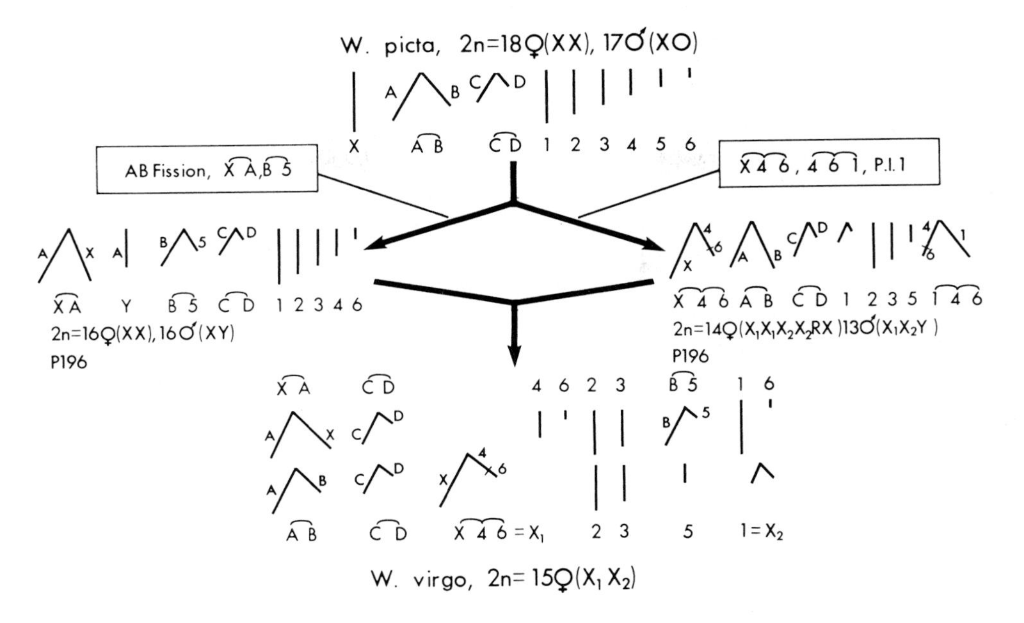

FIGURE 39. Presumed origin of the parthenogenetic morabine grasshopper *Warramaba* (formerly *Moraba*) *virgo*.

ever way the cross is made, they yield more female progeny than male. Male F_1s usually die as embryos. Even if they succeed in hatching they have low viability. Depending on the direction of the cross they have a Y chromosome from either P169 or P196. However, only male individuals of the former category have been reared to adulthood. These have testes of reduced size but normal histology. Their meiosis is highly abnormal though most primary spermatocytes are diploid.[154] A majority of meiocytes have zero, one, or two bivalents. Some of these are composed of partially homologous chromosomes, e.g., the B5 of P169 and the AB of P196. Others are the result of terminal associations between nonhomologs. Irregularities of distribution, resulting from the extreme univalency, lead to abnormal sperm. Nevertheless, such males certainly do copulate, though unsuccessfully, with P169 and female *virgo*.

Females of both P169 and P196 can, if isolated from males, be forced to reproduce parthenogenetically, but the embryos which result from them are invariably inviable haplo/diploid mosaics. This strongly suggests that the premeiotic doubling mechanism that characterizes *virgo* must have arisen as a consequence of hybridization (compare *Caledia*).

Virgo females also cross in the laboratory with males of both P169 and P196. The offspring in all cases are triploid, having a haploid karyotype of the relevant bisexual parent as well as a full *virgo* karyotype. Such triploids are male if they carry a Y of either of the bisexual species but female if they lack one. The viability of all four kinds of triploid is good, much better than that of diploid hybrids between P169 and P196. Female triploids have a much lower fecundity than normal *virgo* females but, even so, it is much higher than that of synthetic *virgo* females. It is possible to maintain triploid laboratory stocks over a number of parthenogenetic generations. Although triploid females will copulate with males of either P169 or P196 no tetraploid offspring have ever resulted. Oocytes of triploid hybrid females show a *virgo*-type meiosis and all the chromosomes, including those derived from the paternal parent, undergo premeiotic doubling.

Triploid males have a very reduced testis in which the spermatogonia degenerate so that only rarely are spermatocytes found. In the few that are, chromosomes of P169 form bivalents with *virgo* chromosomes derived from P169 and so provide further confirmation for the hybrid nature of *virgo.*

Very considerable differences exist between northern and southern populations of P196 in western Australia in size, in fecundity, in C-band features of the karyotype, and in the structure of the X chromosome. All the diagnostic cytogenetic features of the southern race of P196 are present in the Boulder and Zanthus clones of *virgo* from western Australia but not in any other. White and Contreras[155] have interpreted this to imply that these two clones are derived from a hybridization event involving the southern race of P196, whereas the other clones are descended from one or more hybridizations involving the northern race of P196. To what extent the differences between the Boulder and Zanthus clones, on the one hand, and the others were already present in the parent species is not clear. Evidently those translocations and fusions that involve one chromosome from P169 and the other from P196 must have arisen under parthenogenetic conditions, i.e., after hybridization. Such rearrangements are known in at least six clones, namely in Monia Gap and Yatpool from New South Wales and Spargoville, Hayes Hill, Kambalda, and Woolibar from western Australia.

(2) Although the Indian stick insect, *Carausius morosus*, reproduces by obligatory parthenogenesis, impaternate males appear regularly with a low frequency ($<1\%$). Their meiotic divisions, though abnormal, are, nevertheless, very revealing. The endomeiotic DNA replication which characterizes the female is sometimes present, sometimes partial, and sometimes absent in the rare males. When the pairing behavior of primary spermatocytes lacking the endomeiotic replication is analyzed it is possible to show the presence in the male of both structural heterozygosity and aneuploidy.[156]

The females have $2n = 64$ with three metacentric X chromosomes (1 to 3), four metacentric autosomes (4 to 7), and fifty-seven acrocentric autosomes. Of the seven metacentrics in the female only two, 2 and 3, are similar in size. The males have $2n = 63$ with only two X chromosomes. At first metaphase of male meiosis there are univalents, bivalents — both equal and unequal — and multiple associations of from three to six members (Table 57). Autosomes 4 to 7 are either univalent or else form unequal bivalents with an acrocentric or participate with one or more acrocentrics in multiples. The two sex chromosomes of the male are always univalent.

Bergerard[157,158] and Pijnacker[159] initially regarded *C. morosus* as a tetraploid. The number of sex chromosomes, however, is indicative of triploidy. Even so, the 61 autosomes, the apparent monosomy of autosomes 4 to 7, and the variation in morphology and numbers of first metaphase associations in the rare males all tend to conceal the presence of triploidy.

During metaphase I of male meiosis the hexa- and penta-valent multiples seen at earlier stages are no longer present.[156] They apparently break down to smaller elements because of the weak connections between the minute arms of the acrocentrics involved. Tri- and tetra-valent associations are, however, still present. Univalents either move undivided to the poles at first anaphase or else lag. Regular segregation is, thus, disturbed through the asynchronous separation of the various chromosome associations as well as lagging, nondisjunction, and dicentric bridge formation, a consequence of U-type nonsister chromatid exchange. Anaphase II is even more irregular.

The hatchability of parthenogenetically produced eggs is at least 80 to 90% which indicates that associations equivalent to those seen in the rare males do not form during female meiosis. This, of course, is not surprising since the endomeiotic DNA duplication process necessarily obscures the structural and the numerical heterozygosity of the complement. Thus, the aneuploidy, the structural heterozygosity, and the polyploid nature of this parthenogen are both maintained and masked by the endomeiotic duplication in the female. The rare mutant males indicate the complexity of the kary-

MEAN NUMBERS OF CHROMOSOME CONFIGURATIONS AT FIRST METAPHASE OF MEIOSIS IN FOUR IMPATERNATE MALES OF *CARAUSIUS MOROSUS*. THESE MALES HAVE $2n = 63$[156]

Individual no.	No. cells scored	Mean frequency (± SD)		
		Univalents	Bivalents	Multiples
1	13	32.1 ± 8.3	12.5 ± 3.8	4.3 ± 2.1
2	6	44.5 ± 6.7	7.2 ± 3.0	1.2 ± 1.5
3	12	33.7 ± 7.5	10.6 ± 2.9	2.1 ± 2.1
4	15	39.5 ± 6.8	8.7 ± 2.7	1.6 ± 1.3

otype in this species, which suggests that it, like *W. virgo*, may also owe its origin to hybridization since it is best interpreted as an allotriploid of the type ABB.

A genetic system in which a hybrid phenotype, with a concomitantly high level of heterozygosity, can be maintained without a segregational load has many adaptive potentialities and is found both in *W. virgo* and *C. morosus*. A similar situation may also obtain in Italian mole-crickets. While *Gryllotalpa gryllotalpa* ($2n = 22XY,XX$) is known in Italy, probably as a recent invader, it is found mainly in mountain regions. A second group of five distinct cytological forms is also known in Italy. Two of these, *G. sedecim* ($2n = 16$) and *G. octodecim* ($2n = 18$) occur as mixed populations in the Po River region, in Sardinia, and probably in Corsica. In the mixed populations all female hybrids are present with $2n = 17$,[160] but as yet their meiotic behavior is unknown.

VI. THE BASES OF CHROMOSOME CHANGE IN ORTHOPTEROIDS

The evidence dealt with in this review makes it clear that there is extensive intraspecific variability in orthopteroid insects which takes the form of morphs, clines, and geographical races. There are also considerable differences in the karyotypes of related species. Some are content to define evolution in terms of the changes that occur in the genome. Others, more properly, regard it as the consequence of these changes. In this final section, therefore, some consideration is given to the significance to be attached to the impressive body of chromosome variation witnessed in the preceding chapters.

Arguments relating to the significance of chromosome change in evolution fall essentially into four categories.

A. Category 1

Chromosome changes play no role and their presence must be ascribed simply to their lack of any effect, positive or negative, or else to genetic drift or meiotic drive.

As long ago as 1940, Muller[98] argued that, quite apart from any selective advantages or disadvantages associated with it, any genetic change which does not lead to inviable zygotes must be subject to the same process of drift as are gene mutations. In this way it would be possible for a chromosome variant of indifferent survival value to gain a foothold in a population. Few now doubt that chance factors can and do operate on natural populations and that genetic drift represents one process by which some chromosome changes may have become incorporated into natural populations. It is often assumed that random drift depends wholly on accidents of sampling. Wright,[161] however, mentions that fluctuations in selection intensity may also be responsible for random drift. Moreover, while drift may counteract selection it may also sometimes increase its effectiveness.

Meiotic drive offers an alternative mechanism for preserving chromosome changes of indifferent or even negative survival value within a population and is undoubtedly implicated in some sense in certain of the B-chromosome polymorphisms in orthopteroid populations (see Section II.C.1).

Part of the difficulty in assessing the real role of drift and drive stems from the fact that the largely theoretical approaches to the problem have tended to treat all forms of chromosome change in a common category. It is evident that the conditions which lead to the development of a polymorphism or a polytypism are not necessarily those that lead to the fixation of a chromosome variant within an entire species. Where a chromosome difference between two related species is assumed to play a causal role in speciation, it can do so only if it leads to lowered heterokaryotic fertility. In a stable polymorphism it does not usually do so (but see Section II.B). Defining the theoretical conditions which lead to the fixation of a chromosome change with lowered heterokaryotic fertility may, therefore, give no indication about the origin of chromosome polymorphisms. This failure to define which system is under discussion has in part led to much of the current confusion.[161,163]

Hedrick[164] has shown how, in theory, meiotic drive, a combination of drive with drift, or a combination of drift with inbreeding can lead to a substantial increase in the probability of fixation of a chromosomal variant with reduced heterokaryotypic fertility. But there is no clear case where the observed changes have resulted from the action of these factors. While neutral mutants may indeed be introduced into and spread within a population, such mutants will be expected to fluctuate in frequency over time as a consequence of random drift. They will, therefore, increase or decrease fortuitously. In the course of these fluctuations the overwhelming majority will be lost by chance. Only a minority can be expected to become fixed in a population. Yet Muller[98] was of the opinion that "every translocation, each shift which has become established in a species denotes the occurrence of a bottleneck in the past history when drift has outweighed selection," a view which is still shared by some.

Given that a new chromosome mutant behaves efficiently at mitosis and meiosis, the question of whether it has any positive fitness effects is irrelevant to its immediate survival. It is, thus, possible for a chromosome variant of indifferent survival value to gain a foothold in a population. Some of the chromosome variants found in natural populations are, thus, essentially transients without any evolutionary significance. This is likely to be the case with at least certain of the supernumerary segment systems or the supernumerary chromosome systems found only infrequently in orthopteroid populations. On the other hand, widely distributed chromosome polymorphisms are not easily explained on a neutralist argument.

The difficulty in dealing with natural populations is that no genotype of karyotype can be regarded as unconditionally advantageous or disadvantageous. These concepts have a meaning only in relation to the environment and this can, and does, change. Equally, no chromosome change is likely to be unconditionally neutral if we wait, as nature does, to assess its long-term effects.

B. Category 2

Chromosome changes are associated with position effects which lead to a modification of gene action.

There are two quite different aspects to the position effect argument which, in any event, can apply only to structural chromosome changes. The first supposes that those newly occurring chromosome rearrangements which do not survive and which constitute the majority are necessarily associated with deleterious position effects. This view-

Table 58
THE FREQUENCY OF SPONTANEOUSLY OCCURRING HETEROZYGOUS STRUCTURAL CHROMOSOME CHANGES IN THREE PLANT GENERA BELONGING TO THE TRIBE ALOINEAE[167]

	Paracentric inversions		Pericentric inversions			Interchanges		
Genus	No. plants	No. inversions	No. plants	No. inversions	No. different inversions	No. plants	No. interchanges	No. different interchanges
Aloe	123	8(6.5%)	1027	2(0.2%)	2(0.2%)	785	16(2.0%)	4(0.5%)
Gasteria	48	2(4.2%)	132	0	0	132	2(1.5%)	2(1.5%)
Haworthia	72	0	1046	7(0.7%)	2(0.2%)	597	23(3.9%)	8(1.3%)

point has been consistently maintained by White, although I know of no evidence in support of it. Even in *Drosophila*, where position effects are certainly known, they are not often lethal. The only other cases where there is even a suggestion of a chromosome mutation having possible deleterious effects on the phenotype is in man, and even here the evidence is equivocal. Sunmitt et al.,[165] in reviewing the literature on female carriers of balanced X-A translocations, noted that where the breakpoint falls between bands Xq 13 and Xq 27, females are liable to be infertile, whereas carriers with X breakpoints at other locations are fertile. Jacobs,[166] on the other hand, found an excess of apparently balanced *de novo* structural rearrangements among the mentally subnormal. In both cases a possible position effect has been mooted as the basis for the deleterious effect associated with the balanced heterozygous translocations but, in fact, the real cause is not known.

From somatic chromosome analysis in plants of wild origin belonging to the tribe Aloineae, Brandham[167] has shown that structural changes are by no means infrequent and that a fair proportion of them are genuinely novel (Table 58). Despite this, the basic karyotype of all three genera is remarkably conservative and, coupled with this, not one case of homozygosity for any of the observed changes was found. Additionally, none of the plants carrying individually distinctive mutants gave any overt sign of position effects, let alone deleterious ones. In this case, as in many of the spontaneous chromosome mutants described in orthopteroids, the reason for failure to survive has possibly more to do with their meiotic performance and its consequences than with deleterious position effects.

The other side of the argument rests on the assumption that position effects stemming from chromosome rearrangements produce adaptive phenotypic effects. Here the actual evidence is no better. It has been claimed that chromosome mutations, by rearranging blocks of euchromatin, generate new patterns of gene regulation and temporal expression and, as such, offer the most reasonable mechanism for providing for a major adaptive shift through a change in some critical regulatory pathway.[168] This is certainly an attractive proposition but is still speculative. Bush,[168] however, concludes that from the point of view of such adaptive shifts "the most profound effects of chromosome arrangements are at the developmental level" and that "they can play an important role in repatterning developmental pathways that lead to striking phenotypic change." But no evidence for such striking phenotypic change in relation to chromosome rearrangements has been found. Thus, in none of the polymorphic or polytypic populations of orthopteroids referred to earlier is there any indication that the differences in chromosome constitution which characterize them are associated with any phenotypic change, let alone of a striking nature. Likewise the sibling species in *Caledia* and *Eyprepocnemis* show that chromosome differences between species

Table 59
CHIASMA FREQUENCY VARIATION IN 14 SPECIES OF ACRIDID GRASSHOPPERS WITH A $2n = 23$(XO) ALL-ROD COMPLEMENT

Species	Cell range	Mean individual range
Stethophyma grossum		
♂	11—13	11.0—11.7
♀	12—20	14.0—17.8
Phaulacridium vittatum	11—17	11.8—15.7
Pezotettix giorni	11—17	13.3—15.3
Oedaleonotus enigma	13—18	14.1—16.4
Trimerotropis topanga	14—20	15.3—18.7
Conozoa carinata	14—20	16.4—17.3
Bootettix punctatus	14—21	16.6—18.9
Calliptamus wattenwylianus	15—22	16.4—19.1
Buforania species 6	13—23	16.1—19.8
Encotophus subgracilis	15—22	17.1—19.5
Podisma pedestris	14—22	16.0—21.8
Derotemema haydenii	14—23	16.1—21.8
Spianacris deserticola	19—25	22.0—22.2
Chimerocephalus pacifica	20—25	22.7—23.7

Note: With the exception of *Stethophyma* all scores are from spermatocytes.

need not lead to any apparent phenotypic change. The same is true in the Eumastacidae where, although karyotypes vary greatly from species to species, they may be barely distinguishable in external morphology.

It is true that a consistent correlation of certain plumage and behavioral characters with a specific pericentric inversion has been identified in the white-throated sparrow,[169] but whether this association is indeed the result of a position effect again remains to be clarified. Until a great deal more is known about the relationship between chromosome organization and gene action we will be unlikely to resolve satisfactorily the issue of adaptive position effects, but as things now stand the evidence is not compelling.

C. Category 3

Chromosome changes may, by virtue of their effects on recombination, give an indirect selective advantage which, in some cases, may even be strong enough to counteract a fertility disadvantage associated with the mutant state.

Recombination at meiosis is recognized as a major source of heritable variation between phenotypes. As such, adjustments in the amount of recombination may assume an important adaptive function in regulating the amount of variability both within and between populations. As Rees and Dale[68] have elegantly shown, significant correlations can be demonstrated between chiasma distribution and phenotypic variability. The nature of recombination in orthopteroids and the way it changes in the presence of variant chromosome types are interesting phenomena. In most orthopteroids meiosis is chiasmate in character. Achiasmate meiosis is, however, known in at least 14 genera of mantids, including *Callimantis, Humbertiella, Promiopteryx, Pseudomiopteryx, Amorphoscelis, Leptomantis, Haldwania, Holaptilon, Harpagomantis, Bolbe, Kongobatha, Nanomantis, Glabromantis,* and *Cliomantis,* as well as in a few species of thericleine eumastacid grasshoppers,[1,8] though in both groups only in the male. Even in chiasmate forms there is a wide range of mean cell chiasma frequencies in different species. For example, in acridid grasshoppers with $2n = 23♂$ (11 II + X), it extends from a low of 11 to a high of 23 (Table 59, Figure 42). As White[170] has emphasized, however, few species show chiasma frequencies more than twice the haploid number and, especially when rod chromosomes only are present in the complement, it is more usual to find lower frequencies. This follows from the fact that individual bivalents ordinarily form either one or two chiasmata, rarely more. Higher bivalent chiasma frequencies are sometimes, though not invariably,[171] found in metacentric chromosomes in orthopteroids (Figure 40).

In *Myrmeleotettix maculatus* ($2n = 8$ II + X), B chromosomes produce a quite dramatic increase in mean cell chiasma frequency (Section II.C.1). Whereas the male value does not exceed 16.9 in individuals with a basic karyotype it can rise to as high as 19.8 in the presence of B chromosomes.[44] Supernumerary segments have an equivalent effect in *Chorthippus parallelus* where the mean cell chiasma frequency is normally below 16.0, whereas in the presence of segments it rises to a high of 18.8.[172] In both these cases the rise is achieved by an increase in the number of bivalents with higher chiasma frequencies which inevitably means more interstitial chiasmata, especially in the large metacentric bivalents where as many as five to six chiasmata are now present in a single bivalent.

Chiasma frequency tends to be lowest in species with strict localization and especially where that localization is proximal (Figure 41, Table 60). Frequency and pattern are, thus, to some extent interrelated. Many species show a combination of proximal and distal chiasma distributions within the members of the same complement (Figure 42). Differences between males and females of the same species in respect of either chiasma frequency or chiasma distribution patterns are also evident (Tables 60 and 61). Finally, differences of this type also exist between related species in the same genus or species group (Table 62).

Hillel et al.[179] have emphasized the difference between interstitial chiasmata and those localized proximally or distally in respect of the release of variability. They contend that localized chiasmata must largely serve a mechanical function in securing segregation. Certainly in many orthopteroid species there are considerable regions of the genome which appear to be exempt from recombination. The fact, therefore, that almost every category of chromosome change is capable of influencing either chiasma frequency, chiasma distribution, or both these parameters, offers considerable potential for modifying the amount of variation available to a population. Moreover, the same type of change can produce different effects in different species. Thus, a compensatory increase in the mean cell chiasma frequency of pericentric inversion heterozygotes occurs in *Trimerotropis suffusa*, *T. helferi*, and *T. pseudofasciata* but not in *T. pallidipennis* where the mean chiasma frequency per individual is negatively correlated with the number of heterozygous chromosomes per individual. In *Stethophyma*, on the other hand, the influence of supernumerary material may vary not only between species but also between populations of the same species. Thus, in *S. grossum* there are two independent segment systems on the S11 sited at different locations but both produce a significant increase in chiasma frequency.

Since the principal effects of many chromosome changes relate to recombination, so, too, may their evolutionary significance. While future conditions cannot affect present survival, the genetic composition of future generations must largely determine their capacity to survive future conditions. No one doubts the adverse effects which chromosome changes determine in respect of recombination when these lead to embryonic death of the progeny. White,[180] for example, in discussing the recombination argument in relation to polymorphic systems, has recently stated that "it is simply not

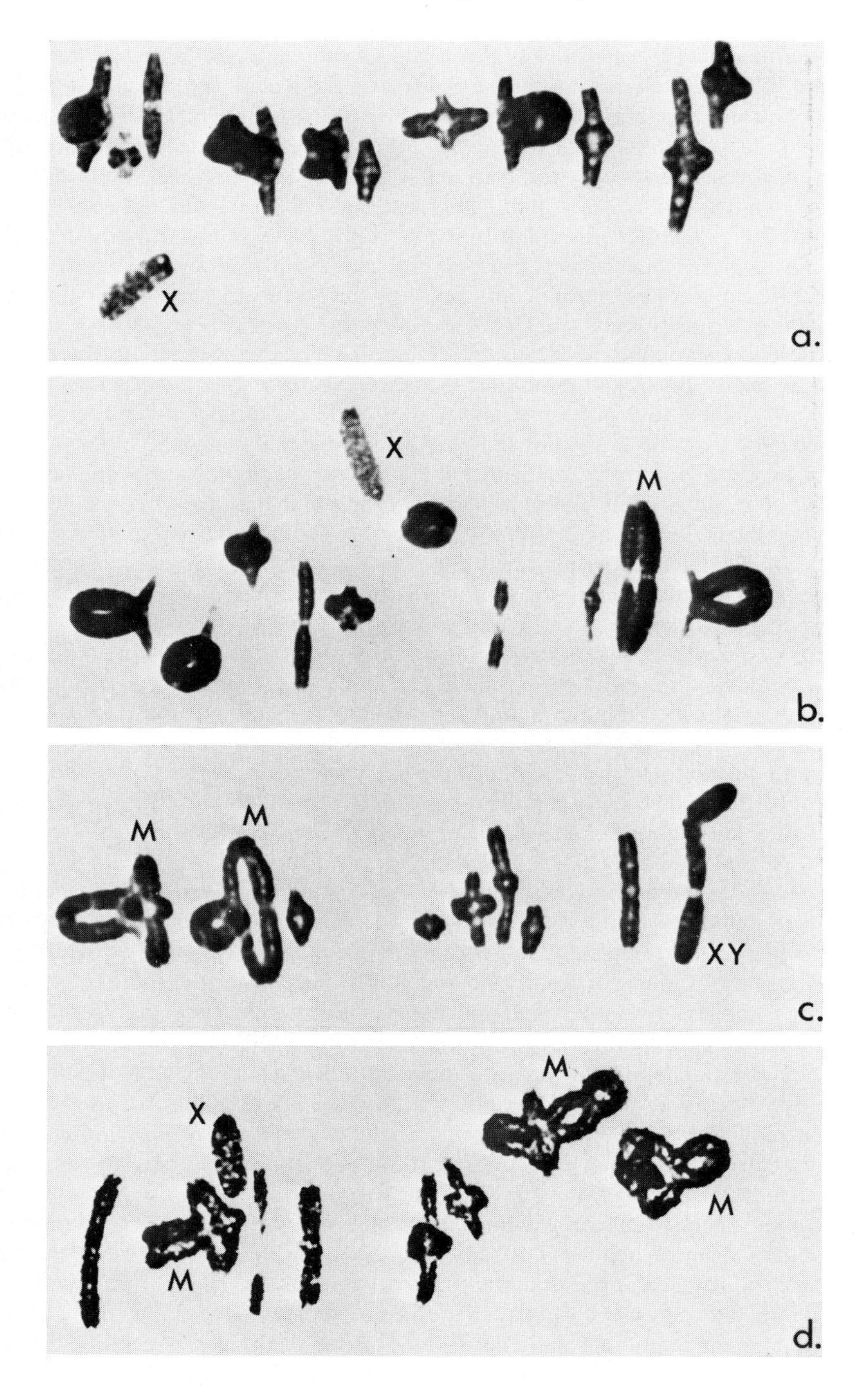

FIGURE 40. Metaphase I of male meiosis in four acridid grasshoppers with zero, one, two, and three metacentric bivalents. (a) *Cuparessa testacea* ($2n$ = 23♂,XO = 11 II + X), (b) *Percassa rugifrons* ($2n$ = 21♂,XO = 10 II + X), (c) *Tolgadia bivittata* ($2n$ = 18♂,XY = 8 II + neo XY II), and (d) *Omocestus viridulus* ($2n$ = 17♂,XO = 8 II + X).

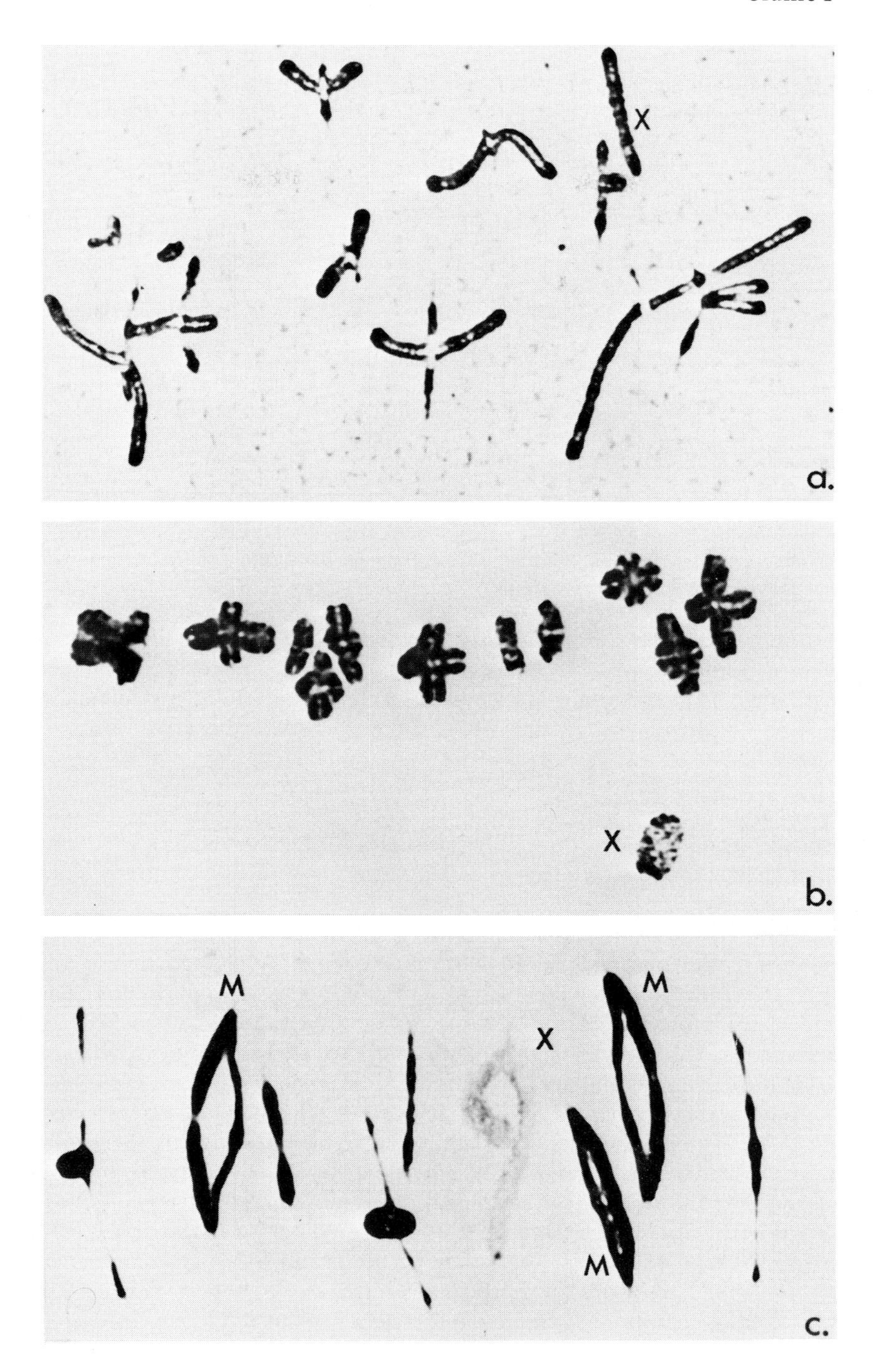

FIGURE 41. Metaphase I of male meiosis in three acridid grasshoppers to illustrate variation in chiasma distribution patterns. (a) *Stethophyma grossum* (2*n* = 23♂,XO) with predominantly proximal chiasma localization, (b) *Chortoicetes terminifera* (2*n* = 23♂,XO) with a nonlocalized distribution, and (c) *Euthystira brachyptera* (2*n* = 17♂,XO) with predominantly distal localization; M denotes metacentric bivalents.

Table 60
SEX DIFFERENCES IN MEAN CELL CHIASMA FREQUENCY ($\overline{X}a$) AND CHIASMA DISTRIBUTION IN ACRIDID GRASSHOPPERS

		Male		Female	
Species	2n ♂/♀	$\overline{X}a$	Xa distribution	$\overline{X}a$	Xa distribution
Chortoicetes terminifera	23/24	13.07	Random	13.02(12.02)	Random
Parapleurus alliaceus		12.32	Proterminal	14.47(12.87)	Proterminal
Stethophyma grossum		11.28	Predominantly extreme proximal	14.98	Distal/ interstitial
Chrysochroan dispar	17/18	12.56	Mainly proterminal	13.57(12.10)	Random
Chorthippus brunneus		13.57	Random	14.23(13.23)	Random

Note: The values in parenthesis for the female $\overline{X}a$ are adjusted means obtained by subtracting the X-bivalent contribution from the total. These values are significantly different from that of the corresponding male in all cases except *Chorthippus brunneus.*[173-175]

credible that there should simultaneously be selection for high and low levels of recombination in a population.'' The existence of a chiasmate meiosis in one sex and an achiasmate meiosis in the other is, of course, a clear example of just such a situation. So, too, is the occurrence of a significant difference in recombination level between the two sexes.

D. Category 4

Chromosome changes introduce segregational problems at meiosis which reduce fertility. Such changes may thus serve to restrict gene flow and so facilitate genetic divergence leading to specific differentiation.

The fact that closely related species do often differ in karyotype suggests, at the very least, that the conditions associated with speciation may also favor the fixation of karyotypic differences. In trimerotropine grasshoppers (Section II.A.2), apart from existing in a polymorphic state, some pericentric inversions had reached fixation in particular species, but clearly in this instance these fixed differences between species do not constitute isolating barriers. The study of interspecies hybrids between different trimerotropine species indicates that the important chromosome differences between them and those which generate infertility in the hybrid are fusions (Section IV.A.2).

Fixed differences between karyotypes can also arise under polytypic conditions (Section III.B) so that should such geographical races subsequently become reproductively isolated by other means, these fixed differences would persist in the new specific types. There is good reason to suspect that some of the examples of so-called chromosome speciation may well be more correctly interpreted as polytypic systems. Key,[181] for example, has long maintained this to be the case for those parapatrically distributed and chromosomally distinct taxa of the *Vandiemenella* complex which White regards as species. It is often a matter of convenience when we decide to consider two taxa as separate species rather than subspecies or races since in nature the transition cannot always be abrupt. But for chromosome speciation in the strict sense the demarcation is expected to be much more clear-cut. Thus, the only cases where fixed chromosome differences can be seriously considered as causative factors in speciation are those where in the hybrid between the two parents, it is the structural rearrangement in respect of which the parents are differentiated that leads to consistent malbehavior. On

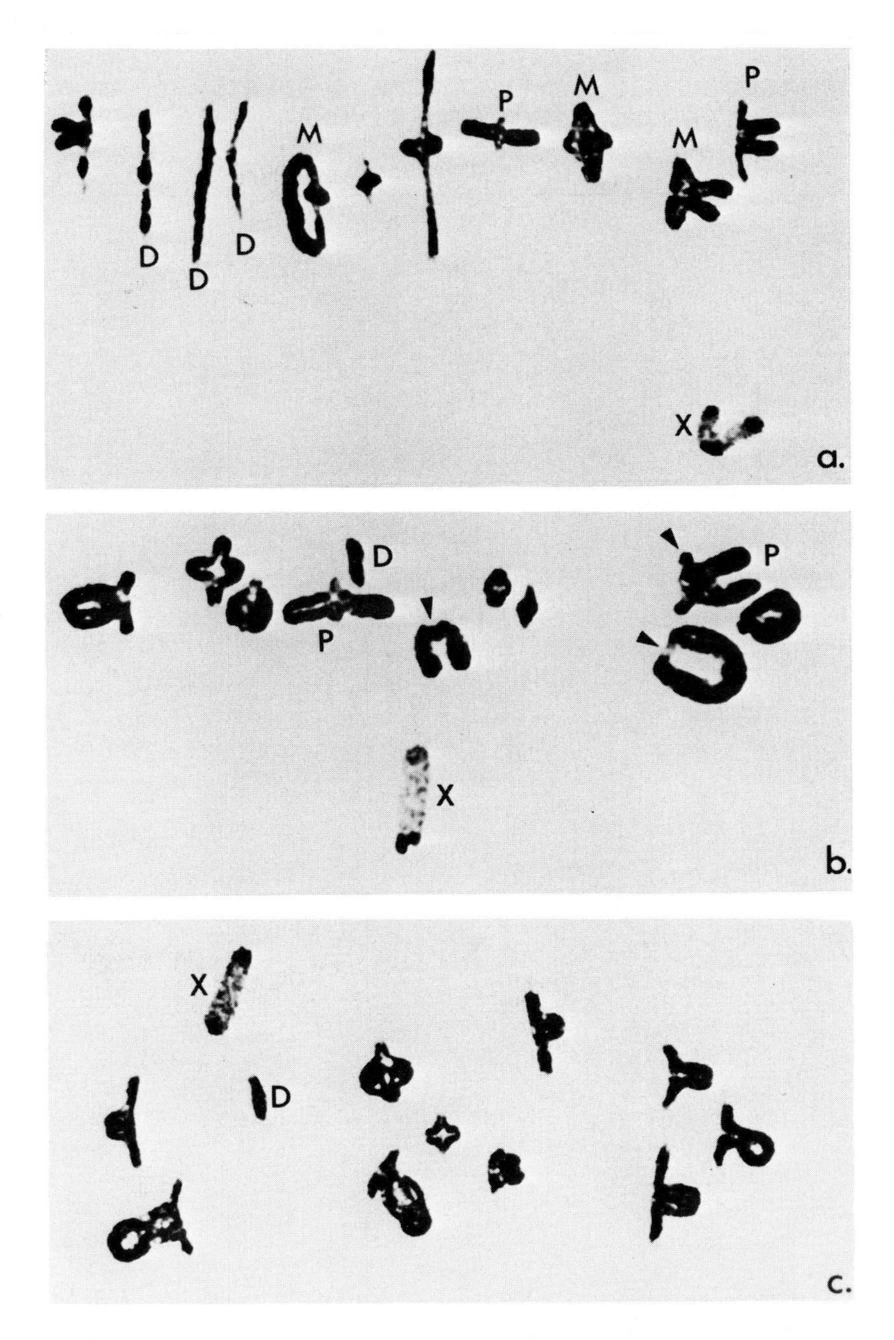

FIGURE 42. Metaphase I of male meiosis in three acridid grasshoppers, all with $2n = 23♂$,XO, to illustrate variation in chiasma frequency and chiasma location. P = proximal and D = distal chiasmata while M = metacentric bivalent. (a) *Trimerotropis pseudofasciata* with three M bivalents and a total of 14 chiasmata; (b) *Platypygius platypygius* with a total of 15 chiasmata; note three bivalents have clear short arms (▲) which in two cases form terminal associations; (c) *Oedipoda germanica* with a total of 22 chiasmata.

Table 61
CHIASMA DISTRIBUTION PATTERNS IN MALES[a] AND FEMALES[b] OF *STETHOPHYMA GROSSUM*[175]

Percent chiasmata	Bivalent types							
	L_1L_2	L_3M_4	M_5	M_6	M_7M_8	M_9	S_{10}	S_{11}
Males								
Proximal	99.84	100.00	97.05	8.68	98.51	7.58	16.62	9.70
Interstitial	0.00	0.00	1.77	89.82	0.15	19.95	38.48	48.82
Distal	0.16	0.00	1.18	1.50	1.35	72.47	44.90	48.48
Females								
Proximal	3.93			6.64			0.00	0.00
Interstitial	50.71			35.51			4.55	0.00
Distal	45.36			58.05			95.45	100.00

[a] 330 Spermatocytes from 22♂s, 11 II + XI.
[b] 65 Oocytes from 10♀s, 11 II + XII.

Table 62
VARIATION IN MEAN MALE CELL CHIASMA FREQUENCY ($\overline{X}a$) AND CHIASMA DISTRIBUTION IN RELATED SPECIES WITHIN THREE DISTINCT GROUPS OF MORABINE GRASSHOPPERS[176-178]

Species	♂ 2*n*	A/B C/D systems	Male sex system	Mean Xa frequency	Xa distribution pattern
Moraba	17	$\overset{\frown}{AB}$, $\overset{\frown}{CD}$	XO	10.46	Moderate distal localization
walkeri					
darwiniensis	19	A, B, $\overset{\frown}{CD}$	XO	10.00	Pronounced proximal localization
serricornis	19	A, B, $\overset{\frown}{CD}$	XO	11.34	Random
Culmacris					
archaica	16	$\overset{\frown}{AB}$, $\overset{\frown}{CD}$	XY	8.16	Strict distal localization
orientalis	16	$\overset{\frown}{AB}$, $\overset{\frown}{CD}$	XY	10.00	Predominantly distal
curvicerus	15	$\overset{\frown}{AB}$, $\overset{\frown}{CD}$	X_1X_2Y	11.99	Little localization
diversa	17	$\overset{\frown}{AB}$, $\overset{\frown}{CD}$	X_1X_2Y	9.00	Strong distal localization
	15	$\overset{\frown}{AB}$, $\overset{\frown}{CD}$	X_1X_2Y	10.00	Strong distal localization
	19	A, B, $\overset{\frown}{CD}$	X_1X_2Y	14.00	A—D form proximal distal rings, rest show distal localization
P 52a	16	$\overset{\frown}{AB}$, $\overset{\frown}{CD}$	XY	11.40	Random
85	16	$\overset{\frown}{AB}$, $\overset{\frown}{CD}$	XY	11.00	Distal localization
53	16	$\overset{\frown}{AB}$, $\overset{\frown}{CD}$	XY	9.6	Random
52b	15	$\overset{\frown}{AB}$, $\overset{\frown}{CD}$	X_1X_2Y	8.8	Extreme distal localization

a priori grounds this is especially likely when several members of the parental karyotypes are characterized by a similar structural change so that multiple differences appear in F_1 hybrids, as in *Eyprepocnemis* (Section IV.A.3).

Unfortunately, many of the arguments concerning speciation are based on circumstantial evidence rather than direct observation and, as such, are not amenable to conventional forms of proof or disproof. This is particularly so in the controversy as to whether or not complete geographical separation of populations is an essential precondition for speciation, the allopatric hypothesis. It clearly cannot be a prerequisite in speciation by alloploidy which, however, in orthopteroids has occurred only very exceptionally and then in conjunction with parthenogenesis. Most authors, nevertheless, believe that, this situation aside, species otherwise originate in allopatry. Genetic isolation is, however, likely to be much more effective than geographical isolation which is capricious. Moreover, present-day geographic isolation may have been determined by a prior genetic discontinuity that is no longer apparent. Equally, of course, genetic isolation may relate to an ecological discontinuity which is also not apparent.

Chromosome rearrangements which impair meiotic pairing or segregation both lead to infertility. The most common cause of such infertility in orthopteroid hybrids is the production of univalent chromosomes. These may arise either as a result of structural differences between the chromosomes of the parents involved in the production of the hybrid or from genotypic differences between them. In several of the hybrids in Section IV.A the infertility appeared to be determined as much by differences in genotype as by chromosome differences per se. This was true in *Caledia* and *Teleogryllus*. In *Trimerotropis* and some of the *Vandiemenella* hybrids the chromosome differences between the parents evidently do make a contribution to infertility though, in the former at least, genotypic differences again play a role. This is not surprising when one recalls that the reproductive isolation in many species depends on two or more mutually reinforcing mechanisms. In trimerotropine grasshoppers, for example, distinct crepidation patterns occur which serve to attract females to males of the same species, but evidently this does not preclude successful mating between distinct species. Chromosome and genotype differences, however, ensure that no progeny ensue from such matings.

VII. CONCLUSION

The contributions made by the study of orthopteroid chromosomes to evolution leave us with more questions than answers. In attempting to provide missing answers six facts should be remembered:

1. The seemingly diverse chromosome systems in nature sometimes offer different ways of achieving much the same end result. The meiotic consequences of inversion hybridity, for example, are in many ways similar to those resulting from chiasma localization. Different species may, therefore, follow different evolutionary pathways to arrive at essentially the same goal.
2. The population structure of most orthopteroids is as yet unknown. The definition of a population as a sympatric group of individuals interbreeding at random is an ideal concept for model builders and theoreticians. Random breeding, unquestionably, is one of the great illusions of theoretical population biology.
3. Muller[98] argued, and many have accepted his argument, that unlike allelic mutations which recur, each chromosome mutation constitutes a unique event. The implication is that in any population or any species all identical chromosome variants must be descended from a single original mutant. This applies no matter how widely scattered the descendants may now be. By contrast allelic mutants can be polyphyletic. There seems little doubt that this view is wrong. Some variants such as B chromosomes may well also recur. So, too, must some of the mutant chromosome conditions found in man. The key question is whether two rearrangements which are morphologically and indeed functionally indistinguishable do, in fact, share identical break points.
4. Several authors have attempted to correlate karyotype stability or lability with life characters, arguing that stability will be favored by factors such as high mobility, uniform environment, and low reproductive rate. Alternatively, karyotype lability might be expected in species with low vagility, a varied environment, and a high reproductive rate. While there may be some truth to this argument it is not without its complications. White,[108] for example, relates the extensive chromosome changes in the Australian morabine grasshoppers to the fact that they

constitute a group of apterous insects with low vagility. However, as he himself points out,[88] the two African subfamilies of eumastacids — the Thericleinae and the Euschmiditiinae — are also flightless and seem to have essentially the same population structure and restricted geographical ranges as the morabines but have nothing like the cytogenetic diversity. Even within the Morabinae there are genera, such as *Geckomima* and *Callita,* in which the karyotype is uniform. Yet here, too, one finds restricted vagility, small populations, complete lack of sound production, and an association with a specific group of host plants.[182] Moreover, much depends on what one means by karyotypic stability. The B chromosome of *Chortoicetes terminifera,* for example, is unique among orthopteroids in responding to the G-banding technique. Using this technique it is possible to identify no less than 23 variant forms of the chromosome.[183] All of them, however, retain a relatively constant external morphology. Thus, in the absence of banding one would have concluded that the B was invariant and stable whereas, in reality, it has evidently undergone considerable diversification since its origin.

5. In eukaryotes there is an important distinction between the soma and the germ line. The soma is, in reality, a transient entity serving essentially to sustain and support the germ line which is set aside from it at an early embryonic stage. While evolutionary change in eukaryotes appears to have been concerned solely with the elaboration and expansion of the soma and is conventionally discussed exclusively in these terms, these somatic complications serve ultimately to ensure efficient germ line transfer. The soma is a mortal entity which the germ line builds anew in each generation. The germ line itself provides the only continuity between generations. The consequences of meiosis in the germ line, thus, has no meaning for the soma of the individual that houses it. But it has a meaning for the soma of the progeny of that germ line since the composition of the gametes of one generation determines the structure of the soma in the next.
6. The new insights which DNA sequencing studies are providing on the architecture of the genome must ultimately affect many perceptions of chromosome evolution. They have already confirmed that only a small proportion of the total genome codes for structural genes. This, in turn, has led to a widespread belief that it is the repetitive component of the genome rather than its unique sequence component which provides the basis for much of evolutionary change. As yet, however, it has not been possible to graft the new information in any meaningful sense onto the present evolutionary framework. Consequently no attempt has been made in this article to add to the speculations currently available.

REFERENCES

1. **White, M. J. D.,** *Blattodea, Mantodea, Isoptera, Grylloblattodea, Phasmatodea, Dermaptera and Embioptera. Animal Cytogenetics 3, Insecta 2,* Gebrüder Borntraeger, Stuttgart, 1976.
2. **Hewitt, G. M.,** *Orthoptera. Animal Cytogenetics 3, Insecta 1,* Gebrüder Borntraeger, Stuttgart, 1979.
3. **Haines, R. L., Roberts, P. A., and Lattin, J. D.,** Paracentric inversion polymorphism in the grasshopper *Boonacris alticola, Chromosoma,* 65, 185, 1978.
4. **Nur, U.,** Synapsis and crossing over within a paracentric inversion in the grasshopper *Camnulla pellucida, Chromosoma,* 25, 198, 1968.
5. **Weissman, D. B.,** Geographical variability in the pericentric inversion system of the grasshopper *Trimerotropis pseudofasciata, Chromosoma,* 55, 325, 1976.
6. **Jackson, R. C.,** Chromosomal evolution in *Haplopappus gracilis:* a centric transposition race, *Evolution,* 27, 243, 1973.

7. Schroeter, G. L., Pericentric Inversion Polymorphism in *Trimerotropis helferi*, Ph.D. thesis, University of California, Davis, 1968.
8. White, M. J. D., *Animal Cytology and Evolution*, Cambridge University Press, New York, 1973.
9. de Vaio, E. S., Goñi, B., and Rey, C., Chromosome polymorphism in populations of the grasshopper *Trimerotropis pallidipennis* from southern Argentina, *Chromosoma*, 71, 371, 1979.
10. White, M. J. D., A cytological survey of wild populations of *Trimerotropis* and *Circotettix* (Orthoptera: Acrididae). II. Racial differentiation in *T. sparsa*, *Genetics*, 36, 31, 1951.
11. White, M. J. D., Structural heterozygosity in natural populations of the grasshopper *Trimerotropis sparsa*, *Evolution*, 5, 376, 1951.
12. White, M. J. D. and Morley, F. H. W., Effects of pericentric rearrangements on recombination in grasshopper chromosomes, *Genetics*, 40, 604, 1955.
13. White, M. J. D., Adaptive chromosomal polymorphism in an Australian grasshopper, *Evolution*, 10, 298, 1956.
14. White, M. J. D., Cytogenetics of the grasshopper *Moraba scurra*. II. Heterotic systems and their interaction, *Aust. J. Zool.*, 5, 305, 1957.
15. Lewontin, R. C. and White, M. J. D., Interaction between inversion polymorphisms of two chromosome pairs in the grasshopper *Moraba scurra*, *Evolution*, 14, 116, 1960.
16. White, M. J. D., Lewontin, R. C., and Andrew, L. E., Cytogenetics of the grasshopper *Moraba scurra*. VII. Geographical variation of adaptive properties of inversions, *Evolution*, 17, 147, 1963.
17. Colgan, D. J. and Cheney, J., The inversion polymorphism of *Keyacris scurra* and the adaptive topography, *Evolution*, 34, 181, 1980.
18. White, M. J. D. and Andrew, L. E., Cytogenetics of the grasshopper *Moraba scurra*. V. Biometric effects of chromosomal inversions, *Evolution*, 14, 284, 1960.
19. White, M. J. D. and Key, K. H. L., A cytotaxonomic study of the pulsilla group of species in the genus *Austroicetes* Uv. (Orthoptera: Acrididae), *Aust. J. Zool.*, 5, 56, 1957.
20. Nankivell, R. N., Interactions between inversion polymorphisms and the colour pattern polymorphism in the grasshopper *Austroicetes interioris* (White and Key), *Acrida*, 3, 93, 1974.
21. Ohmachi, F. and Ueshima, N., A study of local variation of chromosome complements in *Scapsipedus aspersus* Chopard (Orthoptera: Gryllidea), *Bull. Fac. Agric. Mie Univ.*, 14, 43, 1957.
22. Lewis, K. R. and John, B., Studies on *Periplaneta americana*. II. Interchange heterozygosity in isolated populations, *Heredity*, 11, 11, 1957.
23. John, B. and Lewis, K. R., Studies on *Periplaneta americana*. III. Selection for heterozygosity, *Heredity*, 12, 185, 1958.
24. John, B. and Quraishi, H. B., Studies on *Periplaneta americana*. IV. Pakistani populations, *Heredity*, 19, 147, 1964.
25. John, B. and Lewis, K. R., Selection for interchange heterozygosity in an inbred culture of *Blaberus discoidalis* (Serville), *Genetics*, 44, 251, 1959.
26. Lewis, K. R. and John, B., *Chromosome Marker*, Churchill, London, 1963, 66.
27. Wise, D. and Rickards, G. K., A quadrivalent studied in living and fixed grasshopper spermatocytes, *Chromosoma*, 63, 305, 1977.
28. Lewis, K. R. and John, B., Spontaneous interchange in *Chorthippus brunneus*, *Chromosoma*, 14, 618, 1963.
29. John, B. and Hewitt, G. M., A spontaneous interchange in *Chorthippus* with extensive chiasma formation in an interstitial segment, *Chromosoma*, 14, 638, 1963.
30. James, S. H., Complex hybridity in *Isotoma petraea*. II. Components and operation of a possible evolutionary mechanism, *Heredity*, 25, 53, 1970.
31. Ross, M. H. and Cochran, D. G., The German cockroach *Blatella germanica*, in *Handbook of Genetics*, Vol. 3, King, R. C., Ed., Plenum Press, New York, 1975, 35.
32. Ross, M. H. and Cochran, D. G., Two new reciprocal translocations in the German cockroach: cytology and genetics of T(3,12) and T(7,12), *J. Heredity*, 66, 79, 1975.
33. Cochran, D. G., Patterns of disjunction frequencies in heterozygous reciprocal translocations from the German cockroach, *Chromosoma*, 62, 191, 1977.
34. Ross, M. H. and Cochran, D. G., Properties of a 3-chromosome double translocation heterozygote in the German cockroach, *J. Heredity*, 70, 259, 1979.
35. Ross, M. H. and Cochran, D. G., Synthesis and properties of a double translocation heterozygote involving a stable ring of six interchange in the German cockroach, *J. Heredity*, 72, 39, 1981.
36. Ueshima, N., Chromosomal polymorphism in a species of field cricket, *Anaxipha pallidula* (Orthoptera, Gryllodea), *J. Nagoya Jogakuin Coll.*, 4, 78, 1957.
37. Hewitt, G. M. and Schroeter, G., Population cytology of *Oedaleonotus*. I. The karyotypic facies of *Oedaleonotus enigma* (Scudder), *Chromosoma*, 25, 121, 1968.
38. Kayano, H. and Nakamura, K., Chiasma studies in structural hybrids. V. Heterozygotes for a centric fusion and for a translocation in *Acrida lata*, *Cytologia*, 25, 476, 1960.

39. **Lewis, K. R. and John, B.,** Breakdown and restoration of chromosome stability following inbreeding in a locust, *Chromosoma,* 10, 589, 1959.
40. **Sharma, G. P., Parshad, R., and Bedi, T. S.,** Breakdown of the meiotic stability in *Crotogonus trachypterus* (Blanchard) (Orthoptera: Acridoidea: Pyrgomorphidae), *Res. Bull. (N.S.) Panjab Univ.,* 13, 281, 1962.
41. **Peters, G. B.,** Germ line polysomy in the grasshopper *Atractomorpha similis, Chromosoma,* 81, 593, 1981.
42. **Hewitt, G. M. and John, B.,** Parallel polymorphism for supernumerary segments in *Chorthippus parallelus* (Zetterstedt). I. British populations, *Chromosoma,* 25, 319, 1968.
43. **Nur, U.,** Mitotic instability leading to an accumulation of B-chromosomes in grasshoppers, *Chromosoma,* 27, 1, 1969.
44. **Hewitt, G. M. and John, B.,** The B-chromosome system of *Myrmeleotettix maculatus* (Thunb.). IV. The dynamics, *Evolution,* 24, 169, 1970.
45. **Camacho, J. P. M., Carballo, A. R., and Cabrero, J.,** The B-chromosome system of the grasshopper *Eyprepocnemis plorans* subspecies *plorans* (Charpentier), *Chromosoma,* 80, 163, 1980.
46. **Bergerard, J., Carton, Y., and Lespinasse, R.,** Analyse électrophorétique en gel de polyacrylamide des protéines de l'hémolymphe de *Locusta migratoria* L. Influence des chromosomes surnuméraires dans la sous-espêce migratoroides, Reiche et Fairmaire, *C.R. Acad. Sci. (Paris),* 275, 783, 1972.
47. **Nur, U.,** A mitotically unstable supernumerary chromosome with an accumulation mechanism in a grasshopper, *Chromosoma,* 14, 407, 1963.
48. **Kayano, H.,** Accumulation of B chromosomes in the germ line of *Locusta migratoria, Heredity,* 27, 119, 1971.
49. **Nur, U.,** Maintenance of 'parasitic' B chromosome in the grasshopper *Melanoplus femur-rubrum, Genetics,* 87, 499, 1977.
50. **Lucov, Z. and Nur, U.,** Accumulation of B chromosomes by preferential segregation in females of the grasshopper *Melanoplus femur-rubrum, Chromosoma,* 42, 289, 1973.
51. **Kimura, M.,** A suggestion on the experimental approach to the origin of supernumerary chromosomes, *Am. Nat.,* 96, 319, 1962.
52. **Jackson, W. D. and Cheung, D. S. M.,** Distortional meiotic segregation of a supernumerary chromosome producing differential frequencies in the sexes in the short horned grasshopper *Phaulacridium vittatum, Chromosoma,* 23, 24, 1967.
53. **John, B. and Freeman, M.,** B chromosome behavior in *Phaulacridium vittatum, Chromosoma,* 46, 181, 1974.
54. **John, B. and Freeman, M.,** The cytogenetic structure of Tasmanian populations of *Phaulacridium vittatum, Chromosoma,* 53, 283, 1975.
55. **Rowe, H. J. and Westerman, M.,** Population cytology of the genus *Phaulacridium.* V. *Phaulacridium vittatum* (Sjöst): Australian mainland populations, *Chromosoma,* 46, 197, 1974.
56. **Fontana, P. G. and Vickery, V. R.,** Segregation — distortion in the B chromosome system of *Tettigidea lateralis* (Say) (Orthoptera: Tetrigidae), *Chromosoma,* 43, 75, 1973.
57. **Fontana, P. G. and Vickery, V. R.,** The B chromosome system of *Tettigidea lateralis.* II. New karyomorphs, patterns of pycnosity and giemsa banding, *Chromosoma,* 50, 371, 1975.
58. **Kayano, H., Sannomiya, M., and Nakamura, K.,** Cytogenetic studies on natural populations of *Acrida lata, Heredity,* 25, 113, 1970.
59. **Sannomiya, M.,** Cytogenetic studies on natural populations of grasshoppers with special reference to B chromosomes. I. *Gonista bicolor, Heredity,* 32, 251, 1974.
60. **Hewitt, G. M.,** Meiotic drive for B chromosomes in the primary oocytes of *Myrmeleotettix maculatus* (Orthoptera: Acrididae), *Chromosoma,* 56, 381, 1976.
61. **Hewitt, G. M.,** Variable transmission rates of a B chromosome in *Myrmeleotettix maculatus* (Thunb.) (Acrididae: Orthoptera), *Chromosoma,* 40, 83, 1973.
62. **Hewitt, G. M. and John, B.,** The B chromosome system of *Myrmeleotettix maculatus. III. The statistics, Chromosoma,* 21, 140, 1967.
63. **Ramel, C.,** A B chromosome system of *Myrmeleotettix maculatus* (Thunb.) (Orthoptera: Acrididae) in Sweden, *Hereditas,* 92, 309, 1980.
64. **Hewitt, G. M. and Brown, F. M.,** The B chromosome system of *Myrmeleotettix maculatus.* V. A steep cline in East Anglia, *Heredity,* 25, 363, 1970.
65. **John, B. and Hewitt, G. M.,** The B-chromosome system of *Myrmeleotettix maculatus.* II. The statics, *Chromosoma,* 17, 121, 1965.
66. **Fletcher, H. L. and Hewitt, G. M.,** Effect of a B-chromosome on chiasma localisation and frequency in male *Euthystira brachyptera, Heredity,* 44, 341, 1980.
67. **Gururaj, M. E. and Rajasekarasetty, M. R.,** Impact of B chromosome polymorphism on the endophenotype of *Acrotylus humbertianus* (Acridoidea: Orthoptera), *J. Cytol. Genet. Congr. Suppl. Proc. 1st Ind. Congr. Cytol. Genet.,* p. 151, 1971.

68. **Rees, H. and Dale, P. J.,** Chiasmata and variability in Lolium and Festuca populations, *Chromosoma,* 47, 335, 1974.
69. **Shaw, D. D.,** The supernumerary segment system of *Stethophyma.* II. Heterochromatin polymorphism and chiasma variation, *Chromosoma,* 34, 19, 1971.
70. **Nur, U.,** Meiotic behaviour of an unequal bivalent in the grasshopper *Calliptamus palaestinensis* (Bdhr.), *Chromosoma,* 12, 272, 1961.
71. **John, B.,** The cytogenetic systems of grasshoppers and locusts. II. The origin and evolution of supernumerary segments, *Chromosoma,* 44, 123, 1973.
72. **John, B. and Hewitt, G. M.,** A polymorphism for heterochromatic supernumerary segments in *Chorthippus parallelus, Chromosoma,* 18, 254, 1966.
73. **Hewitt, G. M. and John, B.,** Parallel polymorphism for supernumerary segments in *Chorthippus parallelus* (Zetterstedt). I. British populations, *Chromosoma,* 25, 319, 1968.
74. **Westerman, M.,** Parallel polymorphism for supernumerary segments in *Chorthippus parallelus* (Zetterstedt). II. French populations, *Chromosoma,* 26, 7, 1969.
75. **Shaw, D. D.,** The supernumerary segment system of *Stethophyma.* I. Structural basis, *Chromosoma,* 30, 326, 1970.
76. **Kayano, H.,** Polymorphism for a supernumerary segment in natural populations of *Mecostethus magister* (Orthoptera), *Jpn. J. Genet.,* 48, 151, 1973.
77. **Miklos, G. L. G. and Nankivell, R. N.,** Telomeric satellite DNA functions in regulating recombination, *Chromosoma,* 56, 143, 1976.
78. **John, B.,** Heterochromatin variation in natural populations, *Chromosomes Today,* 7, 128, 1980.
79. **White, M. J. D.,** Some general problems of chromosomal evolution and speciation in animals, *Survey Biol. Prog.,* 3, 109, 1957.
80. **White, M. J. D.,** Cytogenetics of speciation, *J. Aust. Ent. Soc.,* 9, 1, 1970.
81. **White, M. J. D.,** The value of cytology in taxonomic research on Orthoptera, in *Proc. Int. Study Conf. on Current and Future Problems of Acridology,* Hemming, C. F. and Taylor, T. H. D., Eds., Centre for Overseas Pest Research, London, 1972, 27.
82. **Syren, R. M. and Luykx, P.,** Permanent segmental interchange complex in the termite *Incisitermes schwarzi, Nature,* 266, 167, 1977.
83. **Luykx, P. and Syren, R. M.,** The cytogenetics of *Incisitermes schwarzi* and other Florida termites, *Sociobiology,* 4, 191, 1979.
84. **Syren, R. M. and Luykx, P.,** Geographic variation of sex linked translocation heterozygosity in the termite *Kalotermes approximatus, Chromosoma,* 82, 65, 1981.
85. **Vincke, P. O. and Tilquin, J. P.,** A sex linked ring quadrivalent in Termitidae (Isoptera), *Chromosoma,* 67, 151, 1978.
86. **Fontana, F.,** Interchange complexes in Italian populations of *Reticulitermes lucifugus* Rossi (Isoptera: Rhinotermitidae), *Chromosoma,* 18, 169, 1980.
87. **Mesa, A. and De Mesa, R. S.,** Complex sex determining mechanisms in three species of South American grasshoppers (Orthoptera: Acridoidea), *Chromosoma,* 21, 163, 1967.
88. **White, M. J. D.,** Karyotypes of some members of the grasshopper families Lentulidae and Charilaidae, *Cytologia,* 32, 184, 1967.
89. **John, B. and Hewitt, G. M.,** Interpopulation sex chromosome polymorphism in the grasshopper *Podisma pedestris.* I. Fundamental facts, *Chromosoma,* 31, 291, 1970.
90. **Hewitt, G. M. and John, B.,** Interpopulation sex chromosome polymorphism in the grasshopper *Podisma pedestris.* II. Population parameters, *Chromosoma,* 37, 23, 1972.
91. **Hewitt, G. M.,** A sex chromosome hybrid zone in the grasshopper *Podisma pedestris* (Orthoptera: Acrididae), *Heredity,* 35, 375, 1975.
92. **Hewitt, G. M. and Barton, N. H.,** The structure and maintenance of hybrid zones as exemplified by *Podisma pedestris,* in *Insect Cytogenetics,* Symp. No. 10, Royal Entomological Society of London, 1980, 149.
93. **Barton, N. H.,** The fitness of hybrids between two chromosomal races of the grasshopper *Podisma pedestris, Heredity,* 45, 47, 1980.
94. **Craddock, E. M.,** Chromosomal diversity in the Australian Phasmatodea, *Aust. J. Zool.,* 20, 445, 1972.
95. **Saez, F. A. and Perez-Mosquera, G.,** Citogenética del género *Dichroplus* (Orthoptera: Acrididae), *Recientes Adelantos Biol. 5° Congr. Argent. Ciencias Biol. Buenos Aires,* p. 111, 1971.
96. **Craddock, E. M.,** Chromosomal evolution and speciation in *Didymuria,* in *Genetic Mechanisms of Speciation in Insects,* White, M. J. D., Ed., Australia and New Zealand Book Co., Sydney, 1974, 24.
97. **Craddock, E. M.,** Intraspecific karyotypic differentiation in the Australian phasmatid *Didymuria violescens* (Leach). I. The chromosome races and their structural and evolutionary relationships, *Chromosoma,* 53, 1, 1975.

98. **Muller, H. J.**, Bearing of the *Drosophila* work on systematics in *The New Systematics,* Huxley, J., Ed., Clarendon Press, Oxford, 1940, 185.
99. **Moens, P. B.**, Kinetochores of grasshoppers with Robertsonian fusions, *Chromosoma,* 67, 41, 1978.
100. **Moens, P. B.**, Kinetochore microtubule numbers of different sized chromosomes, *J. Cell Biol.,* 83, 556, 1979.
101. **Sarto, G. E. and Therman, E.**, Replication and inactivation of a dicentric X formed by telomeric fusion, *Am. J. Obstet. Gynaecol.,* 136, 904, 1980.
102. **Holmquist, G. P. and Dancis, B.**, Telomere replication, kinetochore organisers and satellite DNA evolution, *P.N.A.S.,* 76, 4566, 1979.
103. **White, M. J. D.**, Cytogenetics of the grasshopper *Moraba scurra.* I. Meiosis of interracial and interpopulation hybrids, *Aust. J. Zool.,* 5, 285, 1957.
104. **Southern, D. I.**, Stable telocentric chromosomes produced following centric misdivision in *Myrmeleotettix maculatus* (Thunb.), *Chromosoma,* 26, 140, 1969.
105. **Hewitt, G. M. and John, B.**, The cytogenetic systems of grasshoppers and locusts. I. *Chortoicetes terminifera, Chromosoma,* 34, 302, 1971.
106. **Solari, A. J. and Counce, S. J.**, Synaptonemal complex karyotyping in *Melanoplus differentialis, J. Cell Sci.,* 26, 229, 1977.
107. **White, M. J. D.**, Chromosomal rearrangements and speciation in animals, *Ann. Rev. Genet.,* 3, 75, 1969.
108. **White, M. J. D. and Chinnick, L. J.**, Cytogenetics of the grasshopper *Moraba scurra.* III. Distribution of the 15 and 17 chromosome races, *Aust. J. Zool.,* 5, 338, 1957.
109. **White, M. J. D.**, *Modes of Speciation,* Freeman, San Francisco, 1978.
110. **John, B. and King, M.**, Heterochromatin variation in *Cryptobothrus chrysophorus.* I. Chromosome differentiation in natural populations, *Chromosoma,* 64, 219, 1977.
111. **John, B. and King, M.**, Heterochromatin variation in *Cryptobothrus chrysophorus.* II. Patterns of C-banding, *Chromosoma,* 65, 59, 1977.
112. **John, B. and King, M.**, Heterochromatin variation in *Cryptobothrus chrysophorus.* III. Synthetic hybrids, *Chromosoma,* 78, 165, 1980.
113. **John, B. and Lewis, K. R.**, *The Meiotic System, Protoplasmatologia,* VIf1, Springer-Verlag, Vienna, 1965.
114. **Fontana, P. G. and Hogan, T. W.**, Cytogenetic and hybridisation studies of geographic populations of *Teleogryllus commodus* (Walker) and *T. oceanicus* (Le Guillold) (Orthoptera: Gryllidae), *Aust. J. Zool.,* 17, 13, 1969.
115. **Shaw, D. D.**, Population cytogenetics of the genus *Caledia* (Orthoptera: Acridinae). I. Inter and intraspecific karyotype diversity, *Chromosoma,* 54, 221, 1976.
116. **Shaw, D. D., Webb, G. C., and Wilkinson, P.**, Population cytogenetics of the genus *Caledia* (Orthoptera: Acridinae). II. Variation in the pattern of C-banding, *Chromosoma,* 56, 169, 1976.
117. **Moran, C. and Shaw, D. D.**, Population cytogenetics of the genus *Caledia* (Orthoptera: Acridinae). III. Chromosomal polymorphism, racial parapatry and introgression, *Chromosoma,* 63, 181, 1977.
118. **Moran, C.**, The structure of the hybrid zone in *Caledia captiva, Heredity,* 42, 13, 1977.
119. **Shaw, D. D., Moran, C., and Wilkinson, P.**, Chromosomal reorganisation, geographical differentiation and the mechanism of speciation in the genus *Caledia,* in *Insect Cytogenetics,* Symp. No. 10, Royal Entomological Society of London, 1980, 171.
120. **Shaw, D. D. and Wilkinson, P.**, 'Homologies' between non-homologous chromosomes in the grasshopper *Caledia captiva, Chromosoma,* 68, 241, 1978.
121. **Shaw, D. D. and Wilkinson, P.**, Chromosome differentiation, hybrid breakdown and the maintenance of a narrow hybrid zone in *Caledia captiva, Chromosoma,* 80, 1, 1980.
122. **Moran, C.**, Spermatogenesis in natural and experimental hybrids between chromosomally differentiated taxa of *Caledia captiva, Chromosoma,* 81, 579, 1981.
123. **Perdeck, A. C.**, The Isolating Value of Specific Song Patterns in Two Sibling Species of Grasshopper (*Chorthippus brunneus* Thunb. and *Chorthippus biguttulus* L.), E. J. Brill, Leiden, 1957.
124. **Klingstedt, H.**, Failure of anaphase separation in species hybrids, *Nature,* 141, 606, 1938.
125. **Klingstedt, H.**, Taxonomic and cytological studies on grasshopper hybrids. I. Morphology and spermatogenesis of *Chorthippus bicolor* Charp. and *Chorthippus biguttulus* L., *J. Genet.,* 37, 389, 1939.
126. **Darlington, C. D.**, Crossing over and its relationships in *Chorthippus* and *Stauroderus, J. Genet.,* 33, 465, 1936.
127. **Lewis, K. R. and John, B.**, The meiotic consequences of spontaneous chromosome breakage, *Chromosoma,* 18, 287, 1966.
128. **Weissman, D. B. and Rentz, D. C. F.**, Cytological, morphological and crepitational characteristics of the trimerotropines *(Aerochoreutes, Circotettix and Trimerotropis)* grasshoppers (Orthoptera: Oedipodinae), *Trans. Am. Ent. Soc.,* 106, 253, 1980.
129. **John, B. and Weissman, D. B.**, Cytogenetic components of reproductive isolation in *Trimerotropis thalassica* and *Trimerotropis occidentalis, Chromosoma,* 60, 187, 1977.

130. **Helwig, E. R.,** Spermatogenesis in hybrids between *Circotettix verruculatus* and *Trimerotropis suffusa* (Orthoptera: Oedipodidae), *Univ. Colo. Stud. Ser. Biol.*, 10, 49, 1955.
131. **White, M. J. D., Carson, H. L., and Cheney, J.,** Chromosomal races in the Australian grasshopper *Moraba viatica* in a zone of geographic overlap, *Evolution*, 18, 417, 1964.
132. **White, M. J. D., Blackith, R. E., Blackith, R. M., and Cheney, J.,** Cytogenetics of the viatica group of Morabine grasshoppers. I. The coastal species, *Aust. J. Zool.*, 15, 263, 1967.
133. **Mrongrovius, M. J.,** Cytogenetics of the hybrids of three members of the grasshopper genus *Vandiemenella* (Orthoptera: Eumastacidae: Morabinae), *Chromosoma*, 71, 81, 1979.
134. **John, B. and Lewis, K. R.,** Genetic speciation in the grasshopper *Eyprepocnemis plorans*, *Chromosoma*, 16, 308, 1965.
135. **Atchley, W.,** The chromosome karyotype in estimation of lineage relationships, *Syst. Zool.*, 21, 199, 1972.
136. **Kumaraswamy, K. R. and Rajasekaresetty, M. R.,** Chromosome repatterning in Acrididae, *Cytobios*, 19, 21, 1977.
137. **Saez, F. A. and Perez-Mosquera, G.,** Structure, behaviour and evolution of the chromosomes of *Dichroplus silveiraguidoi* (Orthoptera: Acrididae), *Genetica*, 47, 105, 1977.
138. **Fontana, P. G. and Vickery, V. R.,** Cytotaxonomic studies on the genus *Boonacris*. I. The 'eastern' taxa and a comparison with the related genera *Dendrotettix* and *Appalachia* (Orthoptera: Catantopidae: Podismini), *Can. J. Genet. Cytol.*, 18, 625, 1976.
139. **Gosalvez, J., López-Fernandez, C., and Morales Agacino, E.,** Consideraciones cromosomicas sobre algunas especies del grupo Podismini (Orthoptera: Acrididae), *Acrida*, 9, 133, 1980.
140. **Ueshima, N. and Rentz, D. C.,** Chromosome systems in the North American Decticinae with reference to Robertsonian changes (Orthoptera: Tettigoniidae), *Cytologia*, 44, 693, 1979.
141. **Lim, H.-C.,** Note on the chromosomes of the Nemobiinae (Orthoptera: Gryllidae), *Can. J. Zool.*, 49, 391, 1971.
142. **Alicata, P., Messina, A., and Oliveri, S.,** Determino cromosomico del sesso in *Odontura stenoxipha* (Orthoptera, Phaneropteridae): un nuovo caso di neo-XY, *Animalia*, 1, 109, 1974.
143. **Gosalves, J., López-Fernandez, C., and Esponda, P.,** Variability of the DNA content in five orthopteran species, *Caryologia*, 33, 275, 1980.
144. **Kiknadze, I. I. and Vysotskaya, L. V.,** Measurement of DNA mass per nucleus in the grasshopper species with different numbers of chromosomes, *Tsitologiya*, 12, 1100, 1970.
145. **Rees, H., Shaw, D. D., and Wilkinson, P.,** Nuclear DNA variation among acridid grasshoppers, *Proc. R. Soc. London*, B 202, 517, 1978.
146. **King, M. and John, B.,** Regularities and restrictions governing C-band variation in acridoid grasshoppers, *Chromosoma*, 76, 123, 1980.
147. **Bergerard, J. and Seugé, J.,** La parthénogenèse accidentelle chez *Locusta migratoria* L., *Bull. Biol. Fr. Belg.*, 93, 16, 1959.
148. **White, M. J. D., Cheney, J., and Key, K. H. L.,** A parthenogenetic species of grasshopper with complex structural heterozygosity (Orthoptera: Acridoidea), *Aust. J. Zool.*, 11, 1, 1963.
149. **White, M. J. D.,** Further studies on the cytology and distribution of the Australian parthenogenetic grasshopper *Moraba virgo*, *Rev. Suisse Zool.*, 73, 30, 1966.
150. **White, M. J. D. and Webb, G. C.,** Origin and evolution of parthenogenetic reproduction in the grasshopper *Moraba virgo* (Eumastacidae: Morabinae), *Aust. J. Zool.*, 16, 647, 1968.
151. **Hewitt, G. M.,** A new hypothesis for the origin of the parthenogenetic grasshopper *Moraba virgo*, *Heredity*, 34, 117, 1975.
152. **Webb, G. C., White, M. J. D., Contreras, N., and Cheney, J.,** Cytogenetics of the parthenogenetic grasshopper *Warramaba* (formerly *Moraba*) *virgo* and its bisexual relatives. IV. Chromosome banding studies, *Chromosoma*, 67, 309, 1978.
153. **White, M. J. D., Contreras, N., Cheney, J., and Webb, G. C.,** Cytogenetics of the parthenogenetic grasshopper *Warramaba* (formerly *Moraba*) *virgo* and its bisexual relatives. II. Hybridisation studies, *Chromosoma*, 61, 127, 1977.
154. **White, M. J. D. and Contreras, N.,** Cytogenetics of the parthenogenetic grasshopper *Warramaba* (formerly *Moraba*) *virgo* and its bisexual relatives. III. Meiosis of male "synthetic virgo" individuals, *Chromosoma*, 67, 55, 1978.
155. **White, M. J. D. and Contreras, N.,** Chromosome architecture of the parthenogenetic grasshopper *Warramaba virgo* and its bisexual ancestors, *Chromosomes Today*, 7, 165, 1981.
156. **Pijnacker, L. P. and Harbott, J.,** Structural heterozygosity and aneuploidy in the parthenogenetic stick insect *Carausius morosus* Br. (Phasmatodea: Phasmatidae), *Chromosoma*, 76, 165, 1980.
157. **Bergerard, J.,** Intersexualité expérimentale chez *Carausius morosus* Br. (Phasmidae), *Bull. Biol. Fr. Belg.*, 95, 273, 1961.
158. **Bergerard, J.,** Parthenogenesis in the Phasmidae, *Endeavour*, 21, 137, 1962.

159. **Pijnacker, L. P.**, The Cytology, Sex Determination and Parthenogenesis of *Carausius morosus* Br., Dissertation, University of Groningen, Groningen, The Netherlands, 1964.
160. **Baccetti, B. and Capra, F.**, Notulae Orthopterologicae. XXXIV. Le specie Italiane del genere *Gryllotalpa* L., *Redia,* 61, 401, 1978.
161. **Wright, S.**, Genic and organismic selection, *Evolution,* 34, 825, 1980.
162. **Bengtsson, B. O. and Bodmer, W. F.**, On the increase of chromosome mutations under random mating, *Theor. Pop. Biol.,* 9, 260, 1976.
163. **Lande, R.**, Effective deme sizes during long term evolution estimated from rates of chromosomal rearrangement, *Evolution,* 33, 234, 1979.
164. **Hedrick, P. W.**, The establishment of chromosomal variants, *Evolution,* 35, 322, 1981.
165. **Sunmitt, R. L., Tipton, R. E., Wilroy, F. S., Martens, P. R., and Phelan, J. P.**, X-autosome translocations: a review, *Birth Defects,* 14, 219, 1978.
166. **Jacobs, P. A.**, Correlation between euploid structural chromosome rearrangements and mental subnormality in humans, *Nature,* 249, 164, 1974.
167. **Brandham, P. E.**, The frequency of spontaneous chromosome change, in *Current Chromosome Research,* Jones, K. and Brandham, P. E., Eds., Elsevier/North Holland, Amsterdam, 1976, 77.
168. **Bush, G. L.**, Stasipatric speciation and rapid evolution in mammals, in *Evolution and Speciation,* Atchley, W. R. and Woodruff, D. S., Eds., Cambridge University Press, New York, 1981, 201.
169. **Thorneycroft, H. B.**, A cytogenetic study of the white throated sparrow *Zonotrichia albicollis* (Gmelin), *Evolution,* 29, 611, 1976.
170. **White, M. J. D.**, Cytogenetic mechanisms in insect reproduction, in *Insect Reproduction,* Symp. No. 2, Royal Ent. Soc. Lond., 1964, 1.
171. **John, B. and Freeman, M. F.**, Causes and consequences of Robertsonian exchange, *Chromosoma,* 52, 123, 1975.
172. **John, B. and Hewitt, G. M.**, Parallel polymorphism for supernumerary segments in *Chorthippus parallelus* (Zetterstedt). III. The Ashurst population, *Chromosoma,* 28, 73, 1969.
173. **Fletcher, H. L. and Hewitt, G. M.**, A comparison of chiasma frequency and distribution between sexes in three species of grasshoppers, *Chromosoma,* 77, 129, 1980.
174. **Jones, G. H., Stamford, W. K., and Perry, P. E.**, Male and female meiosis in grasshoppers. II. *Chorthippus brunneus, Chromosoma,* 51, 381, 1975.
175. **Perry, P. E. and Jones, G. H.**, Male and female meiosis in grasshoppers. I. *Stethophyma grossum, Chromosoma,* 47, 227, 1974.
176. **White, M. J. D.**, Karyotypes and meiosis of the morabine grasshoppers. I. *Moraba, Spectriforma* and *Filoraba, Aust. J. Zool.,* 25, 567, 1977.
177. **White, M. J. D.**, Karyotypes and meiosis of the morabine grasshoppers. II. The genera *Culmacris* and *Stilletta, Aust. J. Zool.,* 27, 109, 1979.
178. **White, M. J. D. and Cheney, J.**, Cytogenetics of a group of morabine grasshoppers with XY and X_1X_2Y males, *Chromosomes Today,* 3, 177, 1972.
179. **Hillel, J., Feldman, M. W., and Simchen, G.**, Mating systems and population structure in two closely related species of the wheat group, *Heredity,* 31, 1, 1973.
180. **White, M. J. D.**, Rectangularity in allopatric and non-allopatric speciation and the roles of unique and repetitive DNA sequences in speciation, in *Convergno Internazionale sui Meccanisms di Speciazione,* Accad. Naz. dei Lincei, in press.
181. **Key, K. H. L.**, The concept of stasipatric speciation, *Syst. Zool.,* 17, 14, 1968.
182. **White, M. J. D.**, Modes of speciation in orthopteroid insects, *Bull. Biol.,* 47(Suppl.), 83, 1980.
183. **Webb, G. C. and Neuhaus, P.**, Chromosome organisation in the Australian plague locust *Chortoicetes terminifera.* II. Banding variants of the B chromosome, *Chromosoma,* 70, 205, 1979.

Chapter 2

FISH CYTOGENETICS

Yoshio Ojima

TABLE OF CONTENTS

I. INTRODUCTION

The Pisces are very important for studying animal evolution and for understanding the development of vertebrates in general. A reference to the chromosome monograph, the "Fish Chromosome Retrieval System List" (CDR List) published by Ojima,[1] indicates that basic data necessary to resolve relevant taxonomic and evolutionary problems are quite inadequate. Consequently, an accurate cytogenetic survey of the Pisces becomes increasingly important in order to establish chromosomal relation with the teleosts, on the one hand, and to enable one to have a glimpse of the relations between chromosomal evolution and differentiation of vertebrate species, on the other. Because members of Class Pisces are comparatively difficult for cytogenetic studies, not much progress has been made in this field. Among approximately 20,000 species of fish, the chromosome number is known in about 1000 and complete karyotyping has been made in about 800. The development of cytogenetic techniques for fish chromosomes since 1961, particularly a current air-drying method, combined with colchicine treatment to accumulate cells at metaphase and facilitate the spread of chromosomes, has made possible the accurate delineation of the chromosomes in somatic cells.

In addition, fish cell culture, which provides important materials for studies in various fields of cell biology, has been carried out subsequent to the development of mammalian techniques initiated by Wolf et al.[2] and the establishment of the in vitro cell line.[3] Cell culture techniques became available for the use of leukocytes and cells derived from a variety of sources such as blood, fins, scales, ovaries, kidneys, and eyed embryos, and were proven to be applicable to cytogenetic investigations.[4]

The recent development of new differential staining techniques has made possible the demonstration of characteristic banding patterns of fish chromosomes. These new technical approaches have proved effective for obtaining a better understanding of the characterization of individual chromosomes and for standardizing karyotypes.

II. STATISTICAL RESULTS ON FISH CHROMOSOMES

According to Gosline[5] the teleostean fishes are classified into three larger groups: a lower group including Osteoglossiformes, Clupeiformes, Salmoniformes, Cypriniformes, Anguilliformes, and Notacanthiformes; an intermediate group represented by Myctophiformes, Cetomimiformes, Beloniformes, Percopsiformes, Gadiformes, Cyprinodontiformes, Syngnathiformes, Beryciformes, Lampridiformes, and Zaiformes; and a higher group involving Perciformes, Pleuronectiformes, Scorpaeniformes, Gobiesociformes, Tetraodontiformes, Echeneiformes, Mastacembeliformes, Synbranchiformes, Icosteiformes, Pegasiformes, and Lophiiformes.

The author appended the lowest group to Gosline's classification such as Cyclostomi, involving Petromyzontiformes and Myxiniformes; Chondrichthyes, including Lamniformes, Rajiformes, and Chimaeriformes; Choanichthyes, represented by Lepidosireniformes, Chondrostei with Polypteriformes and Acipenseriformes; and Holostei, involving Lepisosteiformes and Amiiformes.

The chromosome numbers and karyotypes show a parallel relation with this classification system (Tables 1 and 2). The numbers in Cyclostomi show variations from $2n = 36$ (*Paramyxine atami*) to $2n = 168$ (*Petromyzon marinus*). There is a diploid-tetraploid relation between the Northern Hemisphere ($2n = 164$ to 166) and the Southern Hemisphere ($2n = 76$) specimens in chromosome numbers of the Petromyzontiformes.

Only 11 among 1500 living species of the Chondrichthyes have been investigated. The minimum chromosome number was $2n = 28$ (*Narcine brasiliensis*) and the maximum $2n = 98$ (*Raja clavate*). Data on this group are insufficient and it remains a largely unexplored domain.

Table 1
STATISTICAL RESULTS ON CHROMOSOME NUMBERS

	Lowest group	Lower group	Intermediate group	Higher group	Total
Total no. of species	48	483	248	359	1142
Chromosome no. (2*n*)					
Mean	81.6	60.0	42.0	45.9	52.5
Min	28	22	18	16	16
Max	239	206	69	78	239
Modal no. (2*n*)					
1	36 (31%)	50 (38%)	48 (40%)	48 (56%)	48 (31%)
2	164 (10%)	48 (9%)	46 (10%)	44 (14%)	50 (17%)
3	—	100 (8%)	40 (7%)	46 (12%)	44 (7%)
Standard deviation	51.2	22.6	8.0	4.9	21.0

Table 2
STATISTICAL RESULTS ON KARYOTYPES

	Lowest group	Lower group	Intermediate group	Higher group	Total
Total no. of species	20	337	97	332	789
Arm no. of chromosomes					
Mean	74.6	89.7	51.0	53.4	69.2
Min	36	40	42	24	24
Max	164	288	78	118	288
Mode					
1	72 (50%)	100 (6.8%)	48 (29%)	48 (29%)	48 (19%)
2	36 (20%)	96,98 (5.3%)	50 (20%)	54 (11%)	50 (9%)
3	—	72,94 (4.2%)	52 (11%)	50 (11%)	54 (5%)
Standard deviation	34.3	34.2	6.6	11.0	30.2
M type					
Mean no. of M	24.2	31.6	5.5	7.5	18.0
Max no. of M	38	96	32	58	96
A type					
Mean no. of A	26.2	26.6	40.0	38.3	33.2
Max no. of A	164	124	52	54	164

In Lepidosireniformes represented by Choanichtyes, only one species, *Lepidosiren paradoxa*, with $2n = 38$ and $2n = 68$, is cytologically known. The chromosome numbers of Chondrostei in Teleostomi vary from $2n = 36$ (*Polypterus palmas*), involving specially large chromosomes, to $2n = 239$ (*Acipenser naccari*). In Holostei of Teleostomi, only two species, *Amia calva* ($2n = 46$) and *Lepisosteus productus*, spotted gar ($2n = 68$), have been reported.

Among the lowest group, the chromosome numbers have a wide distribution from $2n = 28$ (*Narcine brasiliensis*) to $2n = 239$ (*Acipenser maccari*) with a standard deviation (SD) of 51.2. Two peaks occur at $2n = 36$ (31%) and $2n = 164$ (10%) (Figure 1).

The mean value of chromosomal arm number is 74.6, similar to that of the lower group. The mean number of M-type (meta- and submetacentrics) and A-type (subtelo- and acrocentrics) per metaphase plate is 24.2 and 26.2.

The chromosome numbers of the lower group vary from $2n = 22$ (*Umbra pygmaea*) to $4n = 206$ (*Carassius auratus langsdorfii*). The wide distribution (SD = 22.6) with three low peaks at $2n = 50$ (38%), $2n = 48$ (9%), and $2n = 100$ (8%) is characteristic of this group. Another feature of the distribution is the spread toward larger numbers

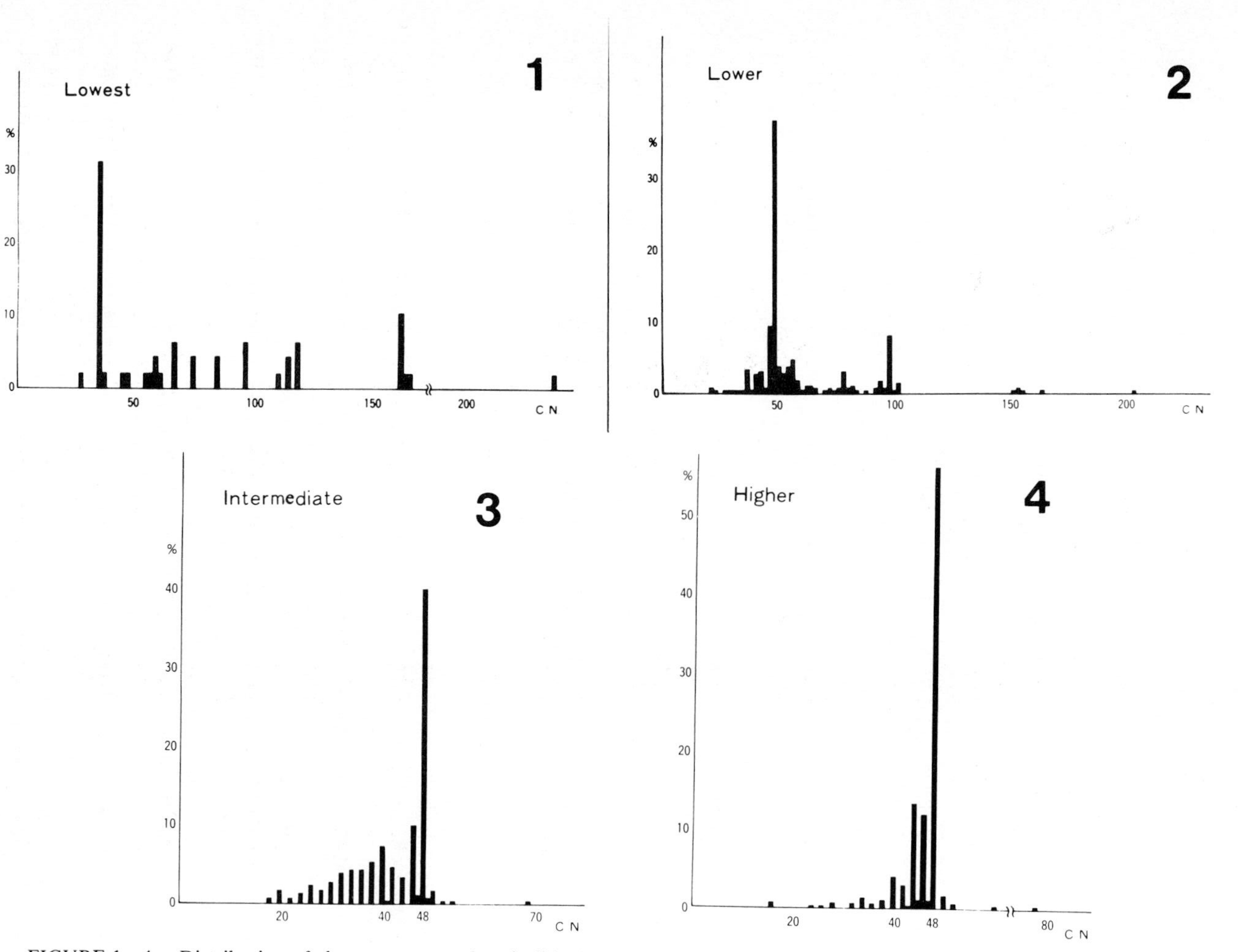

FIGURE 1—4. Distribution of chromosome numbers in fish classification groups. Figure 1: lowest group; Figure 2: lower group; Figure 3: intermediate group; Figure 4: higher group.

than $2n = 50$ ($>2n = 50$) (Figure 2). The arm number of the lower group is higher (89.7) than that of the intermediate (51.0) and the higher (53.4) groups. The mean number of M-type chromosomes for a metaphase plate is 31.6 and of A-type is 26.6.

The chromosome numbers of the intermediate group vary from $2n = 18$ (*Aphyosemion christyi*) to $2n = 69$ (*Poecilia formosa*) with a peak at $2n = 48$ (40%) having a gradual concentration (SD = 8.0) into $2n = 48$. Some evidence is expected to be discovered to establish that the distribution of the chromosome numbers spread towards numbers lower than $2n = 48$ ($<2n = 48$) (Figure 3). The mean arm number of this group is 51.0. A mean karyotypic plate is constituted by 5.5 M-type chromosomes and 40.0 A-type chromosomes.

The chromosome numbers of the higher group vary from $2n = 16$ (*Sphaerichthyo osphromonides*) to $2n = 78$ (*Channa gachua*) with a mode of $2n = 48$ (56%). The SD was 4.9, showing a concentration into $2n = 48$ (Figure 4). The arm number is 53.4. The mean number of M-type chromosomes in a metaphase plate is 7.5 and of A-type 38.3.

The above results indicate that there is a wide distribution in chromosome numbers with a large number of M-type chromosomes, especially in the lower groups. This is a consequence of chromosomal polyploidization and gene duplication. In fact, it is closely related to the evolutional changes in vertebrates. But there is a great gap between the lower and intermediate groups. In the latter, the chromosome number is concentrated at $2n = 48$ where there are many A-type and very few M-type chromosomes. There is a close similarity between higher and intermediate groups.

On the basis of these statistical results, it is hypothesized that there existed fishes with various chromosome numbers around the central figure of $2n = 48$, and that chromosome polyploidization or gene duplications must have occurred in ancient times. These processes contributed to vertebrate evolution. During the stages of evolution, only the lowest or lower group of fishes remained as the remnants of fish evolution. Those with $2n = 48$ constituted of A-type chromosomes became the evolutionary step toward the intermediate and higher groups.

III. POLYPLOIDY IN THE LOWER GROUP

Polyploidy, both spontaneous and induced, has been studied by various authors, particularly in plant systems.[6-11] Svardson[12] hypothesized on the significance of polyploidization in the karyotypic differentiation of European salmon, *Osmerus, Salmo,* and *Thymallus.* He suggested that the ancestral basic salmonids with ten chromosomes in the haploid set differentiated into the recent salmonids as a result of repeated chromosomal duplication by intraspecies hybridization.

Ohno[13] clearly interpreted examples of intraspecific chromosomal polymorphism involving Robertsonian fusion which were found in the Pacific northern anchovy (*Engraulis mordax*), the rainbow trout (*Salmo irideus*), the European brown trout (*Salmo trutta*), and the Pacific coho salmo (*Onchorhynchus kisutch*). Kobayashi[14-16] reported three chromosomal polymorphisms characterized by diploidy, triploidy, and tetraploidy in *Carassius auratus langsdorfii.* Diploid and triploid forms have also been noted in *C. a. buergeri.* Ojima et al.[17] have shown that F_1 females from the cross *C. a. cuvieri* ♀ × *Cyprinus carpio* ♂ lay eggs when backcrossed to male carp, and that the fries thus produced develop without observable abnormality. One half of these backcross offspring possess a $3n$-range chromosome complement. Noticeable is the fact that most of these individuals are females.

Gold and Avise[18] defined a single triploid individual ($3n = 75$) in the Californian loach, *Hesperoleucus symmetrieus,* not associated with a unisexual mode of reproduction and not induced artificially. Ueno and Ojima[19] presented some karyological fea-

Table 3
CHROMOSOME COMPLEMENT AND ARM NUMBER IN THREE SPECIES OF COBITID FISHES

Species	2*n*	Complement	Arm no.
Cobitis biwae	48	20M + 22SM, ST + 6A	90
	48	16M + 24SM, ST + 8A	88
	48	16M + 22SM, ST + 10A	86
	96	32M + 54SM, ST + 10A	182
C. taenia taenia	50	12M + 4SM, ST + 34A	66
	86	32M + 32SM, ST + 22A	150
	94	26M + 32SM, ST + 36A	152
C. taenia striata	50	12M + 4SM, ST + 34A	66
	98	20M + 22SM, ST + 56A	140

From Ueno, K. and Ojima, Y., *Proc. Jpn. Acad.*, 52, 446, 1976. With permission.

tures of the genus *Cobitis* (Cobitidae, Cyprinida) with special regard to the diploid-tetraploid complex, and Ojima and Takai[20] found spontaneous polyploidy in the Japanese common loach, *Misgurnus anguillicaudatus.*

Generally speaking, polyploidy seems to be a characteristic feature of the lowest and lower groups of fishes.

A. Diploid-Tetraploid Complexes

Ueno and Ojima[19] and Ueno et al.[21] have shown that three species of Cobitid fishes in Japan, *Cobitis biwae, C. taenia taenia,* and *C. t. striata,* are phenotypically indistinguishable from each other at various levels. The color patterns on the body side and the lamina circularis of the pectoral fins have been employed as major characters for classification. But the color patterns show considerable aberrations, and the lamina circularis of the pectoral fins appears only in adult males.

Chromosome complement and arm number in three species of Cobitid fishes are shown in Table 3. Three kinds of karyotypes with 48 chromosomes and one with 96 chromosomes were obtained in *Cobitis biwae.* The karyotypes of the specimens with 48 chromosomes can be divided into three groups: one with 10 pairs of metacentrics, 11 pairs of submeta- or subtelocentrics, and 3 pairs of acrocentrics (Figure 5); another with 8 pairs of metacentrics, 12 pairs of submeta- or subtelocentrics, and 4 pairs of acrocentrics (Figure 6); and a third with 8 pairs of metacentrics, 12 pairs of submeta- or subtelocentrics, and 5 pairs of acrocentrics (Figure 7). Specimens collected in the waters of the Kinki district and adjacent regions are diploid ($2n = 48$) (Figure 9).

On the other hand, tetraploid specimens were found in the waters of western Honshu (Figure 9). They were characterized by 96 chromosomes comprising 16 pairs of metacentrics, 27 submeta- or subtelocentrics, and 5 pairs of acrocentrics (Figure 8).

Three types of chromosome number, i.e., $2n = 50$, $2n = 86$, and $2n = 94$, were found in *Cobitis taenia taenia.* Specimens with 50 chromosomes were collected in the lower reaches of a river (Kyushu district). There were 6 pairs of metacentrics, 2 pairs of submetacentrics, and 17 pairs of acrocentrics (Table 3). Tetraploid specimens having $2n = 86$ came from the midstream of the river. The karyotype consisted of 16 pairs of metacentrics, 16 pairs of submeta- or subtelocentrics, and 11 pairs of acrocentrics (Table 3). In most cases, one pair of the small acrocentrics had satellites.

Another tetraploid form with $2n = 94$ was obtained in the northern part of Kyushu

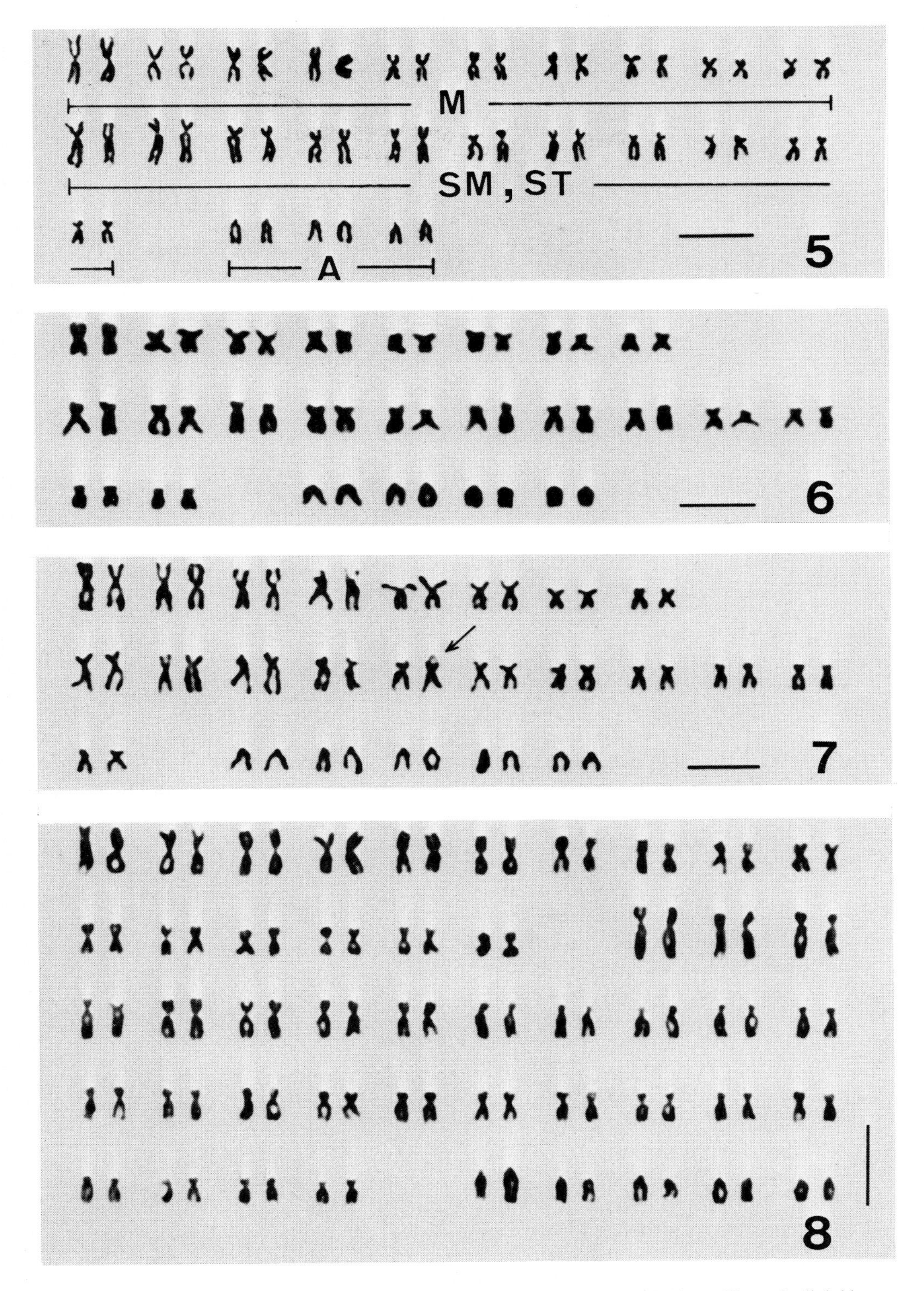

FIGURE 5—8. Karyotype of the diploid and tetraploid forms in *Cobitis biwae*. Figure 5: diploid, $2n$ = 48 (20M + 22SM,ST + 6A); Figure 6: diploid, $2n$ = 48 (16M + 24SM,ST + 8A); Figure 7: diploid, $2n$ = 48 (16M + 22SM,ST + 10A); Figure 8: tetraploid, $2n$ = 96 (32M + 54SM,ST + 10A). The arrow indicates the satellite chromosome. (From Ueno, K. and Ojima, Y., *Proc. Jpn. Acad.*, 52, 446, 1976. With permission.)

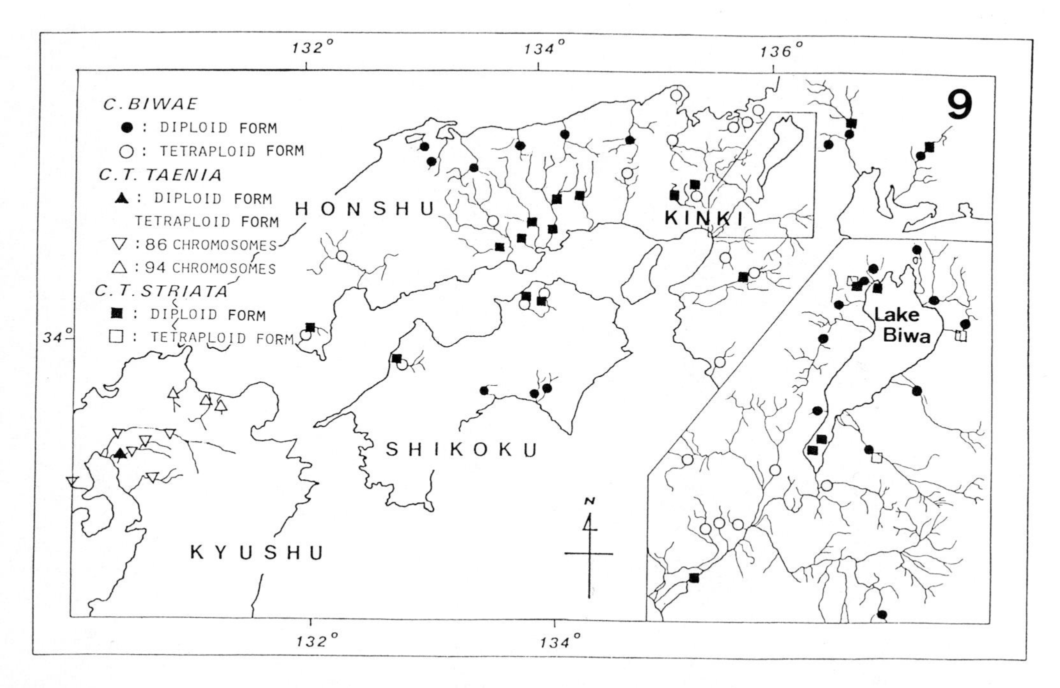

FIGURE 9. Distribution of diploid and tetraploid forms of the genus *Cobitis* in Japan. (From Ueno, K., Iwai, S., and Ojima, Y., *Bull. Jpn. Soc. Sci. Fish.*, 48, 9, 1980. With permission.)

district (Figure 9). The complement comprised 13 pairs of metacentrics, 16 pairs of submetacentrics, and 18 pairs of acrocentrics (Table 3). The AN value in the $4n$ forms was over twice that of the $2n$. It appears that a diploid-tetraploid relation might have existed in these three species. These facts indicate that diploidized tetraploidy becomes endowed with a far greater possibility than its diploid ancestor in accelerating a large evolutionary change.[22]

B. Spontaneous Polyploidy

The loach, *Misgurnus anguillicaudatus*, is a common fresh-water teleost having a wide distribution in Japan. It easily breeds in captivity, and is largely domesticated in small ponds for food. Ojima and Hitotsumachi[23] and Hitotsumachi et al.[24] reported on the chromosomal morphology of this species together with some cytophotometric data on the amount of DNA. It was shown that the chromosome number was $2n = 50$, consisting of 7 pairs of meta- and submetacentrics and 18 pairs of telo- and subtelocentrics.

Recently Ojima and Takai[20] recorded polyploidy in specimens in Japanese common loaches obtained in the market and the field. Of the 80 specimens karyotyped, 4(♀ 3, ♂ 1) were tetraploid with $4n = 100$, while 1(♀) was triploid with $3n = 75$. The diploid chromosome complement was 50 consisting of 5 pairs of metacentrics, 2 pairs of submetacentrics, and 18 pairs of telo- and subtelocentrics (Figure 10). A polymorphism was found in the population of diploid specimens (Figure 11). This chromosome change seemed to result from Robertsonian fusion; the largest metacentrics were created by the fusion of the third and fourth telocentrics. This indicates that this type of intraspecific polymorphism exists in this group.

As expected, the karyotype of triploid individuals comprised 5 triple metacentrics, 2 triple submetacentrics, and 18 triple telo- and subtelocentrics (Figure 12).

The tetraploid karyotype with 100 chromosomes consisted of 10 pairs of metacentrics, 4 pairs of submetacentrics, and 36 pairs of telo- and subtelocentrics (Figure 13).

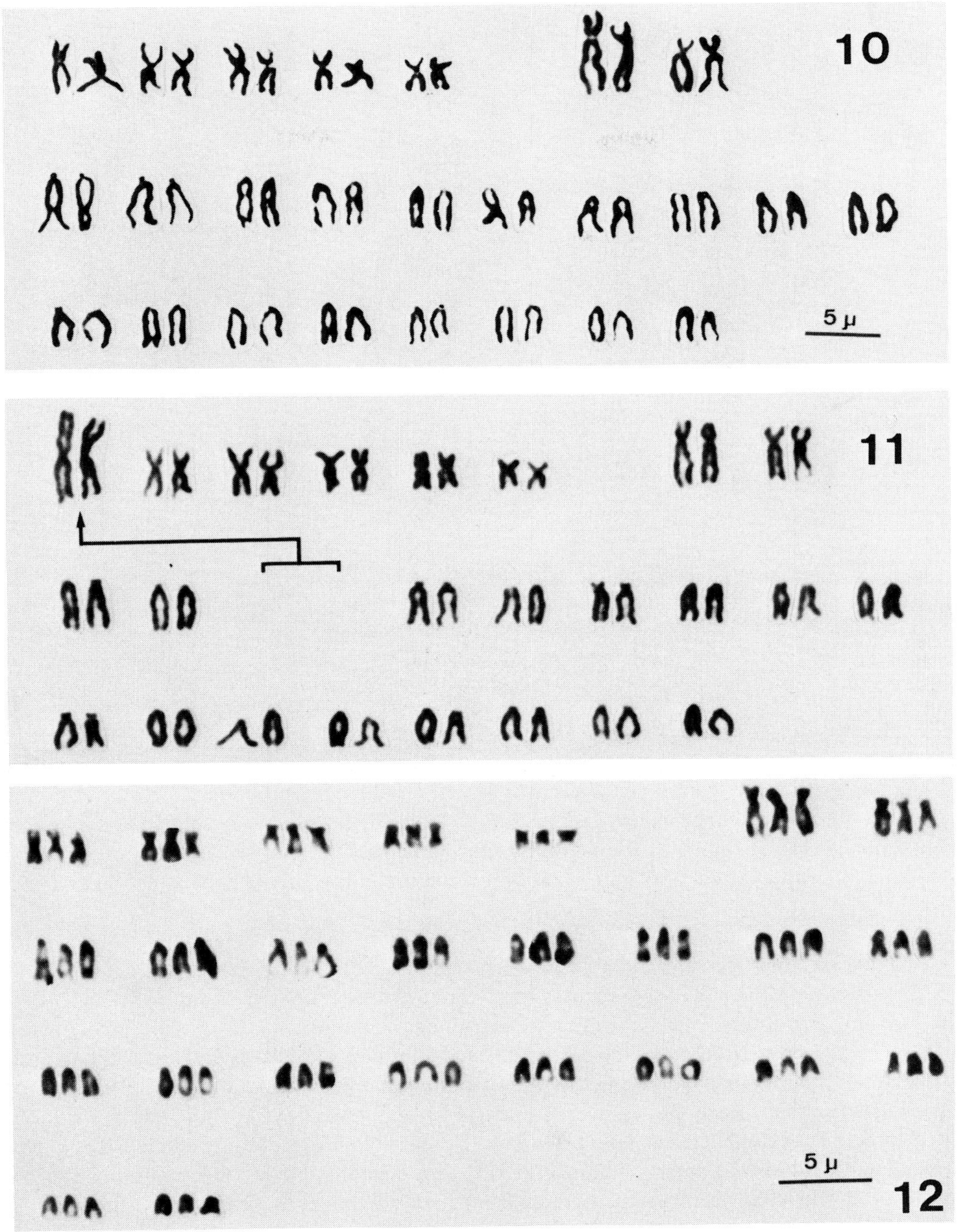

FIGURE 10—13. Serial alignment of somatic chromosomes of Japanese loach, *Misgurnus anguicaudatus.* Figure 10: diploid, $2n = 50$ (10M + 4SM + 36T,ST); Figure 11: Robertsonian fusion in diploid, $2n = 48$ (12M + 4SM + 32T,ST); Figure 12: triploid, $3n = 75$ (15M + 6SM + 54T,ST); Figure 13: tetraploid, $2n = 100$ (20M + 8SM + 72T,ST). (From Ojima, Y. and Takai, A., *Proc. Jpn. Acad.*, 55, 487, 1979. With permission.)

The results of DNA measurements in individual nuclei of liver cells from diploid, triploid, and tetraploid specimens are presented in Figure 14: the tetraploid DNA is about twice the diploid value, and the DNA value of triploidy is intermediate.

Hitotsumachi et al.[24] showed that four loach species, *Misgurnus anguillicaudatus,*

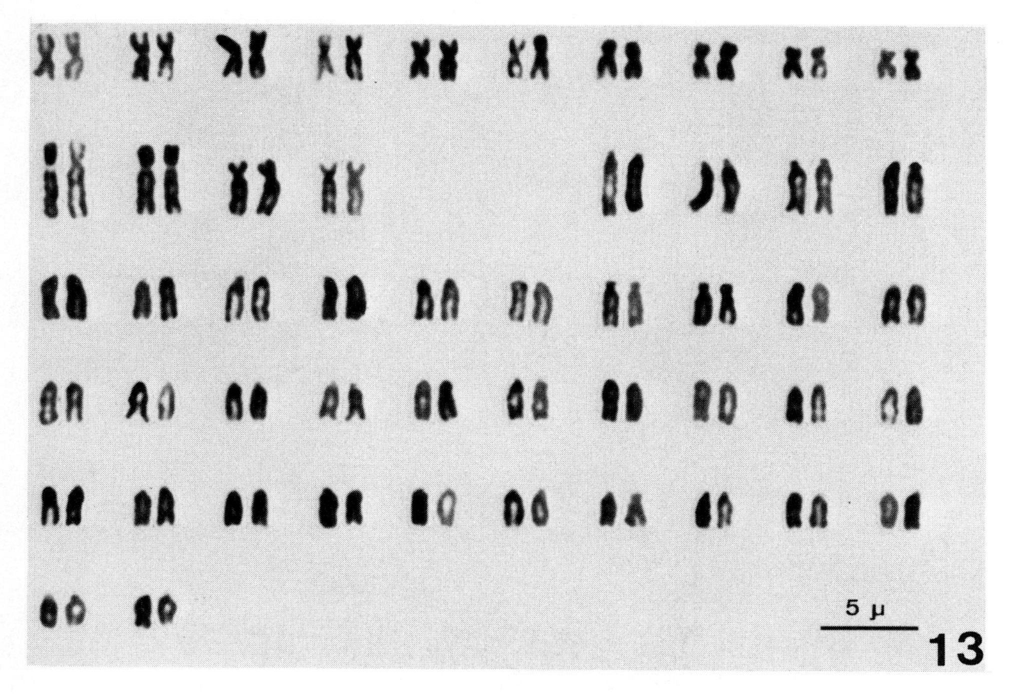

FIGURE 13

Barbatula toni, Lifua nikkonis, and *Cobitis delicata,* despite having the same chromosome number ($2n = 50$), are characterized by specific karyotypes distinguishable from each other. Nogusa[25] reported that one of the Japanese loaches, *Lefua echigonia,* showed 50 diploid chromosomes. The Khulli loach of Southeast Asia also has 50 chromosomes.[26] It appears that the fundamental chromosome number of the family Cobitidae is 50, and that certain structural chromosome components characterize each species.

C. Artificial Polyploidy

Polyploidy in the lower vertebrates has been recorded from time to time.[27-34]

Makino and Ojima[35] presented some cytological evidence showing that refrigeration of carp eggs prevented completion of the second meiotic division and caused chromosome duplication in treated eggs as a consequence of retention of the haploid set of chromosomes normally going into the second polar body; the formation of a diploid egg nucleus resulted. It was suggested that union of the diploid egg nucleus thus produced with a haploid sperm would give rise to a triploid zygote. Ojima and Makino[36] presented some cytological features that confirmed the above suggestion in the carp egg.

Generally in the carp, as in other fishes, the eggs undergo the first meiotic division in the ovary prior to ovulation, and the second meiotic division is arrested at metaphase during ovulation until entry of the spermatozoon occurs. At the time of fertilization, the eggs are at the metaphase stage of the second division and the spindle is present (Figure 15). About 25 to 30 min after insemination, the second division is completed with the formation of the second polar body.

Regularly, at about 10 min after fertilization, the second meiotic division of the egg advances from metaphase to anaphase (Figure 16). The eggs were exposed to a low temperature kept at 0°C for 10 min. At this stage the second meiotic division of the eggs was passing from metaphase to anaphase.

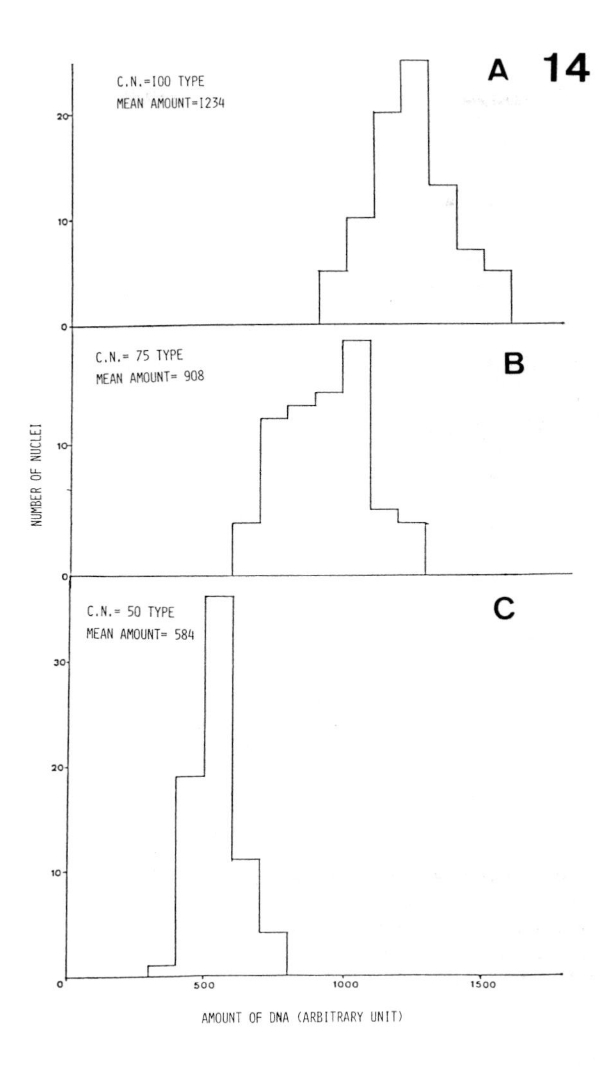

FIGURE 14. Histograms showing relative amounts of DNA in liver cells of diploid (A), triploid (B), and tetraploid loach specimens. (From Ojima, Y. and Takai, A., *Proc. Jpn. Acad.*, 55, 487, 1979. With permission.)

In the nonrefrigerated eggs, the chromosomes of the second meiotic metaphase were separated into two sister groups, each of which contained the haploid number. In contrast, in the refrigerated eggs the chromosomes were in the course of separation at early anaphase (Figure 17); but the eggs remained in this state without further separation of the chromosomes into entire anaphase. There was a complete failure of extrusion of the chromosomes into the second polar body, due to the inhibition of division (Figure 17). Scattering of the chromosomes then occurred inside the nuclear body (Figure 18) and led to the formation of a nucleus containing two haploid sets or a diploid set of chromosomes, as a result of the coming together of the separated elements. The formation of a diploid egg nucleus was thus established through the duplication of the maternal chromosomes.

These observations demonstrate that cold shock applied just after insemination in the egg suppresses the second meiotic division, results in the failure of extrusion of the second polar body, and causes the production of a diploid egg nucleus. The union

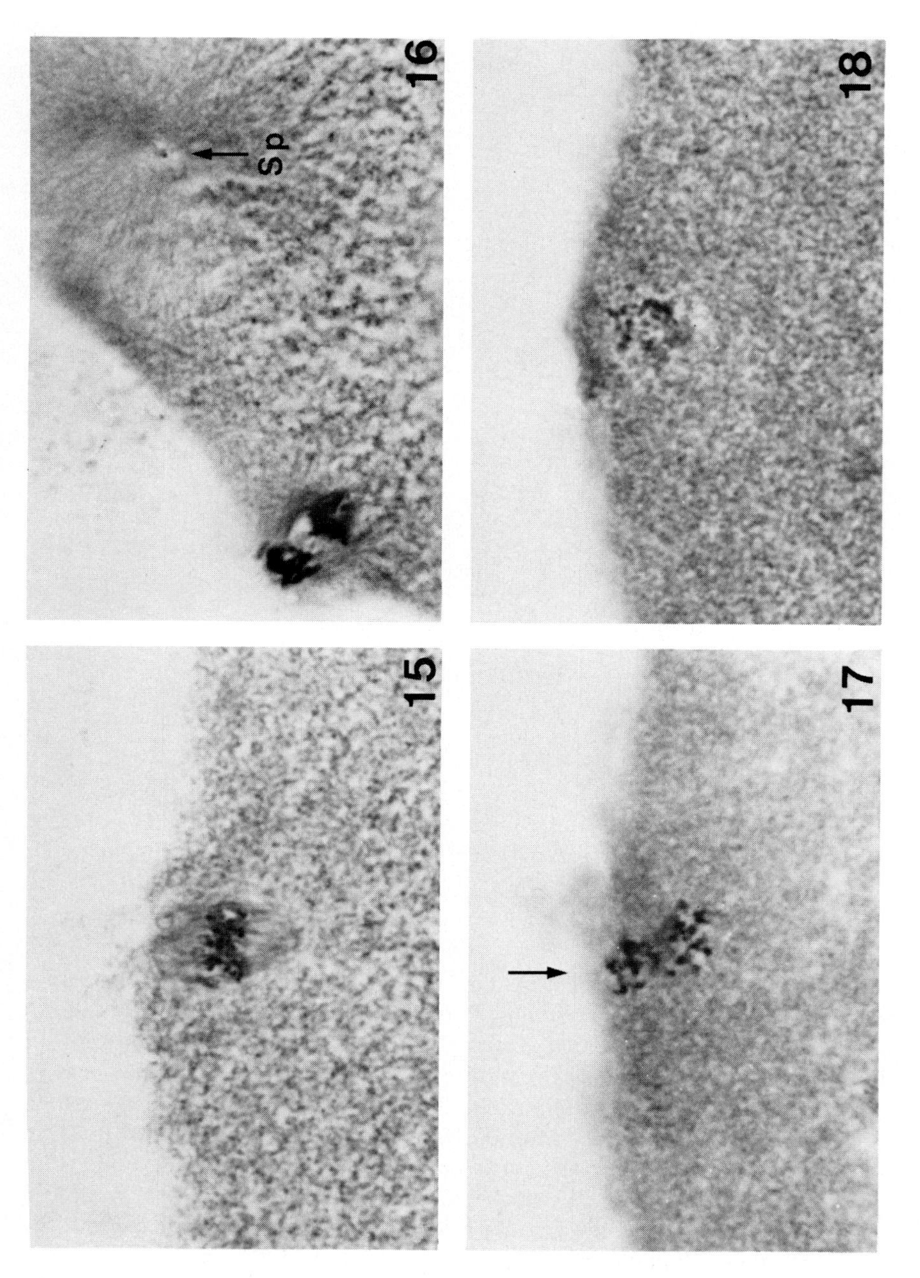

FIGURE 15—18. Photomicrographs from sections of carp eggs. Figure 15: the second meiotic spindle at metaphase, from an egg 8 min after insemination; Figure 16: the second meiotic division at early anaphase from an egg refrigerated for 10 min after insemination. Sp: sperm head with asters; Figure 17: a section of a refrigerated egg showing the failure of extrusion of the second polar body. Two sets of anaphasic chromosomes come together and scatter inside the nuclear membrane; Figure 18: a section of a refrigerated egg showing a nucleus containing two haploid sets of chromosomes. (From Ojima, Y., *Proc. Jpn. Acad.*, 54, 116, 1978. With permission.)

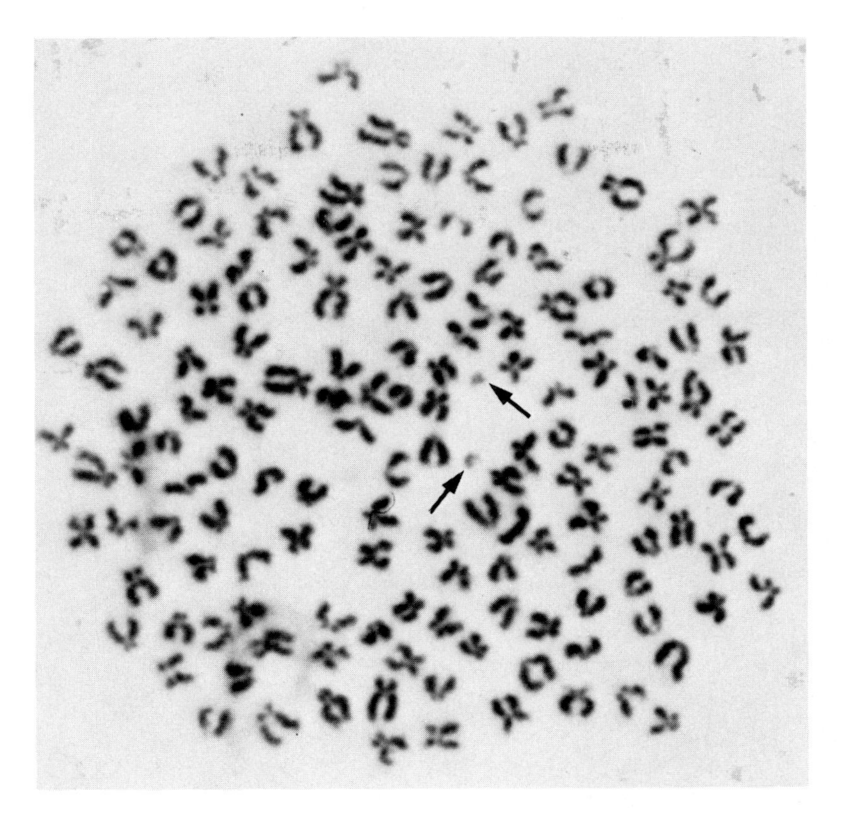

FIGURE 19. A triploid set of chromosomes in an air-dried sample from a juvenile carp derived from one of refrigerated eggs. Arrows indicate microchromosomes. (From Ojima, Y., *Proc. Jpn. Acad.*, 54, 116, 1978. With permission.)

of such a diploid egg with a haploid sperm would result in the formation of a zygote with a triploid chromosome set. This was confirmed experimentally and metaphases with 148 to 154 elements, very close to a triploid set, were counted (Figure 19).

Purdom[37] reported the production of triploids in 10% of plaice and plaice-flounder hybrids by cold shock of the fertilized eggs. The viability of those triploid embryos, larvae, and juveniles appeared normal.

It is likely that the failure of the meiotic division caused by abnormal temperature is based on disturbance of the spindle mechanism resulting from changes in viscosity of the cytoplasm due to changes of temperature.[38]

It has been shown that in mammals overripeness of the eggs may lead to dispermic triploidy, and that dispermy is the major mechanism in the induction of triploidy in man.[39,40]

D. Polyploidy in Hybrids

Although considerable work of both scientific and practical interest has been done on fish hybrids by a number of ichthyologists, knowledge on cytogenetic problems has remained meager. Analyses of genetic systems by means of experimental hybridization proved important for the understanding of cytotaxonomic differences of related forms. In addition, significant correlations between geographical distribution and chromosomal polymorphism or polyploidy have been described in fishes.

Involved in speciation, polyploidy, and reproduction of the genus *Carassius*, the author has been concerned with some cytogenetic investigations in experimentally pro-

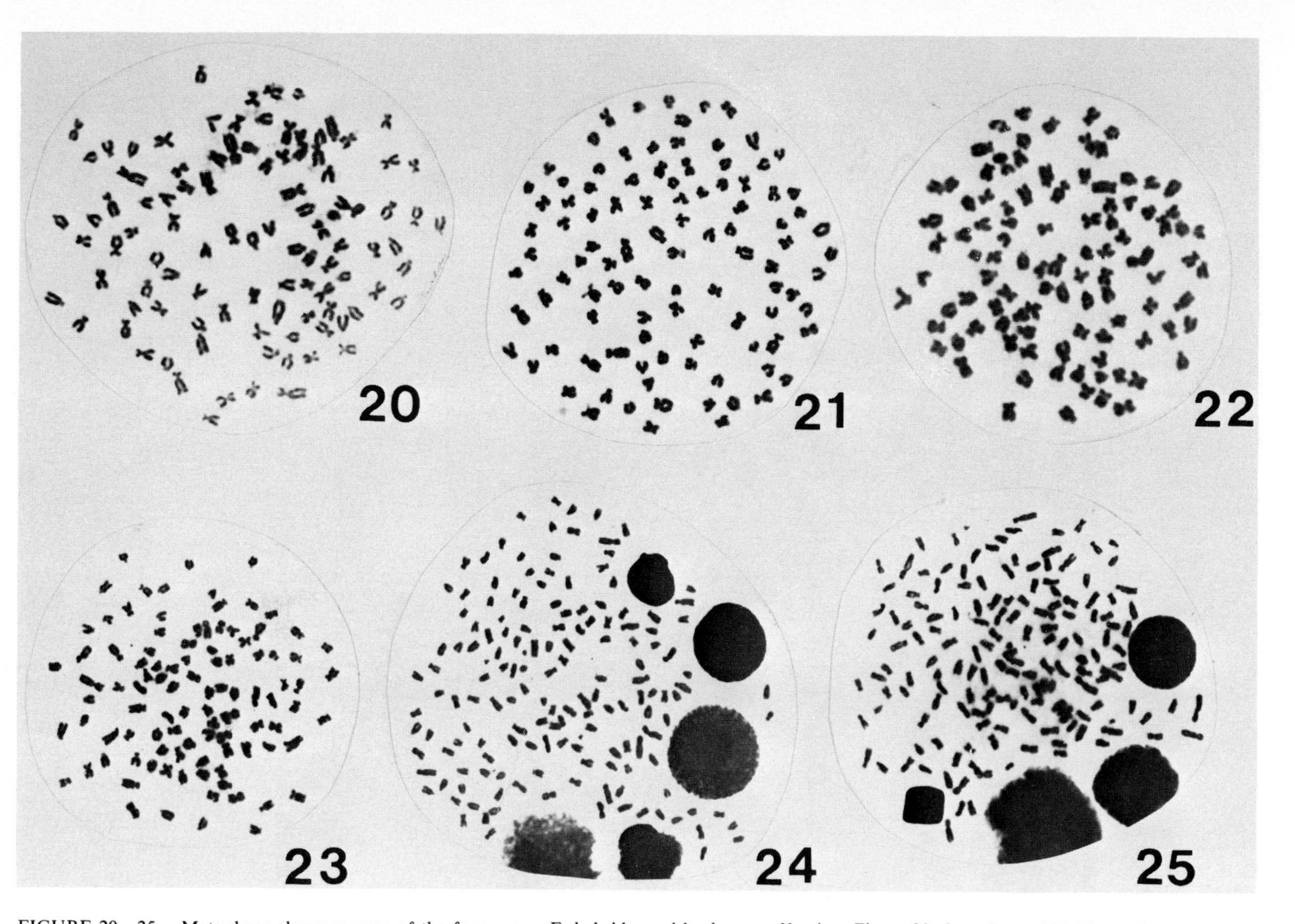

FIGURE 20—25. Metaphase chromosomes of the funa, carp, F_1 hybrids, and backcross offspring. Figure 20: funa $2n = 100$; Figure 21: carp $2n = 100$; Figure 22: F_1 $2n = 100$; Figure 23: diploid backcross individual $2n = 100$; Figure 24: triploid range of chromosomes of a triploid backcross individual; Figure 25: the same of a backcross individual $3n = 155$. (From Ojima, Y., Hayashi, M., and Ueno, K., *Proc. Jpn. Acad.*, 51, 702, 1975. With permission.)

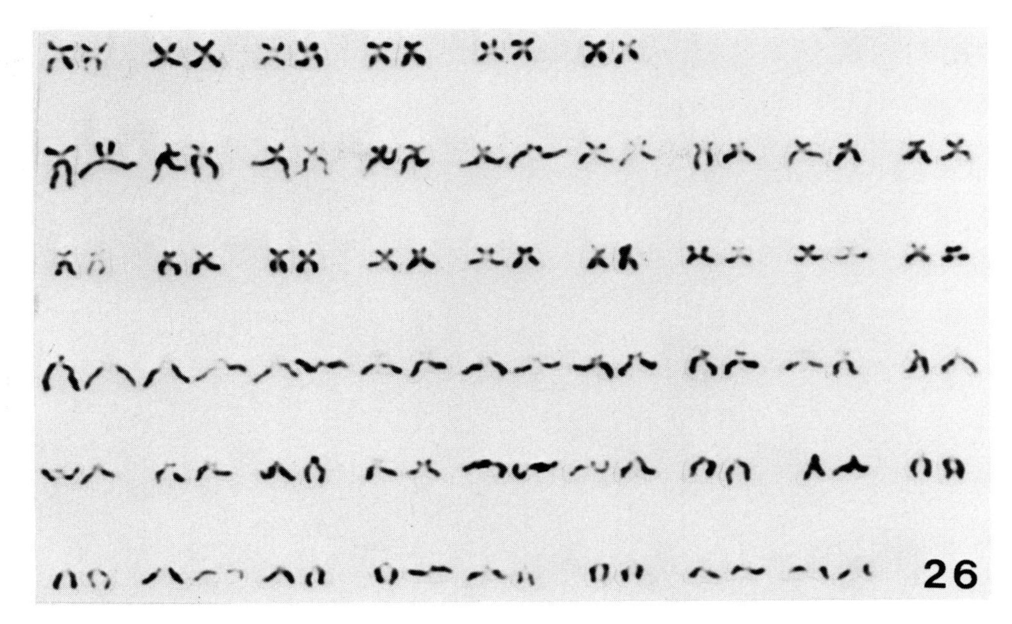

FIGURE 26. Karyotype of the funa, *Carassius auratus cuvieri*, $2n = 100$. (From Ojima, Y., Hitotsumachi, A., and Makino, S., *Proc. Jpn. Acad.*, 42, 62, 1966. With permission.)

duced intergeneric hybrids between the funa* (*Carassius auratus cuvieri*) and the carp (*Cyprinus carpio*).

Ojima et al.[41] and Ojima and Hitotsumachi[42] reported a chromosome number of 100 in the funa and the carp (Figures 20 and 21). The diploid complement was similar in the two forms, consisting of 6 pairs of metacentrics, 18 pairs of submetacentrics, and 26 pairs of acro- or telocentrics (Figures 26 and 27). The chromosome number of the F_1 offspring was 100 (Figure 22). Karyotypically, the F_1 chromosomes did not differ from those of the parent species, except for the largest submetacentrics which originated from the funa (Figure 28).

Breeding experiments by Ojima since 1945 have indicated that the hybrid males were completely sterile. It has recently been shown that F_1 (funa ♀ × carp ♂) females with $2n$ chromosomes laid eggs after backcrossing with the carp, and that these developed into adults without any visible abnormality.

Theoretically, the backcross individuals between the F_2 female with 100 chromosomes and the male carp should have 100 diploid chromosomes. However, backcross offspring having a $3n$-range chromosome complement appeared in one half of the progeny. The chromosome number of the diploid backcross offspring was 100 (Figure 23), comprising 6 pairs of metacentrics, 21 pairs of submetacentrics, and 23 pairs of telo- and acrocentrics.

The chromosomes of the triploid backcross offspring were 145 to 160 in number (Figures 24 and 25). The karyotype of the 156-chromosome cells provided evidence of the trisomic condition for the meta- and submetacentric chromosomes which were morphologically well defined; no informative feature was obtained for the telo- and acrocentric chromosomes because of the paucity of a characteristic configuration for analysis (Figure 29). Noticeable was the occurrence of minute elements, of microchromosomes, in some of the $3n$-range metaphases of the backcross individuals. The individuals with the $3n$-range complement all possessed functional ovaries.

* The name "funa" is commonly used in Japan for the Crucian carp.

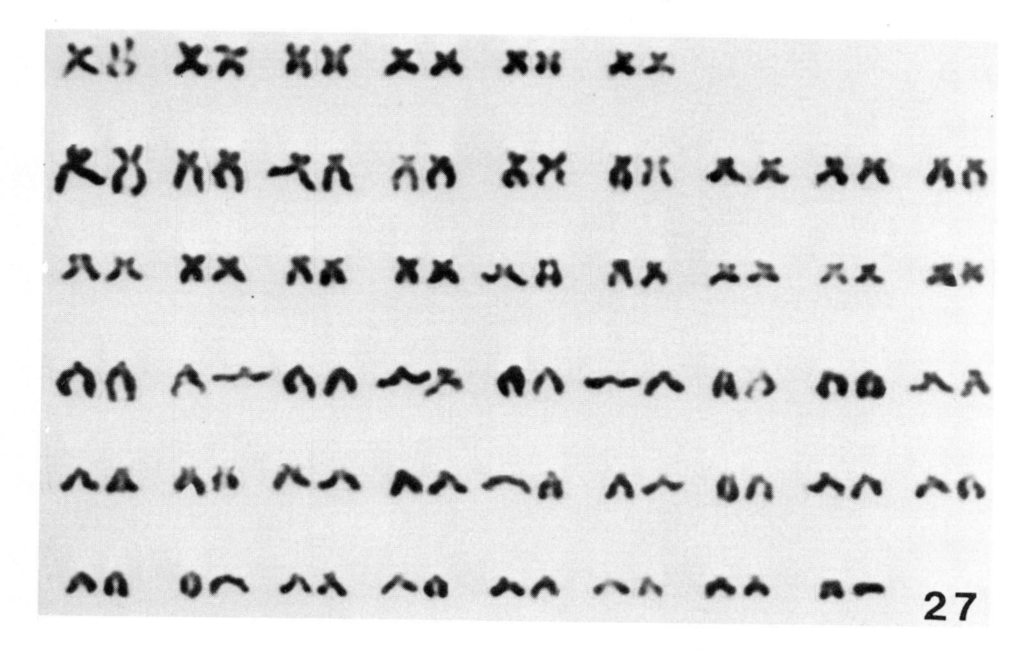

FIGURE 27. Karyotype of the carp, *Cyprinus carpio*, $2n = 100$. (From Ojima, Y. and Hitotsumachi, S., *Jpn. J. Genet.*, 42, 163, 1967. With permission.)

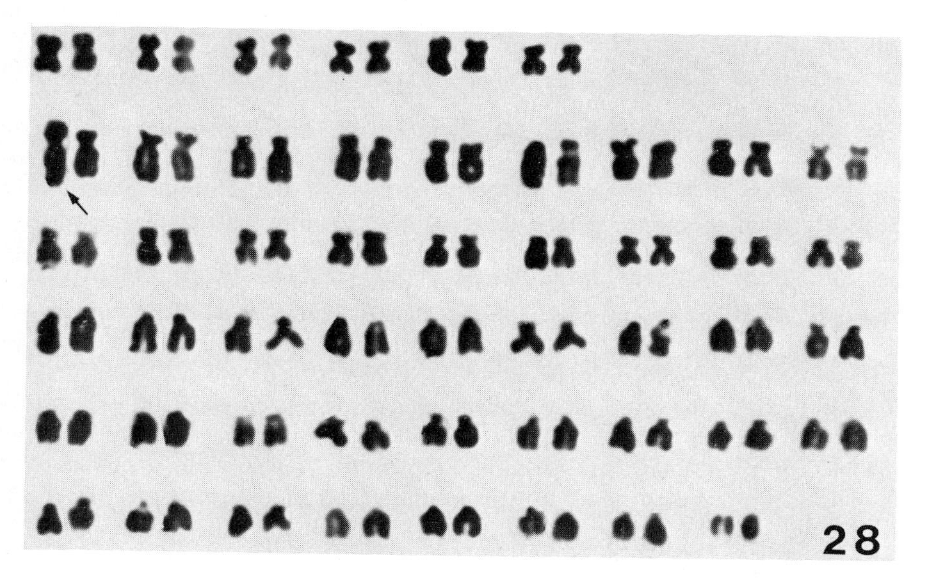

FIGURE 28. Karyotype of the F_1 with 100 chromosomes. Arrow shows the largest submetacentric chromosome originated from the funa. (From Ojima, Y., Hayashi, M., and Ueno, K., *Proc. Jpn. Acad.*, 51, 702, 1975. With permission.)

Kobayashi[43] reported that several triploid individuals with 156 chromosomes consisting of 17 pairs of metacentrics, 31 pairs of submetacentrics, and 30 pairs of acrocentrics appeared among the F_2 hybrids between *Carassius auratus langsdorfii* ($2n$, ♀) and goldfish, Ryukin ($2n$, ♂).

The significance of triploidy in the course of evolution of the funa has remained unclear. The occurrence of $3n$-range backcross specimens seems to throw some light on the differentiation of species in the funa.

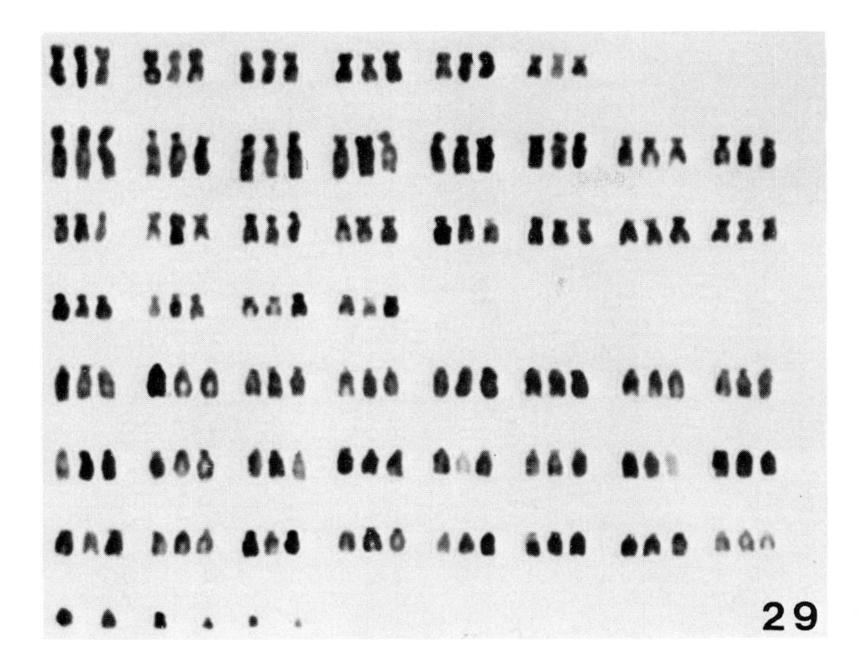

FIGURE 29. Karyotype of a 3*n*-range cell from a triploid backcross specimen. (From Ojima, Y., Hayashi, M., and Ueno, K., *Proc. Jpn. Acad.*, 51, 702, 1975. With permission.)

IV. CHROMOSOME POLYMORPHISMS IN THE HIGHER GROUP

Members of the author's laboratory studied the Labridae and Pomacentridae fishes, with many differentiated species in each, forming colonies inside a coral reef. Though the cytogenetical researches on these marine fishes are still continuing, an outline of the findings is presented here to provide data aimed at gaining more fundamental knowledge about the relationship between chromosomal changes and species differentiation.

A. DNA Duplication

The Japanese Labridae are represented by four subfamilies, consisting of 28 genera and 94 species. These species are distributed in the southern part of Japan, especially inside a coral reef in the Yaeshima group of the Ryukyu Islands. The karyotypes established in the 34 species are described according to their arrangement in three subfamilies in Table 4.

Bodianinae — Three species of the subfamily Bodianinae — *Bodianus axillaris, B. loxozonus,* and *B. mesothorax* — were studied chromosomally; all were found to possess a 2*n* of 48. Although they had the same number of chromosomes, each species had a slightly different karyotype as well as NF, "nombre fundamental". Of interest is the presence of 8 metacentrics in all three species in contrast to different numbers of other types of chromosomes (Figures 30 and 42).

Corinae — The chromosomes of 22 species of Corinae — *Cheilio inermis, Gomphosus varius, Thalassoma quinquevittata, T. lunare, T. cupid, T. lutescens, T. amblycephala, Hemigymus fasciatus, Pseudolabrus japonicus* (— 1 and — 2), *Labroides dimidiatus, Stethojulis bandanensis, S. interrupta, S. strigiventer, Halichoeres trimaculatus, H. centriquadrus, H. tenuispinnis, H. melanochir, H. prosopeion, H. poecilopterus, H. kallochroma, Coris aygula, C. multicolor, C. gaimardi,* and *Holo-*

Table 4
THE CHROMOSOME NUMBERS AND KARYOTYPES ESTABLISHED FOR 34 SPECIES OF LABRIDAE

	CN	M	SM	ST,A	NF
Bodianinae					
Bodianus axillaris	48	8	30	10	86
B. loxozonus	48	8	26	14	82
B. mesothorax	48	8	18	22	74
Corinae					
Cheilio inermis	48	12	12	24	72
Gomphosus varius	48	0	0	48	48
Thalassoma quinquevittata	48	0	0	48	48
T. T. lunare	48	0	0	48	48
T. cupid	48	0	0	48	48
T. lutescens	48	0	0	48	48
T. amblycephala	48	0	0	48	48
Hemigymus fasciatus	48	6	6	36	60
Pseudolabrus japonicus — 1	48	2	2	44	52
Pseudolabrus japonicus — 2	42	20	8	14	70
Labroides dimidiatus	48	0	0	48	48
Stethojulis bandanensis	48	4	0	44	52
S. interrupta	48	0	2	46	50
S. strigiventer	48	2	0	46	50
Halichoeres trimaculatus	48	0	0	48	48
H. centriquadrus	48	0	0	48	48
H. tenuispinnis	48	0	2	46	50
H. melanochir	48	0	2	46	50
H. prosopeion	48	2	0	46	50
H. poecilopterus	48	4	2	42	54
H. kallochroma	48	0	0	48	48
Coris aygula	48	6	6	36	60
C. multicolor	48	6	8	34	62
C. gaimardi	48	2	10	36	60
Hologymnoscus semidiscus	48	2	2	44	52
Cheilininae					
Cirrhilabrus temminckii	34	10	2	22	46
C. cyanopleura	34	10	2	22	46
Hemipteronotus dea	44	0	0	44	44
H. taeniurus	48	0	4	44	52
Epibulus insidiator	48	4	8	36	60
Cheilinus fasciatus	48	0	12	36	60
C. bimaculatus	32	4	2	26	38

From *Proc. Jpn. Acad.*, 56, 328, 1980. With permission.

gymnoscus semidiscus — were studied. All the species, except one variety of *Pseudolabrus japonicus* — 2 (Figures 34 and 46), had 48 chromosomes, though *Cheilio, Pseudolabrus,* and *Coris* were characterized by different karyotypes (Figures 31, 33, 34, 36, 43, 45, 46, and 48). Though the numbers were not exactly the same, the ST or A chromosomal elements were predominant in the complement of almost all species, while the M or SM did not occur or were very low in frequency.

Cheilininae — Seven species — *Cirrhilabrus temminckii, C. cyanopleura, Hemipteronotus dea, H. taeniurus, Epibulus insidiator, Cheilinus fasciatus,* and *C. bimaculatus* — were studied (Figures 37 to 41 and 49 to 53). The variation ranged from 32 to 48 in the $2n$ number, and from 38 to 60 in NF. The variation in karyotypes was also remark-

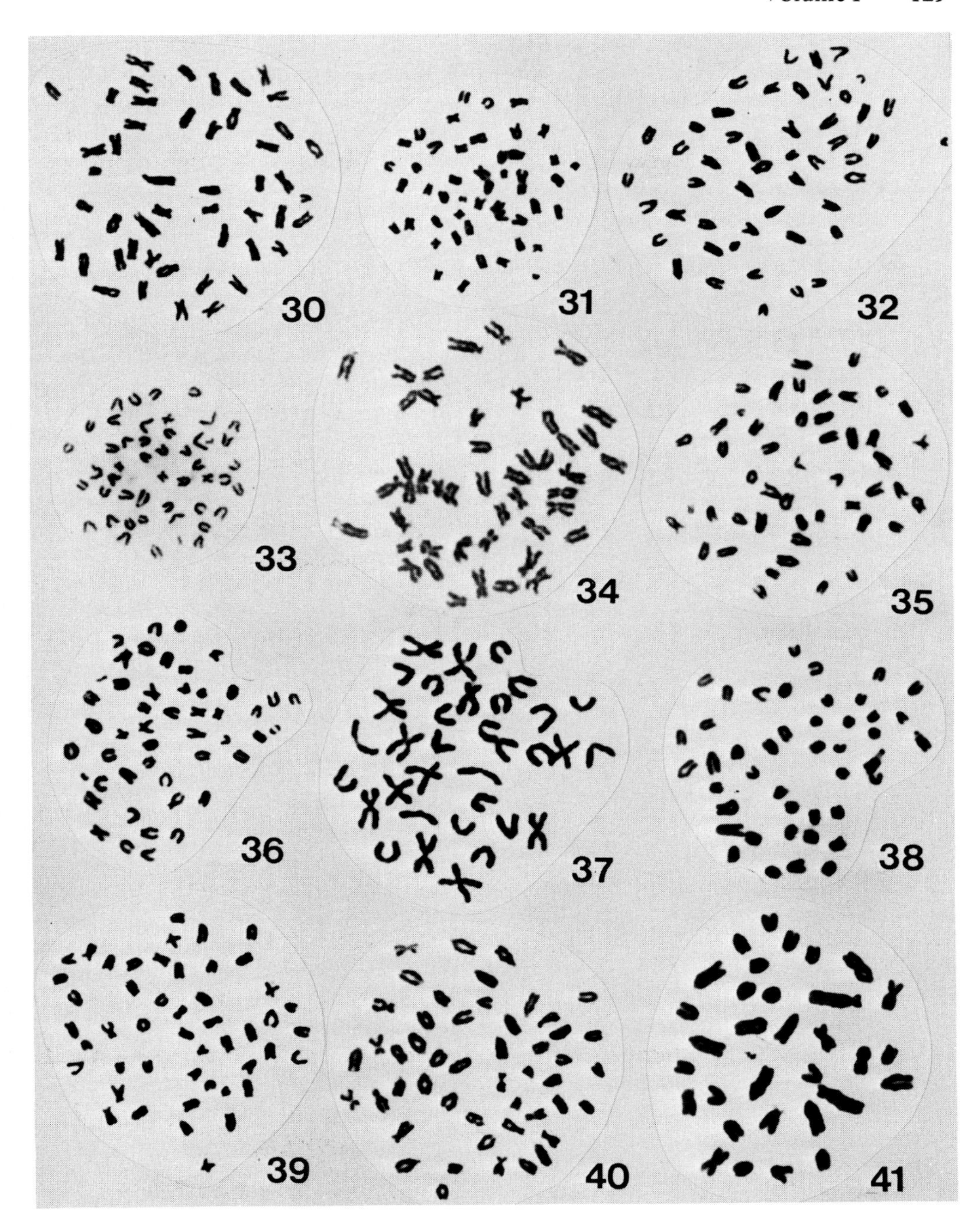

FIGURE 30—41. Photomicrographs of metaphase chromosomes of 12 species of Labrid fishes. Figure 30: *Bodianus axillaris;* Figure 31: *Cheilio inermis;* Figure 32: *Thalassoma lutescens;* Figure 33: *Pseudolabrus japonicus* — 1; Figure 34: *P. japonicus* — 2; Figure 35: *Halichoeres prosopeion;* Figure 36: *Coris multicolor;* Figure 37: *Cirrhilabrus temminckii;* Figure 38: *Hemipteronotus dea;* Figure 39: *Epibulus insidiator;* Figure 40: *Cheilinus fasciatus;* Figure 41: *C. bimaculatus.* (From Ojima, Y. and Kashiwagi, E., *Proc. Jpn. Acad.*, 56B, 328, 1980. With permission.)

able: the number of M, SM, and ST, A chromosomes being different in each species (Table 4 and Figures 42 to 53).

So far as the present investigations are concerned, the members of the Bodianinae, Corinae, except *Pseudolabrus japonicus* — 2, and Cheilininae, except *Hemipteronotus taeniurus, Epibulus insidiator,* and *Cheilinus fasciatus,* were found to possess a chro-

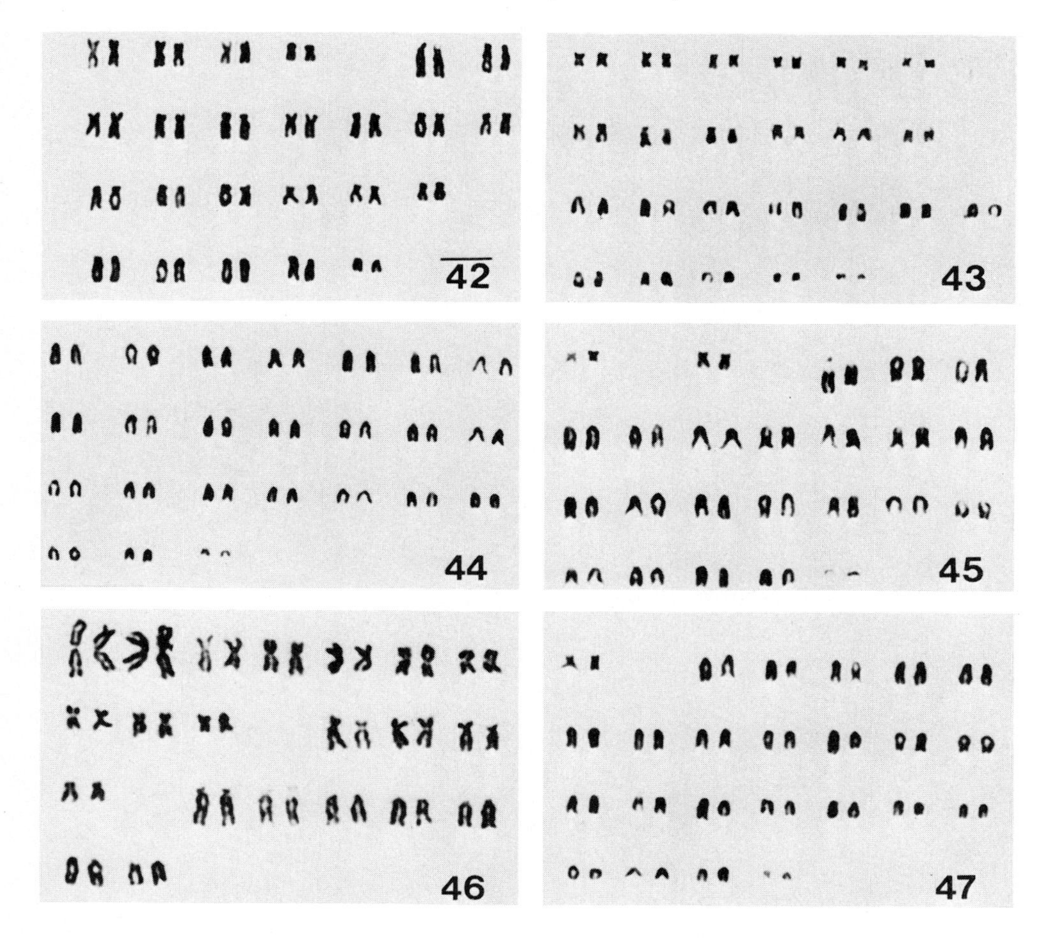

FIGURE 42—53. Serial alignments of somatic chromosomes corresponding to Figures 30 to 41. (From Ojima, Y. and Takai, A., *Proc. Jpn. Acad.*, 55, 487, 1979. With permission.)

mosome number, $2n = 48$, typical of the higher group. Morphologically, the chromosome changes occurring among species in Labridae could be derived from the ancestral type of 48 chromosomes, containing characteristically ST, A chromosomes, through pericentric inversion. In contrast, *P. japonicus* in Corinae and four species in Cheilininae, *Cirrhilabrus temminckii, C. cyanopleura, Hemipteronotus dea*, and *Cheilinus bimaculatus*, are outside the range of the higher group in chromosome number and karyotype.

The comparative amounts of DNA in individual nuclei of liver cells from *Coris multicolor* and *Cirrhilabrus temminckii* are presented in Figure 54. The DNA value of the latter is about twice that of the former. The size of chromosomes in *Cirrhilabrus temminckii, C. cyanopleura,* and *Cheilinus bimaculatus* was very large in contrast to that of the other Corinae (Figures 37 and 41). Several problems arise in relation to chromosomal morphology and DNA value. It seems that Robertsonian rearrangements and tandem duplications occurred continuously during evolution in fishes with $2n =$ 48, consisting of A-type chromosomes, which became the stem of the higher group.

In fact, Tri confirmed anatomically and systematically that *C. fasciatus* and *C. bimaculatus* have a different number of vertebrae than the other Labrid species.[41a] He suggested on the basis of these studies that the Japanese Labridae would include 6 subfamilies, consisting of 34 genera with 110 species.

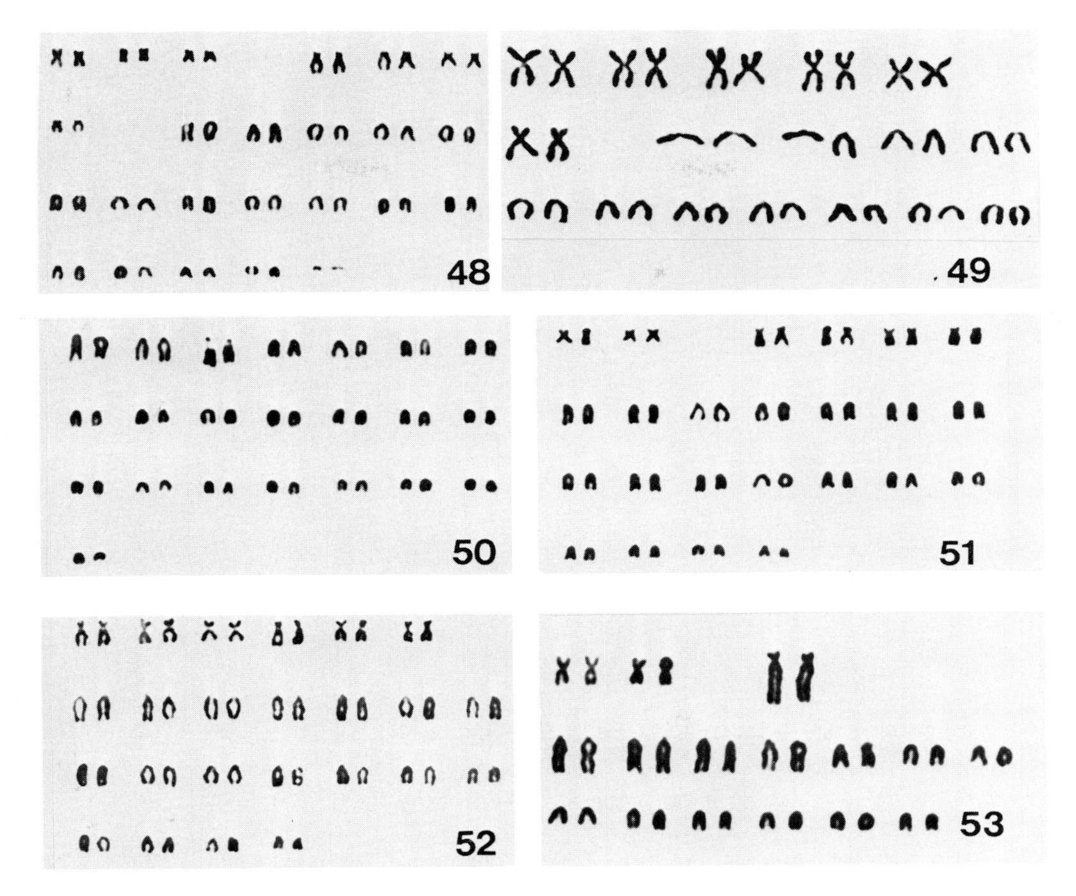

FIGURES 48 to 53

B. Robertsonian Rearrangement

In Japan, 29 species in 4 subfamilies of Amphiprioninae, Chrominae, Pomacentrinae, and Glyphisodontinae in Pomacentridae have been cytogenetically studied by several investigators (Table 5).

It is evident from this table that almost all species possess 48 chromosomes in diploid, the basic karyotype of the higher group. However, there are some intraspecific rearrangements that could be derived through pericentric inversion as were the Labrid chromosomes. The four species *Dascyllus trimaculatus, D. reticulatus, D. auruanus,* and *D. melanurus* in Chrominae showed such characteristic features in chromosome number and karyotype (Table 6). Although this type of chromosomal evolution is known widely in mammals, it is rarely observed in fish karyotypes of the higher group.

D. trimaculatus possesses two chromosome numbers, $2n = 48$ and 47 (Figures 55, 56, 69, 70). *D. reticulatus* shows chromosome polymorphisms of $2n = 37$, 36, 35, and 34 (Figures 57 to 60 and 71 to 74). Polymorphic karyotypes are observed also in *D. auruanus* — 33, 32, 31, 30, 29, 28, and 27 in diploid complexes (Figures 61 to 67 and 75 to 81). Only *D. melanurus* has the $2n = 48$ constituted by A-type chromosomes as a basic type of the higher group (Figures 68 and 82).

Arai and Inoue[44] reported only one pattern of karyotype in *D. trimaculatus* with 47 chromosomes. They suggested that one large heteromorphic metacentric chromosome seems to be the sex chromosome, which could be formed by centric fusion. In investigations here, 15 of 17 individuals of *D. trimaculatus* possessed 48 chromosomes. All species of genus *Dascyllus* perform sex reversal throughout life. Two other closely

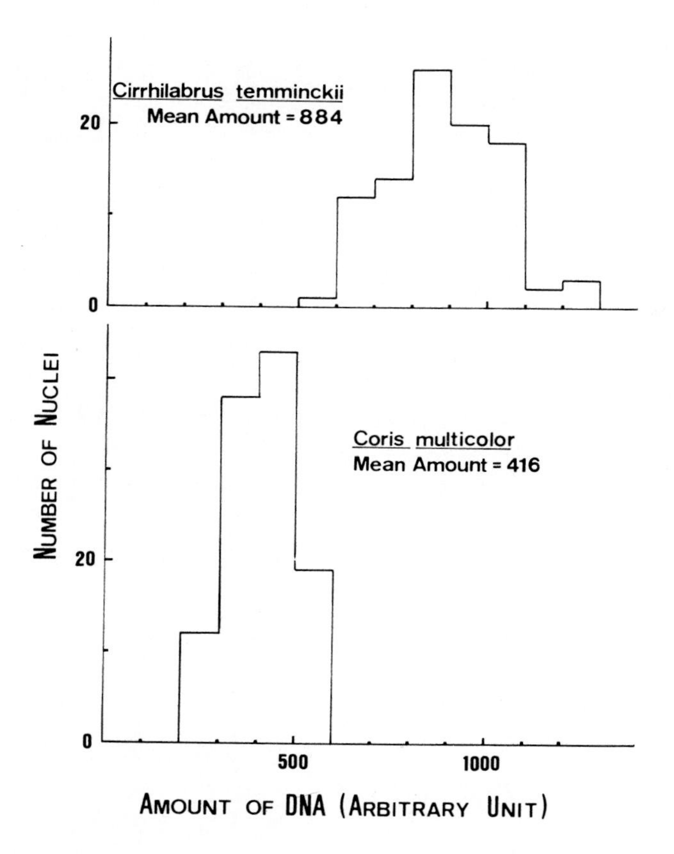

FIGURE 54. Histograms showing relative amount of DNA in liver cells of *Cirrhilabrus temminckii* and *Coris multicolor*.

related species, *D. reticulatus* and *D. auruanus*, showed the same kind of intraspecific chromosomal polymorphisms. It is clear from these facts that Robertsonian rearrangements by centric fusion have occurred in the same species. Furthermore, as shown in Table 6, the same NF value, 48, in all species of genus *Dascyllus* indicates distinctly that the chromosomal polymorphisms were formed by Robertsonian rearrangements throughout evolution. However, it is difficult to determine whether a biarmed chromosome split into two uniarmed chromosomes or, more probably, two uniarmed elements joined to form a biarmed chromosome.

V. SEX CHROMOSOMES

In about 30 teleosts from a thousand species, the existence of possible heteromorphic sex chromosomes has been reported by several investigators (Table 7). An understanding of sex determination in teleosts requires elucidation of cytogenetic mechanisms, morphophysiological processes, and hormonal control. In teleosts, the knowledge of these is very poor in comparison with that of other animals.[46] Svardson and Wickbom[47] state that the cytological identification of the sex chromosomes in teleosts is impossible, even though their existence has been proved by genetic methods.

Generally speaking, teleosts seem to be sexually undifferentiated. The facts of hermaphrodite and sex reversal such as protogyny or protandry through life in many kinds of teleosts are positive proof of an undifferentiated condition. In some cases the heteromorphic constitution is different among closely related species or members of the same species.

Table 5
KARYOTYPE CONSTITUTION OF 29 SPECIES OF THE POMACENTRIDAE STUDIED

Species	CN	M	SM	ST, A	NF
Amphiprioninae					
Amphiprion frenatus	48	14	22	12	84[a]
A. clarkii	48	14	16	18	78[b]
A. ocellaris	48	14	22	12	84[a]
Chrominae					
Chromis chrysura	48	2	0	46	50
C. caerulea	48	0	0	48	48
Dascyllus reticulatus[c]	37—34	*	*	*	48
D. aruanas[c]	33—27	*	*	*	48
D. trimaculatus[c]	48,47	*	*	*	48
D. melanurus[c]	48	0	0	0	48
Pomacentrinae					
Eupomacentrus lividus	48	6	24	18	78
E. nigricans	48	2	2	44	52
Dischistodus prosopotaenia	48	6	10	32	64
Pomacentrus coelestis	48	0	0	48	48[b]
P. moluccensis	48	8	28	12	84
P. amboinensis	48	—	—	—	—
P. rhodonotus	48	8	26	14	82
Pomacentrus sp.	48	10	26	12	84
P. philippinus	48	6	22	20	76
Cheilioprion labitatus	48	2	20	26	70
Glyphisodontinae					
Plectroglyphidodon lacrymatus	48	0	2	46	50
P. leucozonus	48	0	0	48	48[b]
Abudefduf notatus	48	2	0	46	50[b]
A. sordidus	48	2	2	44	52[b]
A. vaigiensis	48	2	2	44	52[b]
Amblyglyphidodon curacao	48	6	22	20	76
Glyphidodontops hemicyaneus	48	0	32	16	80
G. cyaneus	48	0	0	48	48
G. rex	48	8	22	18	78
Paraglyphidodon nigroris	48	8	24	16	80

[a] Arai et al.[45]
[b] Arai and Inoue.[44]
[c] Intraspecific variation of karyotype (Table 6).

There are two cases of heterozygotic chromosomes in which X and Y or Z and W are very nearly or quite equal in size and, on the other hand, Y or W is distinctly smaller than X or Z.

In the author's laboratory, C-banded differential staining was applied to the heteromorphic chromosomes in *Gambusia affinis* with ZW chromosomes[68] and *Stephanolepis cirrhifer* with XY chromosomes.[75] As shown in Figures 83 to 86, no C-banded staining, indicating constitutive heterochromatin, appeared on the heteromorphic chromosomes. The heteromorphic chromosomes of *Apogon notatus* showed a diploid chromosome number of 46, including 2 metacentrics, 4 submetacentrics, 3 large subtelo-, telocentrics, and 37 subtelo-, telocentrics in the male (Figure 87) and 2 M, 4 SM, 2 large ST, T, and 38 ST, T in the female (Figure 88). In this case, the heteromorphic chromosomes in the male are XY. But there is another colony of the same species, *A. notatus*, with homozygotic chromosomes in the male. The karyotypic constitution is 2

Table 6
KARYOTYPE CONSTITUTION OF THE GENUS *DASCYLLUS*

Species	CN	M, SM	ST, A	NF
Dascyllus trimaculatus	48	0	48	48
	47	1	46	48
D. reticulatus	37	11	26	48
	36	12	24	48
	35	13	22	48
	34	14	20	48
D. auruanus	33	15	18	48
	32	16	16	48
	31	17	14	48
	30	18	12	48
	29	19	10	48
	28	20	8	48
	27	21	6	48
D. melanurus	48	0	48	48

From Ojima, Y. and Kashiwagi, E., *Proc. Jpn. Acad.*, 57, 368, 1981. With permission.

metacentrics, 4 submetacentrics, 2 large subtelo-, telocentrics, and 38 subtelo-, telocentrics in the male and female (Figure 89).

Ueda and Ojima[48] and Ojima and Takai[49] have shown that in the Japanese genus *Carassius, C. auratus langsdorfii, C. a.* subsp., *C. a. auratus* (Chinese Funa), and *C. a. aurarus* (goldfish) were characterized by a pair of marker chromosomes with deeply stained C-bands on the short arms of the second largest submetacentric chromosomes. The occurrence of two markers in the female and one in the male (Figures 90 and 91) was notable. These markers could not be demonstrated in other subspecies, *C. a. buergeri, C. a. grandoculis,* and *C. a. cuvieri,* with the same chromosomal constitutions (Figure 92).

These facts indicate that the sex chromosomes in teleostean fishes are still undifferentiated morphologically and functionally.

The cytological evidence unmistakably indicates that the sex chromosomes are fundamentally similar to other chromosomes and that they originally formed a pair of synaptic mates indistinguishable in appearance, behavior, and visible structure. Such a condition, indeed, seems still to exist in many animals (and plants) in which these chromosomes have not yet been identified as such. In this condition they still, for the most part, remain in the homogametic sex, their most striking peculiarities appearing only in the digametic sex. It is probable, therefore, that most of their special characteristics have resulted from their constantly heterozygous condition in one sex, which must have existed from a very ancient period. It is evident further that the X chromosome is a body of complex constitution and that only a part of it, and probably only a very small part, contributes to sex differentiation. This is indicated by the great differences in the size of this chromosome in different species, even in rather closely related ones.[50]

Makino[51] and Galgano[52] are of the opinion that, inasmuch as teleosts occupy a rather low position on the evolutionary ladder, the degree of difference between their sex chromosomes and autosomes is negligible. On the basis of a consensus of estimates through the facts and views mentioned above, it seems that the fish group is still in the process of evolutionary differentiation but that its sexual differentiation is still in an unstable and primitive condition.

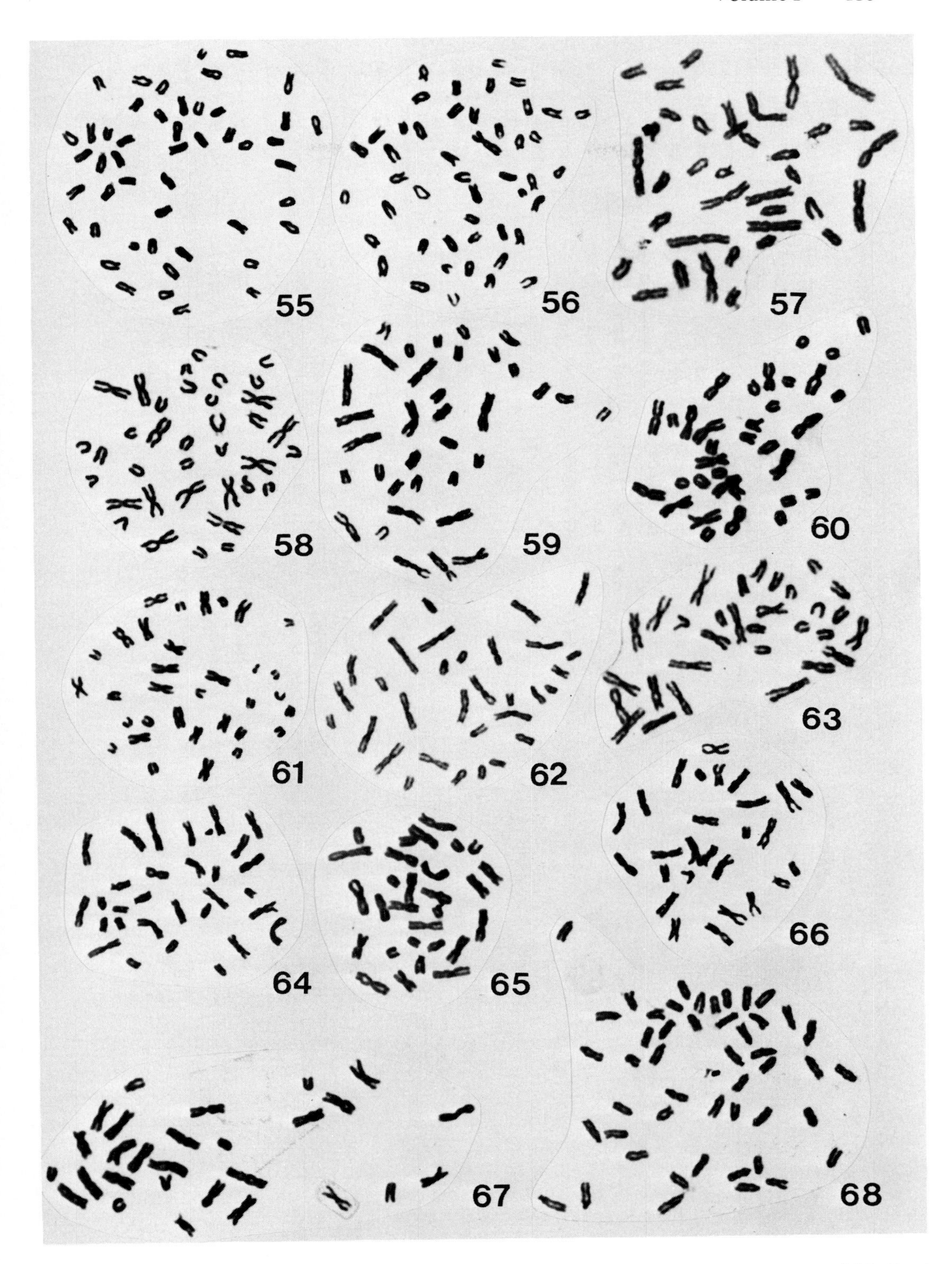

FIGURE 55—68. Photomicrographs of metaphase chromosomes of the genus *Dascyllus*. Figures 55 and 56: *D. trimaculatus;* Figures 57 to 60: *D. reticulatus;* Figures 61 to 67: *D. auruanus;* Figure 68: *D. melanurus.* (Figures 57 to 60 from Ojima, Y. and Kashiwagi, E., *Proc. Jpn. Acad.*, 57, 368, 1981. With permission.)

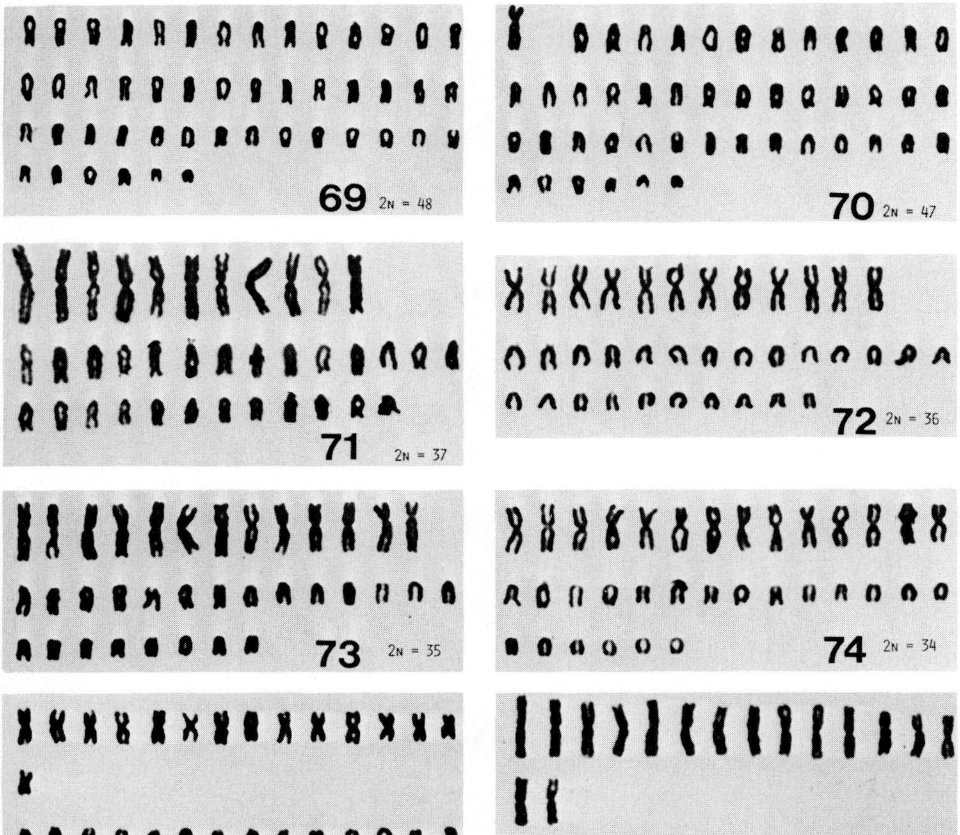

FIGURE 69—82. Serial alignment of somatic chromosomes corresponding to Figures 55 to 68.

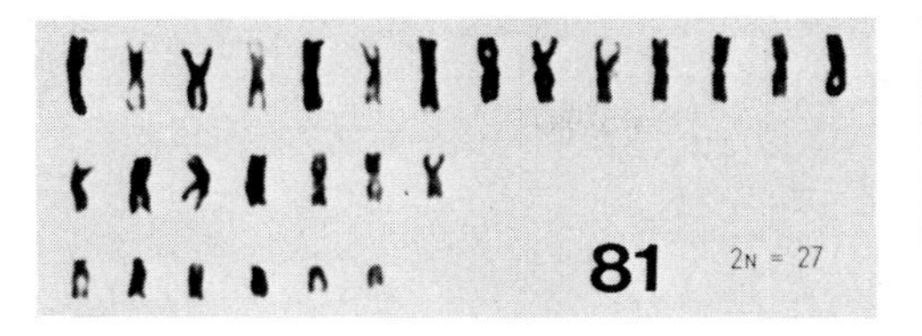

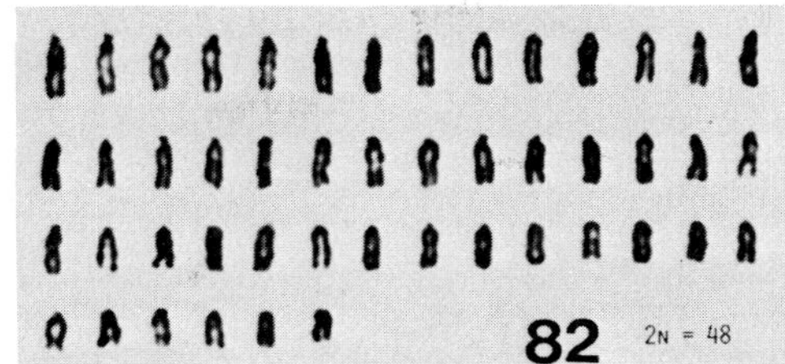

FIGURES 81 to 82

Table 7
HETEROMORPHIC SEX CHROMOSOMES REPORTED IN FISH SPECIES

Lower Teleostean Groups

	Sex chromosomes	Ref.
Salmoniformes		
Bathylagidae		
Bathylagus wesethi	XX-XY	53
B. stilbius	XX-XY	53
B. ochotensis	XX-XY	53
B. milleri	XX-XY	53
Sternoptycidae		
Sternoptyx diphana	XX-XO	53
Salmonidae		
Salmo gairdneri	XX-XY	54
Cypriniformes		
Ictaluridae		
Noturus taylori	XX-XY	55
Bagridae		
Mystus tengara	ZW-ZZ	56
Siluridae		
Callichrous bimaculatus	$X_1X_1X_2X_2$-$X_1X_1X_2$	57
Erythrinidae		
Hoplias lacerdae	XX-XY	58
H. malabaricus	XX-XY	59
Palodontidae		
Apareiodon affinis	ZW_1W_2-ZZ	60
Anostomidae		
Lepolinus lacustris	XX-XY	61
L. silvestrii	ZW-ZZ	61
L. obtusidens	ZW-ZZ	61
Anguilliformes		
Anguillidae		
Anguilla japonica	ZW-ZZ	62
Congridae		
Astroconger myriaster	ZW-ZZ	62

Intermediate Teleostean Groups

	Sex chromosomes	Ref.
Myctophiformes		
Neoscopelidae		
Scopelengys tristis	XX-XO	53
Myctophidae		
Lampanyctus ritteri	XX-XO	53
Parvilux ingens	XX-XO	53
Symbolophorus californiensis	XX-XY	53

Table 7 (continued) HETEROMORPHIC SEX CHROMOSOMES REPORTED IN FISH SPECIES

Intermediate Teleostean Groups

	Sex chromosomes	Ref.
Synodontidae		
Saurida undosquamis	ZW-ZZ	63
Cyprinodontiformes		
Cyprinodontidae		
Fundulus diaphanus	XX-XY	64
F. parvipinnis	XX-XY	64
Cyprinodon sp. (unnamed)	$X_1X_1X_2X_2$-X_1X_2Y	65
Mollienesia sphenops	ZW-ZZ	66
Megupsilon aporus	$X_1X_1X_2X_2$-X_1X_2Y	67
Poeciliidae		
Gambusia affinis	ZW-ZZ	68
G. holbrook	ZW-ZZ	69
Platypoecilus maculatus	ZW-ZZ	70
Syngnathiformes		
Gasterosteidae		
Gasterosteus wheatlandi	XX-XY	64
Apeltes quadracus	ZW-ZZ	64
Beryciformes		
Melamphaeidae		
Melamphase parvus	XX-XY	53
Scopeloberyx robustus	XX-XY	53
S. mizolepis bispinosus	XX-XY	53

Higher Teleostean Groups

	Sex chromosomes	Ref.
Perciformes		
Belontiidae		
Colisa fasciatus	ZW-ZZ	71
Trichogaster fasciatus	ZW-ZZ	71
Gobiidae		
Gobiodon citrinus	XX-XO	73
Pleuronectiformes		
Cynoglossidae		
Symphurus plagiusa	XX-XO	74
Tetraodontiformes		
Aluteridae		
Stephanolepis cirrhifer	$X_1X_1X_2X_2$-X_1X_2Y	75

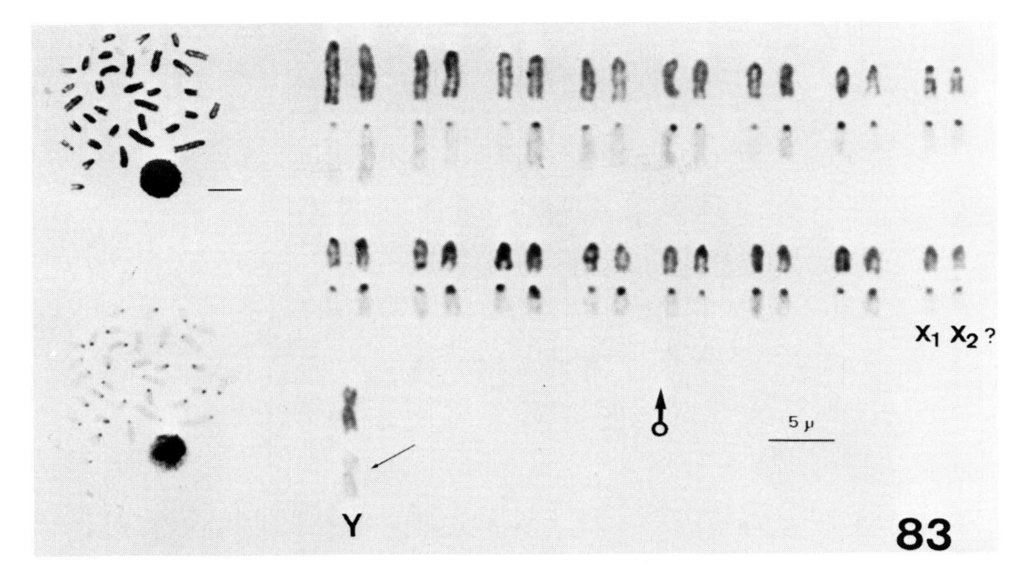

FIGURE 83. C-Banding karyotype of a male of *Stephanolepis cirrhifer* (the lower row) with heteromorphic X_1X_2?Y chromosomes. The arrow indicates Y chromosome without C-banded staining.

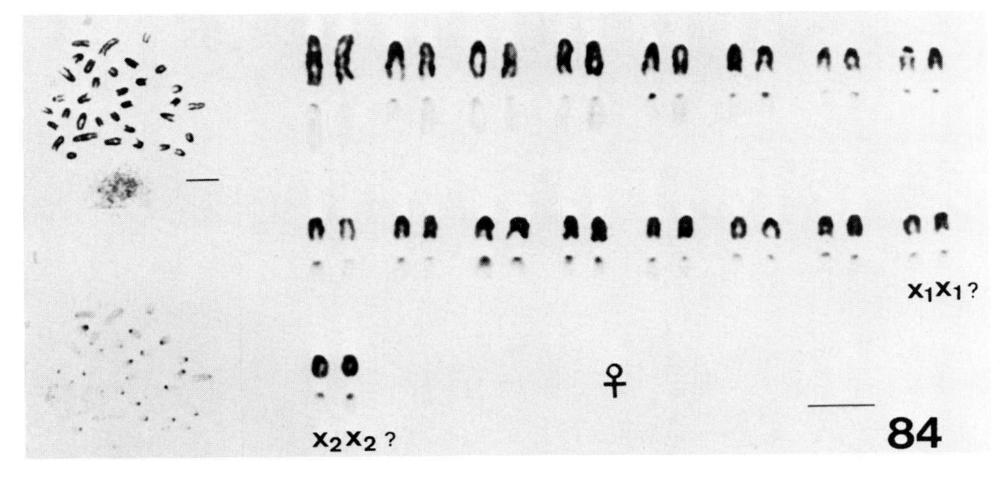

FIGURE 84. C-Banding karyotype of a female of *Stephanolepis cirrhifer* (the lower row) with X_1X_2? chromosomes.

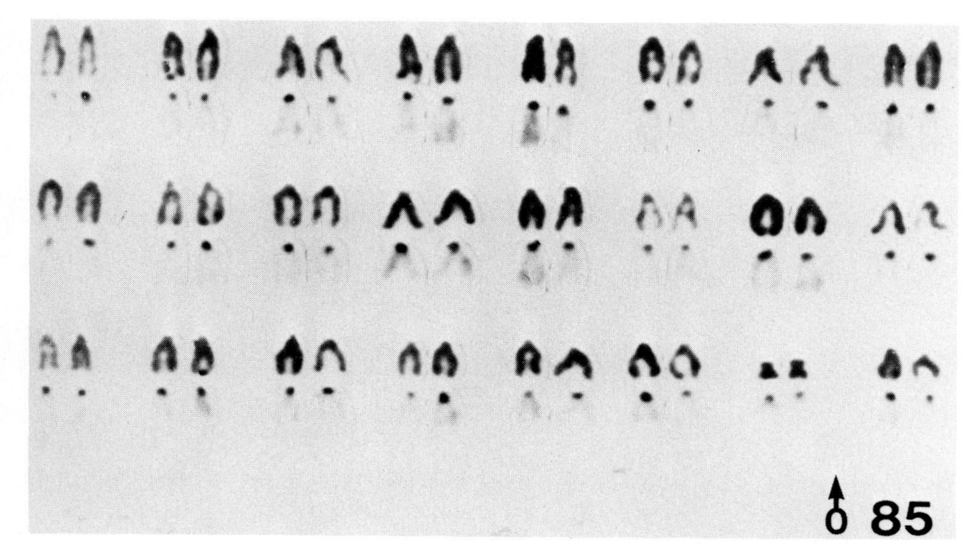

FIGURE 85. C-Banding karyotype of a male of *Gambusia affinis* (the lower row).

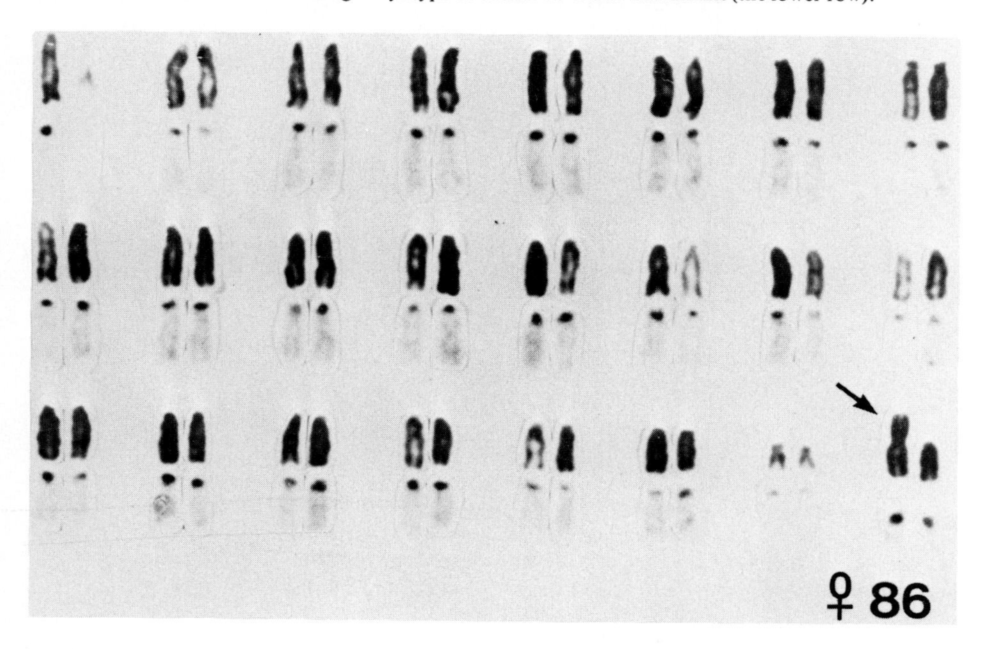

FIGURE 86. C-Banding karyotype of a female of *Gambusia affinis* (the lower row) with heteromorphic ZW chromosomes. The arrow indicates W chromosome without special C-banded staining.

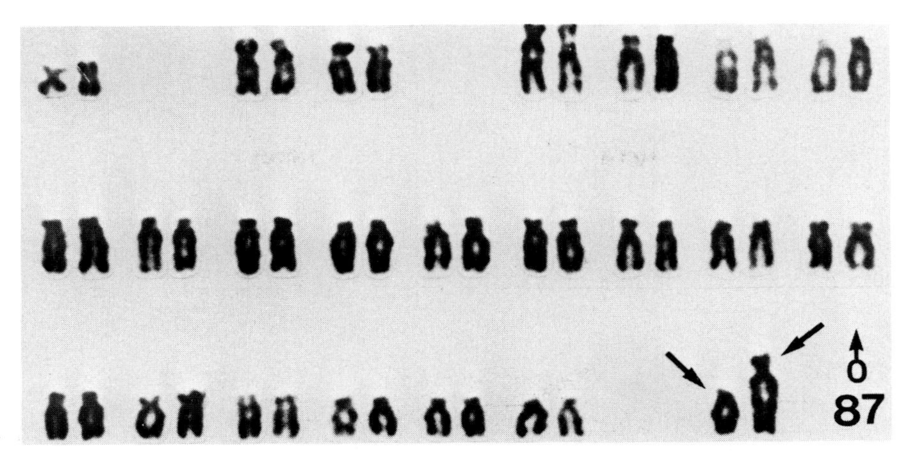

FIGURE 87. Karyotype analysis of a male of *Apogon notatus* with heteromorphic chromosomes (arrow).

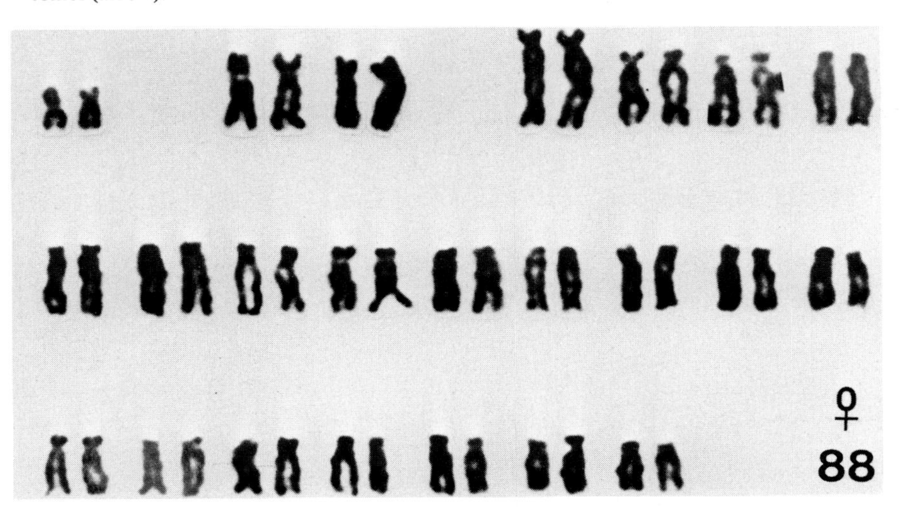

FIGURE 88. Karyotype analysis of a female of *Apogon notatus*.

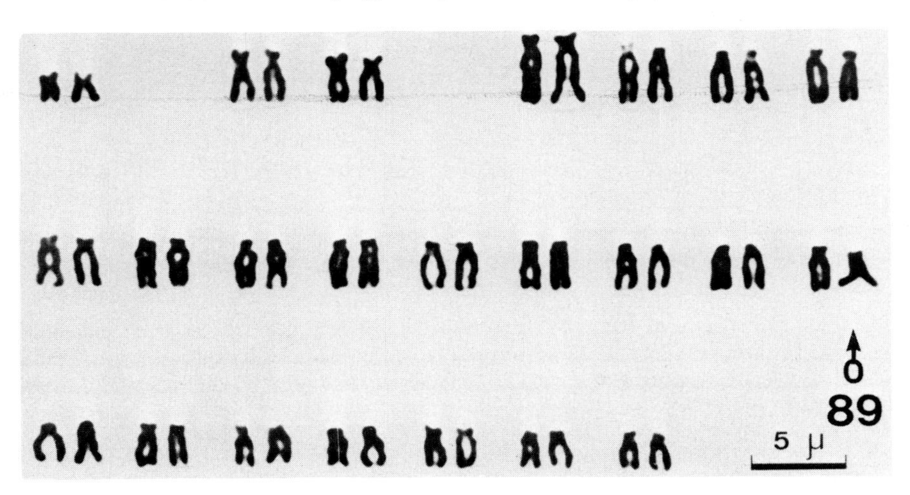

FIGURE 89. Karyotype analysis of a male of *Apogon notatus* without heteromorphic chromosomes.

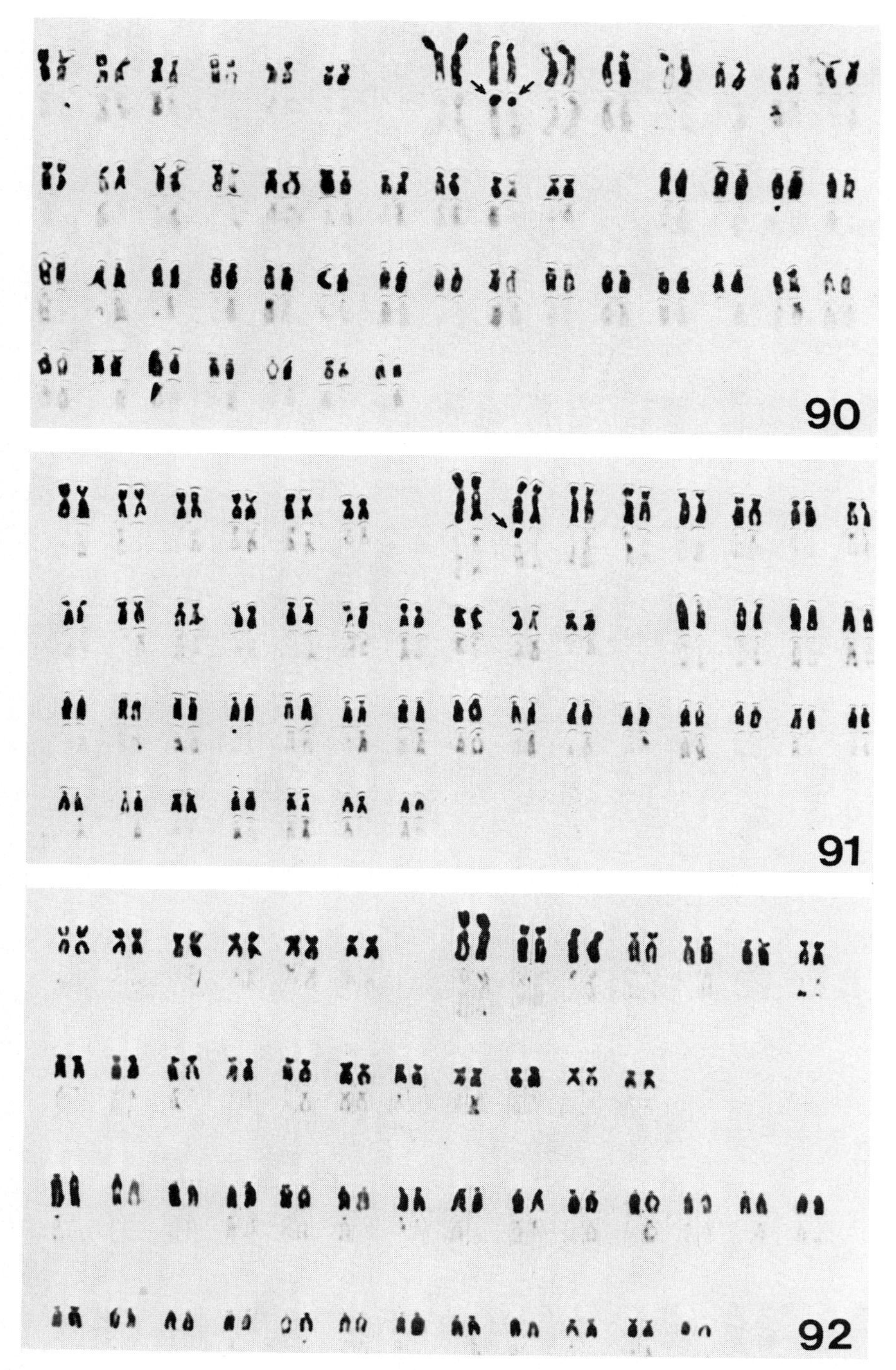

FIGURE 90—92. Figure 90: C-banding karyotype of a female of *Carassius auratus auratus* (the lower row). The arrows denote two large submetacentrics having intensely C-banded short arms; Figure 91: C-banding karyotype of a male of *C. a. auratus* (the lower row). The arrow denotes one large submetacentric having intensely C-banded short arm; Figure 92: C-banding karyotype of *C. a. cuvieri* (the lower row). (Figures 90 and 91 from Ojima, Y. and Takai, A., *Proc. Jpn. Acad.*, 55, 346, 1979. With permission.)

REFERENCES

1. **Ojima, Y.**, Chromosome Data Retrival System List (CDR List), *Rep. Biol. Lab.*, Kwansei Gakuin Univ. Bull. No. 3, Nishinomiya, Japan, 1981.
2. **Wolf, K., Quimby, M. C., Pyle, E. A., and Dexter, R. P.**, Preparation of monolayer cell cultures from tissues of some lower vertebrates, *Science*, 135, 1065, 1960.
3. **Clem, L. W., Moewus, L., and Sigel, M. M.**, Studies with cells from marine fish in tissue culture, *Proc. Soc. Exp. Biol. Med.*, 108, 762, 1961.
4. **Ojima, Y.**, Preparation of cell-cultures for chromosome studies of fishes, *Proc. Jpn. Acad.*, 54, 116, 1978.
5. **Gosline, W. A.**, *Functional Morphology and Classification of Teleostean Fishes*, University Press Hawaii, Honolulu, 1971, 1.
6. **Winkler, H.**, Über die experimentelle Erzeugung von Pflanzen mit abweichenden Chromosomenzahlen, *Zeits. F. Bot.*, 8, 417, 1916.
7. **Rosenberg, O.**, Das Verhalten der Chromosomen in einer hybriden Pflanze, *Ber. D. D. Bot. Gesellsch.*, 21, 110, 1903.
8. **Rosenberg, O.**, Cytologische und morphologishe Studien an Drosera, *Kungl. Svensk. Vetensk. Handl.*, 43, 1, 1909.
9. **Sakamura, T.**, Kurze Mitteilung über die Chromosomenzahlen und die Verwandtschaftsverhältnisse der *Triticum-Arten*, *Bot. Mag. Tokyo*, 32, 150, 1918.
10. **Kihara, H. and Ono, T.**, Chromosomenzahlen und systematishe Gruppierung der *Rumex-Arten*, *Zeits. F. Zellforsch.*, 4, 475, 1926.
11. **Blakeslee, A. F. and Avery, A. G.**, Methods of inducing doubling of chromosomes in plants, *J. Hered.*, 28, 393, 1937.
12. **Svärdson, G.**, Chromosome studies on Salmonidae, *Rep. Inst. Freshwat. Res. Drottningholm*, 23, 1, 1945.
13. **Ohno, S.**, *Evolution by Gene Duplication*, Springer-Verlag, New York, 1970.
14. **Kobayashi, H., Kawashima, Y., and Takeuchi, N.**, Comparative chromosome studies in the genus *Carassius*, especially with a finding of polyploidy in the Ginbuna *(C. auratus langsdorfii)*, *Jpn. J. Ichthyol.*, 17, 153, 1970.
15. **Kobayashi, H. and Ochi, N.**, Chromosome studies of the hybrids, Ginbuna (*Carassius auratus langsdorfii*) × Kinbuna (*C. auratus* subsp.) and Ginbuna × Loach (*Misgurnus anguillicaudatus*), *Zool. Mag. (Tokyo)*, 81, 67, 1972.
16. **Kobayashi, H., Ochi, N., and Takeuchi, N.**, Chromosome studies in the genus *Carassius*: comparison of *C. auratus grandoculis*, *C. auratus buergeri*, and *C. auratus langsdorfii*, *Jpn. J. Ichthyol.*, 20, 7, 1973.
17. **Ojima, Y., Hayashi, M., and Ueno, K.**, Triploidy appeared in the back-cross offsprings from Funa-icarp crossing, *Proc. Jpn. Acad.*, 51, 702, 1975.
18. **Gold, J. R. and Avise, J. C.**, Spontaneous triploidy in the California loach *Hesperoleucus symmetricus* (Pisces: Cyprinidae), *Cytogenet. Cell Genet.*, 17, 144, 1976.
19. **Ueno, K. and Ojima, Y.**, Diploid-tetraploid complexes in the genus *Cobitis* (Cobitidae, Cyprinida), *Proc. Jpn. Acad.*, 52, 446, 1976.
20. **Ojima, Y. and Takai, A.**, The occurrence of spontaneous polyploid in the Japanese common loach, *Misgurnus anguillicaudatus*, *Proc. Jpn. Acad.*, 55, 487, 1979.
21. **Ueno, K., Iwai, S., and Ojima, Y.**, Karyotypes and geographical distribution in the genus *Cobitis* (Cobitidae), *Bull. Jpn. Soc. Sci. Fish.*, 46, 9, 1980.
22. **Ohno, S.**, *Animal Cytogenetics*, Vol. 4, Chordata, 1, Gebrüder Borntraeger, Berlin, 1974.
23. **Ojima, Y. and Hitotsumachi, S.**, Cytogenetical studies in loaches (Pisces, Cobitidae), *Zool. Mag.*, 78, 139, 1969.
24. **Hitotsumachi, S., Sasaki, M., and Ojima, Y.**, Comparative karyotype study in several species of Japanese loaches (Pisces, Cobitidae), *Jpn. J. Genet.*, 44, 157, 1969.
25. **Nogusa, S.**, A comparative study of the chromosomes in fishes with particular considerations on taxonomy and evolution, *Mem. Hyogo Univ. Agr.*, 3(1), (Biological Ser. No. 3), 1960.
26. **Muramoto, J., Ohno, S., and Atkin, N. B.**, On the diploid state of the fish order Ostariopysi, *Chromosoma*, 24, 59, 1968.
27. **Hertwig, G. and Hertwig, P.**, Triploid Froschenlarven, *Arch. Mikr. Anat.*, 84, 34, 1920.
28. **Dalcq, A.**, La formule chromosomiale chez la grenouille, *Ann. Bull. Soc. R. Sci. Med. Nat. Bruxelles*, No. 1-2, 15 to 22, 1930.
29. **Fankhauser, G. and Kaylor, C. T.**, Chromosome numbers in androgenetic embryos of *Triturus viridescens*, *Anat. Rec.*, 64 (Suppl. 1), 41, 1935.
30. **Fankhauser, G.**, Triploidy in the newt, *Triturus viridescens*, *Proc. Am. Phil. Soc.*, 79, 715, 1938.

31. **Fankhauser, G.,** Polyploidy in the salamander, *Eurycea bislineata, J. Hered.,* 30, 379, 1939.
32. **Fankhauser, G. and Griffiths, R. B.,** Induction of triploidy and haploidy in the newt, *Triturus viridescens,* by cold treatment of unsegmented eggs, *Proc. Natl. Acad. Sci.,* 25, 233, 1939.
33. **Griffiths, R. B.,** Triploidy (and haploidy) in the newt, *Triturus viridescens,* induced by refrigeration of fertilized eggs, *Genetics,* 26, 69, 1941.
34. **Kawamura, T.,** Triploid frogs developed from fertilized eggs, *Proc. Imp. Acad.,* 17, 523, 1941.
35. **Makino, S. and Ojima, Y.,** Formation of the diploid egg nucleus due to suppression of the second maturation division, induced by refrigeration of fertilized eggs of the carp, *Cyprinus carpio, Cytologia,* 13, 55, 1943.
36. **Ojima, Y. and Makino, S.,** Triploidy induced by cold shock in fertilized eggs of the Carp. A preliminary study, *Proc. Jpn. Acad.,* 54, 359, 1978.
37. **Purdom, C. E.,** Induced polyploidy in plaice (*Pleuronectes platessa*) and its hybrid with the flounder *(Platichthys flesus), Heredity,* 29, 11, 1973.
38. **Sax, K.,** The experimental production of polyploidy, *J. Arnold Arboretum,* 17, 153, 1936.
39. **Schindler, A. M. and Mikamo, K.,** Triploidy in man. Report of a case and a discussion on etiology, *Cytogenetics,* 9, 116, 1970.
40. **Mikamo, K. and Iffy, L.,** Aging of the ovum, *Obstet. Gynecol. Ann.,* p. 47, 1974.
41. **Ojima, Y., Hitotsumachi, S., and Makino, S.,** Cytogenetic studies in lower vertebrates. I. A preliminary report on the funa (*Carassius auratus*) and gold-fish (a revised study), *Proc. Jpn. Acad.,* 42, 62, 1966.

41a. **Tri, N.,** unpublished.

42. **Ojima, Y. and Hitotsumachi, S.,** Cytogenetic studies in lower vertebrates. IV. A note on the chromosomes of the Carp (*Cyprinus carpio*) in comparison with those of the Funa and the gold-fish *(Carassius auratus), Jpn. J. Genet.,* 42, 163, 1967.
43. **Kobayashi, H.,** The Origin of the Polyploid Funa, *Japan Zoological Meeting, Tokyo,* Oct. 2—4, 1979, and Sapporo, Oct. 1—3, 1980.
44. **Arai, R. and Inoue, M.,** Chromosomes of seven species of Pomacentridae and two species of Acanthuridae from Japan, *Bull. Nat. Sci. Mus.,* 2, 73, 1976.
45. **Arai, R., Inoue, M., and Ida, H.,** Chromosomes of four species of coral fishes from Japan, *Bull. Nat. Sci. Mus.,* 2, 137, 1976.
46. **Vanyakia, E. D.,** *Genetics, Selection, and Hybridization of Fish,* Cherfas, B. I., Ed., Israel Program for Scientific Translations, Jerusalem, 1972, 5.
47. **Svärdson, G. and Wickbom, T.,** The chromosomes of two species of Anabantidae with a new case of sex reversal, *Hereditas,* 28, 1, 1942.
48. **Ueda, T. and Ojima, Y.,** Differential chromosomal characteristics in the Funa subspecies (Carassius), *Proc. Jpn. Acad.,* 54, 283, 1978.
49. **Ojima, Y. and Takai, A.,** Further cytogenetical studies on the origin of the gold-fish, *Proc. Jpn. Acad.,* 55, 346, 1979.
50. **Wilson, E. B.,** *The Cell in Development and Heredity,* Macmillan, New York, 1925.
51. **Makino, S.,** The chromosomes of fishes, *Proc. Jpn. Acad.,* 8, 1, 1932.
52. **Galgano, H.,** Le gametogenesi dell teleostems, *Atti Soc. Ital. Anat.,* 7, 1938.
53. **Chen, T. R.,** Karyological heterogamety of deep-sea fishes, *Postilla,* 130, 1, 1969.
54. **Thorgaard, G. H.,** Heteromorphic sex chromosomes in male rainbow trout, *Science,* 196, 900, 1977.
55. **LeGrande, W. H.,** Chromosomal evolution in North American catfishes (Siluriformes: Ictaluridae) with particular emphasis on the madtoms *Noturus, Copeia,* p. 33, 1981.
56. **Rishi, K. K.,** Somatic karyotypes of three teleosts, *Genen Phaenen,* 16, 101, 1973.
57. **Rishi, K. K.,** Mitotic and meiotic chromosomes of a teleost, *Callichrous bimaculatus* (Bloch) with indications of male heterogamety, *Ciencia Cult.,* 28, 1171, 1976a.
58. **Bertollo, L. A. C., Takahashi, C. S., and Filho, O. M.,** Cytotaxonomic considerations on *Hoplias lacerdae* (Pisces, Erythrinidae), *Rev. Brasil. Genet.,* 1(2), 103, 1978.
59. **Bertollo, L. A. C., Takahashi, C. S., and Filho, O. M.,** Karyotypic studies of two allopatric populations of the genus *Hoplias* (Pisces, Erythrinidae), *Rev. Brasil Genet.,* 2(1), 17, 1979.
60. **Filho, O. M., Bertollo, L. A. C., and Galetti, P. M., Jr.,** Evidences for a multiple sex chromosome system with female heterogamety in *Apareiodon affinis* (Pisces, Parodontidae), *Caryologia,* 33, 83, 1980.
61. **Galetti, P. M., Jr., Foresti, F., Bertollo, L. A. C., and Filho, O. M.,** Heteromorphic sex chromosomes in three species of the genus *Leporinus* (Pisces, Anostomidae), *Cytogenet. Cell Genet.,* 29, 138, 1981.
62. **Park, E. H. and Kang, Y. S.,** Karyological confirmation of conspicuous ZW sex chromosomes in two species of Pacific anguilloid fishes (Anguillioformes: Teleostomi), *Cytogenet. Cell Genet.,* 23, 33, 1979.

63. **Nishikawa, S. and Sakamoto, K.**, Comparative studies on the chromosomes in Japanese fishes. IV. Somatic chromosomes of two lizard fishes, *J. Shimonoseki Univ. Fisheries*, 27, 113, 1978.
64. **Chen, T. R. and Ruddle, H. H.**, A chromosome study of four species and a hybrid of the killifish genus *Fundulus* (Cyprinodontidae), *Chromosoma*, 29, 255, 1970.
65. **Uyeno, T. and Miller, R. R.**, Multiple sex chromosomes in a Mexican cyprinodontid fish, *Nature*, 231, 452, 1971.
66. **Rishi, K. K.**, Cytogenetical female heterogamety in jet-black molly, *Mollienesia* sphenops, *Curr. Sci.*, 45, 669, 1976b.
67. **Miller, R. and Walters, V.**, *Megupsilon aporus* (Cyprinodontiformes), *Chromosome Atlas*, 2(15), 1, 1973.
68. **Chen, T. R. and Ebeling, A. W.**, Karyological evidence of female heterogamety in the mosquitofish *Gambusia affinis*, *Copeia*, 1968, 70.
69. **Yoshida, T. H. and Hayashi, M.**, Preliminary note on karyotype of guppy and topminnow, *Ann. Rep. Nat. Inst. Genet. Jpn.*, 21, 52, 1971.
70. **Anders, A., Anders, F., and Rase, S.**, XY females caused by X-irradiation, *Experientia*, 25, 851, 1969.
70a. **Chen, T. R. and Reisman, H. M.**, A comparative chromosome study of the North American species of sticklebacks (Teleostei: Gasterosteidae), *Cytogenetics*, 9, 321, 1970.
71. **Rishi, K. K.**, Somatic G-banded chromosomes of *Colisa fasciatus* (Perciformes: Belontidae) and confirmation of female heterogamety, *Copeia*, 1979, 146.
72. **Rishi, K. K.**, Somatic and meiotic chromosomes of *Trichogaster fasciatus* (Bl. and Sch.) (Teleostei, Perciformes: Osphronemidae), *Genen Phaenen*, 18, 49, 1975.
73. **Arai, R. and Sawada, Y.**, Chromosomes of Japanese gobioid fishes. I, *Bull. Nat. Sci. Mus. Tokyo*, 17, 97, 1974.
74. **LeGrande, W. H.**, Karyology of six species of Louisiana flatfishes (Pleuronectiformes: Osteichthyes), *Copeia*, 1975, 516.
75. **Murofushi, M., Oikawa, S., Nishikawa, S., and Yoshida, T. H.**, Cytogenetical studies on fishes. III. Multiple sex chromosome mechanism in the filefish, *Stephanolepis cirrhifer*, *Jpn. J. Genet.*, 55, 127, 1980.
76. **Ojima, Y. and Kashiwagi, E.**, Further studies of the chromosomes of the Labridae (Pisces). A preliminary note, *Proc. Jpn. Acad.*, 56B, 328, 1980.
77. **Ojima, Y and Kashiwagi, E.**, Chromosomal evolution associated with Robertsonian fusion in the genus *Dascyllus* (Chrominae, Pisces), *Proc. Jpn. Acad.*, 57B, 368, 1981.

Chapter 3

CHROMOSOME DIFFERENTIATION AND SPECIES EVOLUTION IN RODENTS*

Tosihide H. Yosida

TABLE OF CONTENTS

* Contribution No. 1403 from the National Institute of Genetics, Japan. Supported by a grant-in-aid for scientific research from the Ministry of Education, Science and Culture (Nos. 56010001, 56390019, 56480003).

I. INTRODUCTION

One of the important causes of evolution in organisms is the production of genetic mutations. Mutation occurs at the genic or DNA level and also at the chromosomal or macromolecular level. In the latter, inversion, translocation, deletion, and some other phenomena are known to cause modification of chromosome structure or karyotypes. In this paper a possible relation between the differentiation of chromosome structures and evolution of organisms will be discussed, based mainly on the author's data obtained in the studies of *Rattus* and some other rodents or mammalian species published by several investigators.

II. RELATIONSHIP BETWEEN CHROMOSOME INVERSION AND SPECIES DIFFERENTIATION

Pericentric inversion is often found in rodent species. Relationships between the inversion of chromosomes and species differentiation will be described in the black rat and some other rodent species.

A. Karyotype Evolution by Pericentric Inversion in the Black Rat

Although inversion is common, the karyotype does not change by paracentric inversion. Change of the chromosome morphology is due to pericentric inversion which includes the centromere. This inversion produces a drastic change in the karyotype without any alteration of chromosome number or the number of centromeres.

The black rat, or house rat (*Rattus rattus*), is a very common animal distributed widely in the world (Figure 1). The basic chromosome number of this species is 42 (21 pairs) consisting of 13 acrocentric autosome pairs (1 to 13), 7 metacentric autosome pairs (14 to 20), and acrocentric X and Y (Figure 2). Among 13 acrocentric pairs, nos. 1, 9, and 13 were polymorphic due to acrocentric and subtelocentric members. These chromosomes, in some members, form acrocentric homomorphic pairs (A/A), in some form acrocentric and subtelocentric heteromorphic pairs (A/S), whereas others form subtelocentric homomorphic pairs (S/S).[1-3] Based on measurements of chromosome length between the acrocentric and the subtelocentric, comparison of their G-banding patterns, and also frequencies and distribution patterns of the polymorphic chromosomes, it is suggested that the subtelocentric members were derived by the pericentric inversion of the original acrocentric partners (Figure 3).

An example of chromosome polymorphism of pair no. 1 in black rats distributed in Japan is as follows. All specimens collected from northern and northwestern Japan, where the climate is more severe with heavy snow and cold weather in winter, showed only the monomorphic A/A type, while in southern and southeastern Japan animals with all three karyotypes (A/A, A/S, and S/S) were observed.[2] The break in the distribution of the black rats between those with a monomorphic pair no. 1 and those with polymorphic pair no. 1 closely coincides with the Japanese climate, as Japan is divided into two areas with heavy snowfall (over 50 cm) and light snowfall (below 50 cm). Among 453 rats collected throughout Japan, 75.9% carried the acrocentric homomorphic pair, which is suggested to be a basic type for the animals in Japan (Figure 4).

By laboratory matings between rats with several chromosome types (A/A, A/S, and S/S), it is found that the segregation of these three types approximated the expected values. Although the number of animals with A/S pairs was slightly high, that with S/S pairs was somewhat low. This finding indicated that the rats with an S/S pair no. 1 might be less adaptive to laboratory breeding.[2] The lower adaptability of the rats

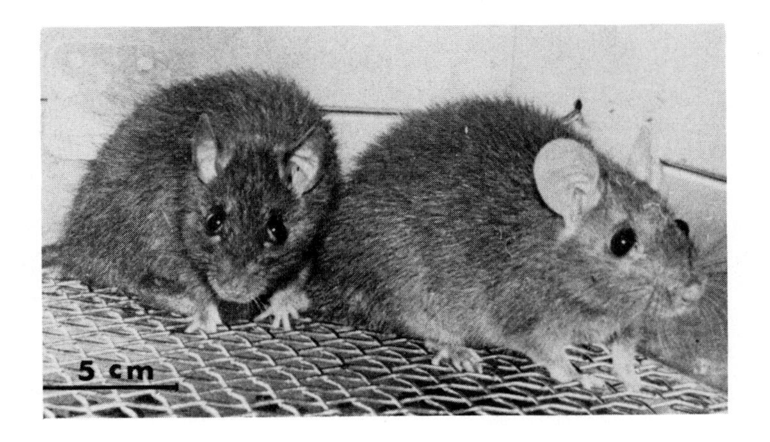

FIGURE 1. Black rat (*Rattus rattus*).

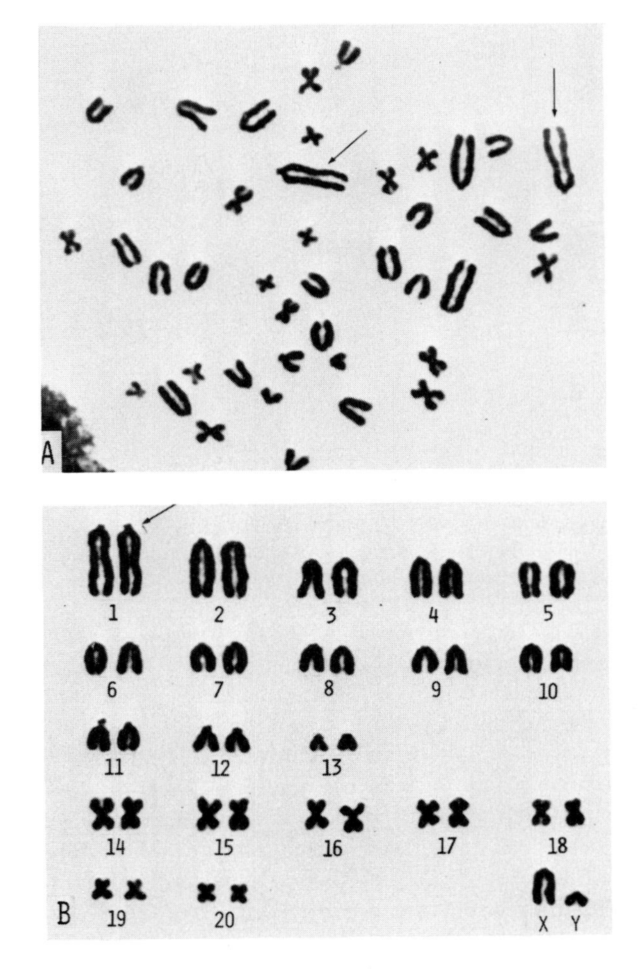

FIGURE 2. Basic karyotype of the black rat, *R. rattus* (Asian type). (A) Metaphase, (B) karyotype.

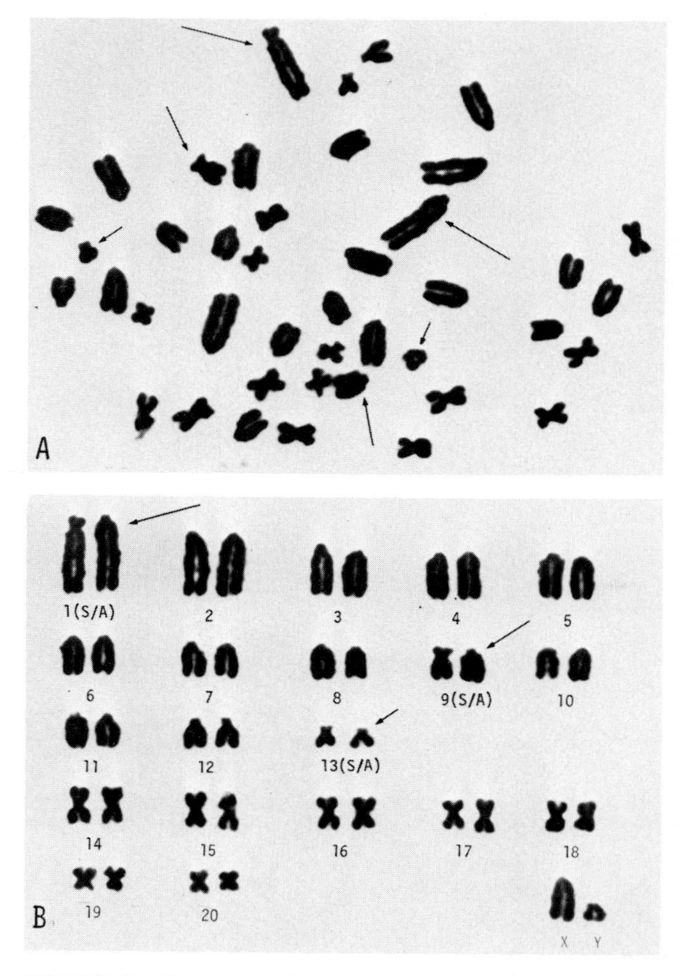

FIGURE 3. Karyotype of an Asian-type black rat with heteromorphic pairs no. 1, 9, and 13. (A) Metaphase, (B) karyotype. Acrocentrics and subtelocentrics are denoted by A and S, respectively. Arrows indicate the heteromorphic pairs.

with A/A pair no. 1 was also confirmed when the rats were bred without any air conditioning in summer and winter.[4]

According to mammalian taxonomists,[5-7] the black rats originated in the Indo-Malayan region and spread around the world after the Middle Ages. They might have been characterized by acrocentric pair no. 1, migrated to Japan from the Asian continent, and eventually spread to all parts of the country. Later, the rats with subtelocentric chromosome no. 1 which arose by pericentric inversion somewhere in Southeast Asia, might have migrated to Japan and mated with the black rats originally living there (Figure 5). The black rats with subtelocentric pair no. 1 might have adapted to the hot climate so that they could not move to northern Japan. In this way a clear boundary between the rat populations arose spontaneously to divide Japan into two parts, Northwest and Southeast. The black rats with acrocentric pair no. 1 or acrocentric and subtelocentric polymorphism have been found in the other countries of East and Southeast Asia, such as Korea, Hong Kong, Taiwan, Thailand, and Malaysia. However, the specimens collected in the Philippines (Luzon, Mindanao) and Indonesia (Java, Celebes) always showed subtelocentric pair no. 1 (Figure 6).[8] By mating the

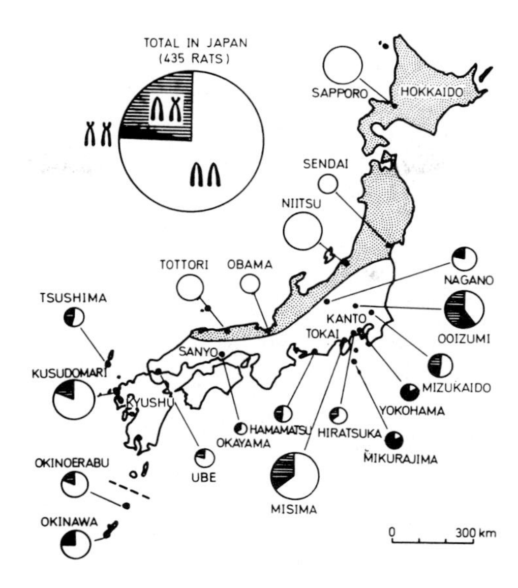

FIGURE 4. Population map of the frequency of no. 1 chromosome polymorphism in the black rat distributed in Japan. Population size is indicated by the dimension of the circles, in which white, shaded, and black areas show the frequency of animals with A/A, A/S, and S/S pairs, respectively. Area with A/A chromosome type is dotted. (From Yosida, T. H., Tsuchiya, K., and Moriwaki, K., *Chromosoma (Berlin)*, 33, 30, 1971. With permission.)

Philippine black rats with the S/S pair no. 1 and Japanese ones with the A/A pair no. 1, the polymorphic chromosomes were successfully established in the laboratory.

Black rats were polymorphic in respect to pairs no. 9 and no. 13 acrocentric and subtelocentric chromosomes. The frequencies of polymorphic pair no. 9 in Japan were similar to that of pair no. 1; namely, in the Japanese population the frequencies of the A/A pair were higher than the A/S and S/S pairs, but those in the islands in Southeast Asia always showed the S/S type. In pair no. 13, however, the S/S and A/S pairs were observed in all parts of Japan. This means that the pericentric inversion of pair no. 13 might have occurred earlier than those of pairs nos. 1 and 9.[9]

B. Species Differentiation in Genus *Rattus* by Pericentric Inversion

By segregation of chromosome polymorphism of pairs 1, 9, and 13 in the black rats, several karyotypes would be produced in a natural population and also in the laboratory. By combination of the homomorphic pairs in these chromosomes, 9 chromosome sets are possible theoretically and are actually observed in the natural population and in the laboratory. A comparison of the karyotypes of the other *Rattus* species with the polymorphic karyotypes of the black rat shows an interesting relationship. The Norway rat, *Rattus norvegicus*, the little house rat, *R. exulans*, the Singapore rat, *R. annandalei*, and the swamp giant rat, *R. muelleri*, all have 42 chromosomes and a karyotype similar to the house rat (Figure 7). They have 13 acrocentric or subtelocentric autosomes and 7 metacentric autosomes, but the numbers of subtelocentric chromosome pairs are different in each species. In the Singapore rat, pair no. 9 only is subtelocentric, but nos. 1 and 13 are acrocentrics. Pair no. 1 and 9 in the little house rat, 1 and 13 in the swamp giant rat, and 1, 9, and 13 in the Norway rat are subtelocentrics. These karyotypes coincide with either one of the polymorphic karyotypes in the black rat.[10] Based on these observations it can be concluded that the karyotypes

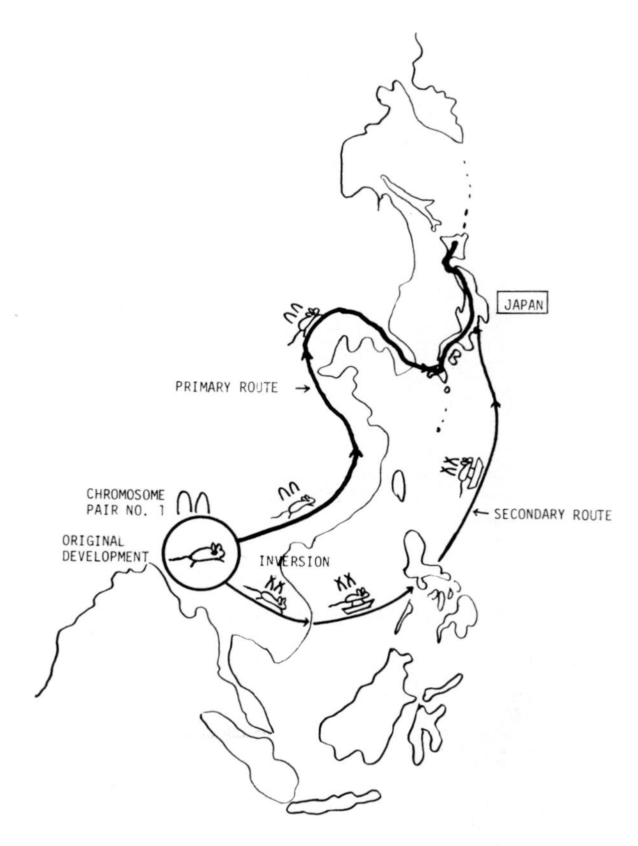

FIGURE 5. Routes of migration of Asian-type black rats from Southeast Asia to East Asia. Primary route is from continent of Southeast Asia through China and Korea. They should have an acrocentric pair no. 1. The secondary route is from islands of Southeast Asia. They might have a subtelocentric pair no. 1.

of the above *Rattus* species are derived from either one of the polymorphic karyotypes of the black rat.

The basic karyotype of *Rattus* species is considered to have 7 small metacentric autosome pairs (14 to 20). In the *Rattus* group, however, a few species appear to have only two or three small metacentrics as found in the long-tailed giant rat, *R. sabanus.* This species has 42 chromosomes, like other *Rattus* species, but 18 autosome pairs (1 to 18) are acrocentric or subtelocentric and only two autosome pairs (19 and 20) are small metacentric, similar to the metacentric pair no. 19 and 20 in the black and other rats (Figure 8). G-banding patterns of all chromosomes in this species are also similar to the other *Rattus* species.[34] These results strongly suggest that the small metacentrics in *Rattus* species could be derived by the inversion of acrocentrics which seem to be the original type of these chromosomes, and the long-tailed giant rat is an ancestral form of the *Rattus* group.[10]

C. Karyotype Evolution in the Norway Rat

The Norway rat (*R. norvegicus*) is a species closely related to the black rat. Although the black rat very often shows chromosome polymorphism or geographical variation, in the Norway rat such karyotype differentiation is very rare in breeding and natural populations. A small inversion of pair no. 3 has been reported.[11] Recently a large

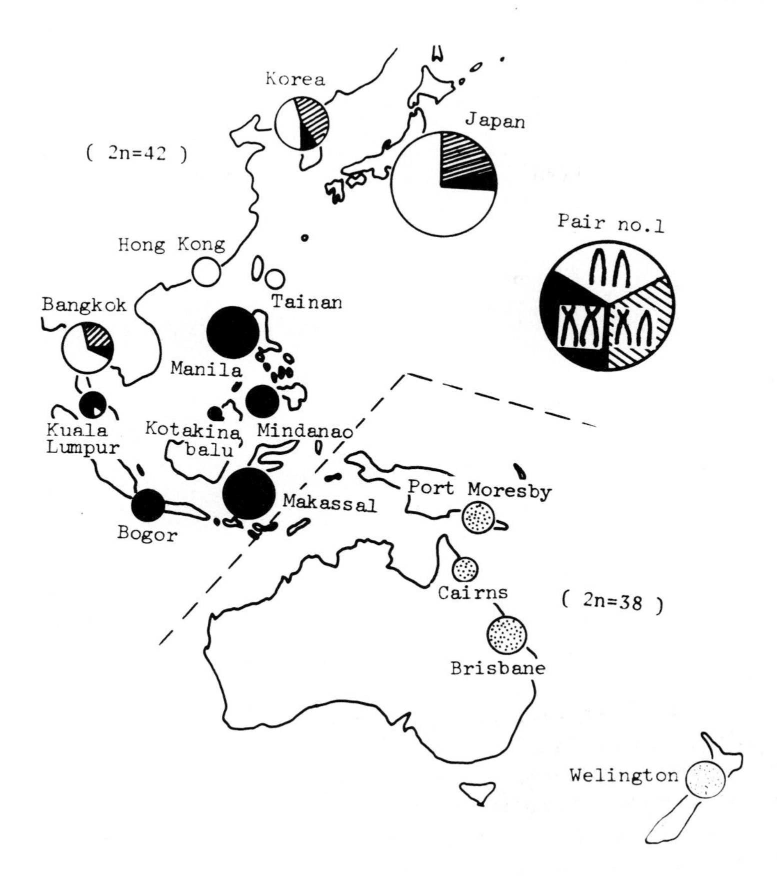

FIGURE 6. Distribution of chromosome polymorphism (pair no. 1) in East and Southeast Asia and Oceania. Refer to Figure 4 in the size of circle and white, shaded, and black areas. They are Asian-type black rats. The Oceanian type ($2n = 38$) is shown by the circles filled with dots.

inversion of pair no. 1 chromosome in the Norway rat was found in laboratory breeding. A history of its occurrence and its propagation into the offspring will be given here.

The chromosomal mutation was first found in a Lewis strain rat. Among six rats (3 ♀:3 ♂) obtained from a breeder in Japan, five had a normal karyotype, but the remaining female was remarkable by having a translocation of chromosomes between nos. 1 and 12 (no. 13 in our arrangement). One member of pair no. 1 of this rat was normal subtelocentric, but the short arm of the other member was constantly longer than the normal partner. On the other hand, pair no. 12 was also heteromorphic with a normal subtelocentric and short partners (Figure 9). By G-banding staining it was found that the broken end of pair no. 12 was translocated to the short arm of one partner of no. 1.[12] By mating the translocation female rat with the normal Lewis male, 17 rats from three litters were obtained. They segregated into nine normal homozygotes and eight translocation heterozygotes as expected. Among the eight heterozygotes, seven rats had the heteromorphic pair no. 1 with the translocation and normal partner, but the remaining one female was exceptional in having a chromosome mosaic with two cell types. One of them had the heteromorphic pair no. 1 with translocation and normal partners, but the other one was conspicuous with the heteromorphic pair no. 1 involving the translocation and inversion partners (Figure 10). The inversion could

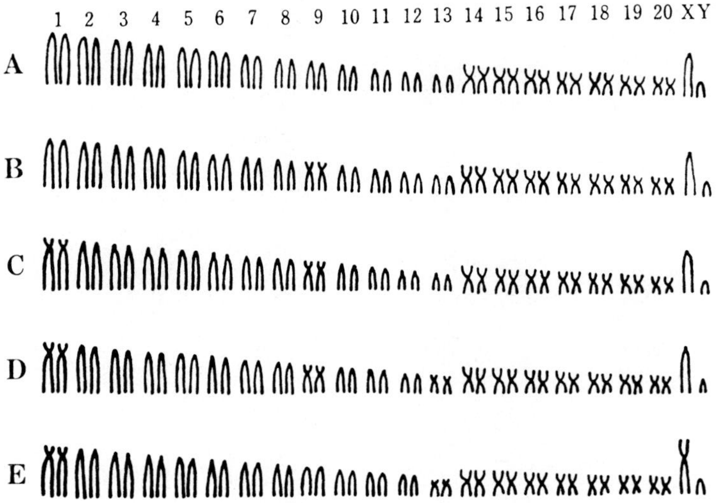

FIGURE 7. Karyotypes of five *Rattus* species with 42 chromosomes similar to the Asian-type black rat. (A) *R. rattus*, (B) *R. annandalei*, (C) *R. exulans*, (D) *R. norvegicus*, (E) *R. muelleri*. They have 42 chromosomes and similar karyotypes, but pairs no. 1, 9, and 13 are either one type, A/A or S/S, which corresponds to either one of the polymorphic karyotypes occurring in the black rat. The X chromosome in *R. muelleri* is differentiated specifically.

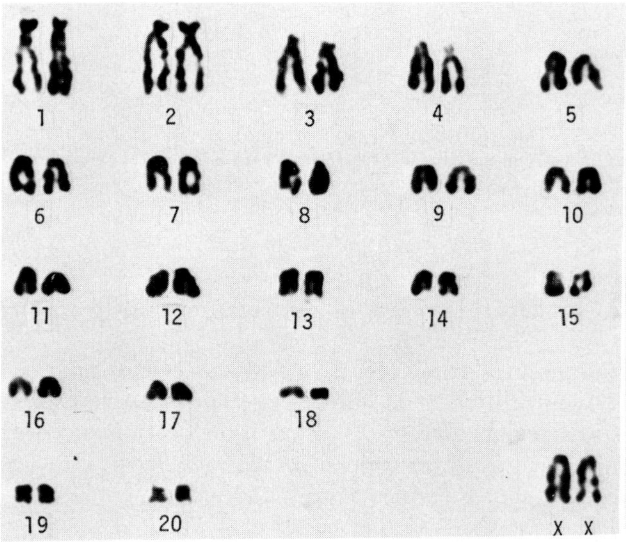

FIGURE 8. Karyotype of *Rattus sabanus* ($2n = 42$). The 18 autosome pairs (nos. 1 to 18) are acrocentrics or subtelocentrics and the other two autosome pairs (nos. 19 and 20) are small metacentrics. X chromosome is similar to the other *Rattus* species.

have occurred in the normal chromosome side in pair no. 1. By pericentric inversion this chromosome changed from subtelocentric to metacentric. This event was also confirmed by G-banding.[13]

The chromosome mosaic was initially found in cells from tail cultures. The fre-

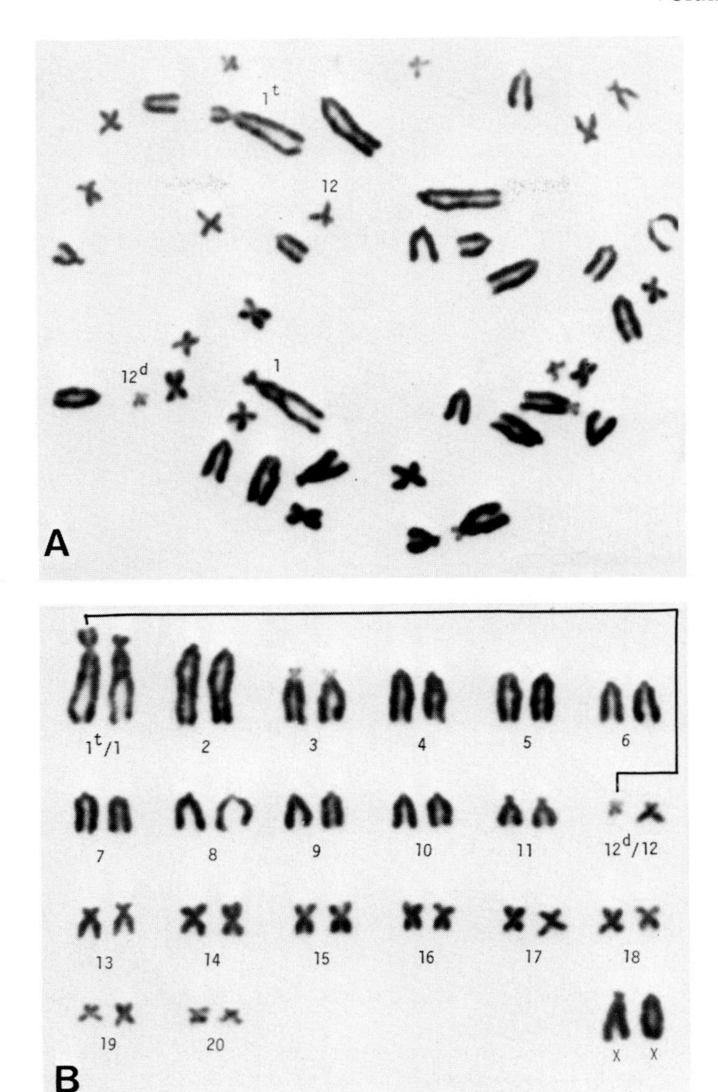

FIGURE 9. Karyotype of a Norway rat with 1/12 translocation. The long arm of one of pair no. 12 is deleted and the broken end is translocated to pair no. 1. The pair no. 1 ($1^t/1$) and no. 12 ($12^d/12$) are heteromorphic. (A) Metaphase, (B) karyotype. (From Yosida, T. H., *Proc. Jpn. Acad.*, 56(Ser. B), 269, 1980. With permission.)

quency of these two types was nearly identical in tail and ear cultures and bone marrow cells. By mating the mosaic female with the normal rat, six offspring (4 ♀:2 ♂) were successfully obtained from three litters. They segregated into three types with normal homomorphic pair no. 1, heteromorphic pair no. 1 with normal and translocation, and the other heteromorphic pair no. 1 with normal and inversion chromosomes, in the expected ratio.

From translocation heterozygotes and inversion heterozygotes it is possible to establish separately translocation homozygote and inversion homozygote rats. Fertility of these rats is quite normal. Litter size of the offspring from both translocation homozygotes was 9.2 ± 3.2 in 16 litters,[14] and from both inversion homozygotes was 7.4 ±

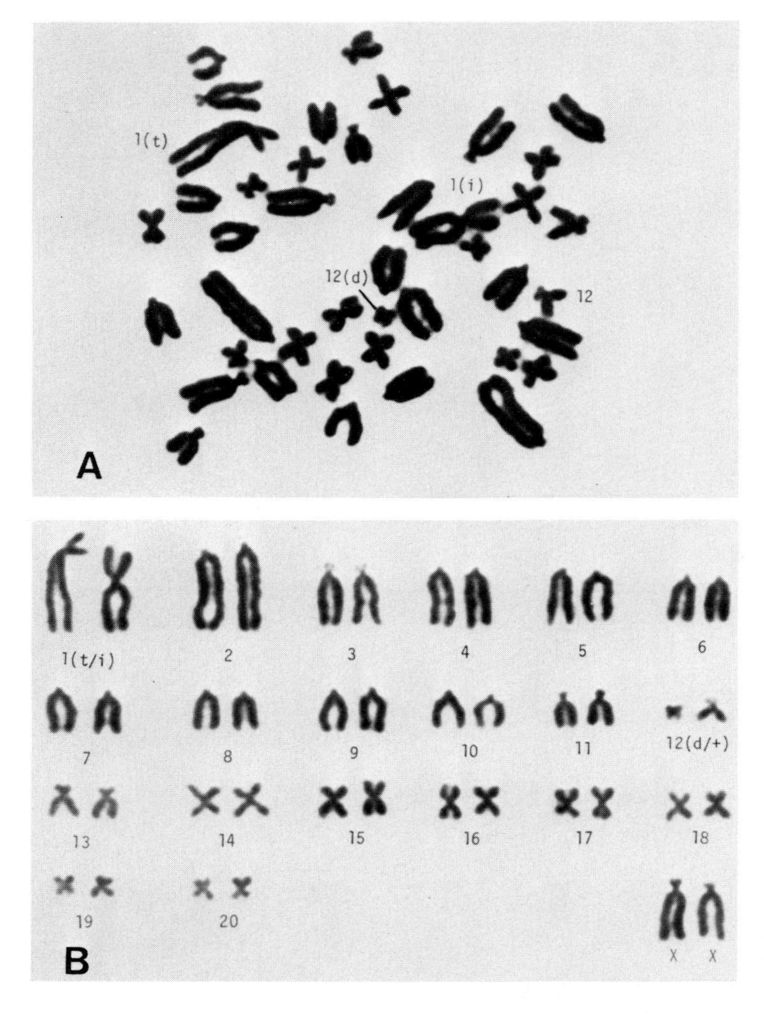

FIGURE 10. Karyotype of a cell with inversion pair no. 1 in the mosaic rat. Pericentric inversion occurs in the normal chromosome in the cell with translocation and normal no. 1 chromosomes. Therefore, the pair no. 12 is heteromorphic. (A) Metaphase, (B) karyotype. (From Yosida, T. H., *Proc. Jpn. Acad.*, 56(Ser. B), 323, 1980. With permission.)

2.8 in 16 litters. In the control with the normal subtelocentric pair no. 1, the litter size was 7.4 ± 2.3.[15] Based on these results, the fertility of the translocation and inversion homozygotes was not much different from the control. (In fact, it was higher than the control.) These results have a significant bearing on karyotype evolution and species differentiation in *Rattus* species.

Another problem in the inversion and translocation rat stocks is that new gene mutations occur successively in the course of breeding. A typical hypotrichosis or hairless mutation occurred in the inversion homozygous stock at the third inbreeding generation (Figure 11), and the other mutant characterized by having yellowish and dusty hair in the young stage occurred in the translocation homozygous stock often in several inbreeding generations. The reason for two chromosomal and two gene mutations occurring sequentially in these rat stocks is not known. Based on these results a speculation may be made that chromosome mutations occurring in an individual may stimulate the induction of the other chromosome or gene mutation due to the unbalanced condition of the genic material.[16]

FIGURE 11. Hairless mutant rat in the inversion mutant stock of the Norway rat.

D. Karyotype Evolution by Pericentric Inversion in Rodents other than *Rattus* Species

The occurrence of polymorphic karyotypes by pericentric inversion has been reported in several mammalian species such as *Mus (Leggada),*[17] *Mastomys,*[18] *Eutamias,*[19] *Peromyscus,*[20,21] *Notomys,*[22] and others. An example of species differentiation by pericentric inversion has been observed in genus *Peromyscus.*[23] Nineteen species belonging to *Peromyscus* had the same chromosome number, $2n = 48$. The number of chromosome arms, however, differed markedly in different species. The difference of arm numbers of chromosomes is due to pericentric inversions. Three species of *Notomys* have the same chromosome number, but the numbers of acrocentric and subtelocentric chromosomes are different. This is also explained by pericentric inversion.[22] Polymorphism of a small autosome in the large naked-soled gerbil (*Tatera indica*) has been reported in specimens collected from India.[25] Chromosome differentiation in this species is due to inversion and other mechanisms.

III. RELATIONSHIP BETWEEN ROBERTSONIAN FUSION AND SPECIES DIFFERENTIATION

According to Robertson,[26] a metacentric chromosome in a certain species, race, or individual of Orthopteran insects may correspond to two acrocentrics in another. This type of karyotype evolution is known widely in plants and animals as the Robertsonian rearrangement. If the two arms of a metacentric are considered to be equivalent to two acrocentrics, the total number of chromosome arms becomes a good guide for evaluation of a phylogenic relation among them. There remains a question, however, whether a bi-armed chromosome arose by fusion of two acrocentrics or vice versa. These two aspects are known as Robertsonian fusion and fission or centromeric fusion or dissociation. Fusion has been reported to be a common phenomenon of karyotype evolution in many animal species.[27-29]

A. Robertsonian Fusion in the Black Rat

As already described, the basic chromosome number of the black rat is 42. They are mostly distributed in Southeast and East Asia. The basic chromosome number has been confirmed by almost all cytogeneticists since the reports by Oguma,[30] Makino,[31] and Matthey.[32] The black rats collected in Oceania, such as Australia, New Zealand, and New Guinea, are predominantly characterized by a diploid chromosome number of 38.[33] Their karyotype is remarkable in having two large metacentric pairs (M_1 and

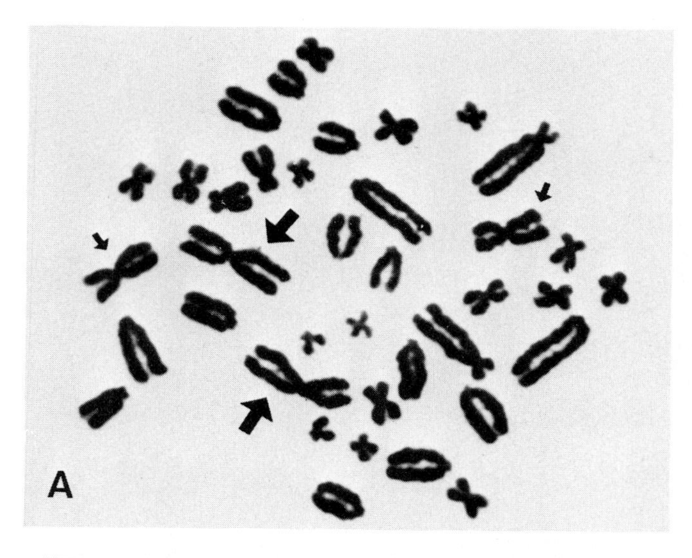

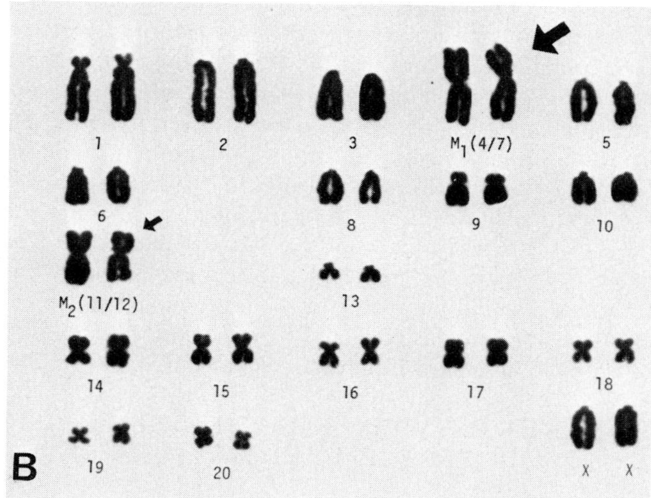

FIGURE 12. Karyotype of the Oceanian-type black rat ($2n = 38$). Two large metacentric pairs, M_1 and M_2, are always included. The M_1 is derived from Robertsonian fusion of acrocentric pairs 4 and 7, and the M_2 from fusion of pairs 11 and 12 included in the Asian-type black rat. (A) Metaphase, (B) karyotype. M_1 and M_2 are denoted by large and small arrows, respectively.

M_2), one (M_1) larger than the other (M_2). The other chromosome pairs are similar to those of the Asian-type rats with 42 chromosomes (Figure 12). A comparison of the idiograms of Asian and Oceanian black rats suggests that the large metacentric M_1 in the Oceanian rat has been derived by Robertsonian fusion of the acrocentric pairs no. 4 and 7 in the Asian-type rat, and the other metacentric M_2 due to the fusion of the acrocentric nos. 11 and 12.[8] This was confirmed by comparative study of the G-banding karyotypes in both black rats.[34] The Oceanian-type black rat with 38 chromosomes was subsequently found throughout the world such as North and South America, Europe, India, Central Asia, Africa, and others except East and Southeast Asia.

The black rat developed originally in the Indo-Malayan region and spread through-

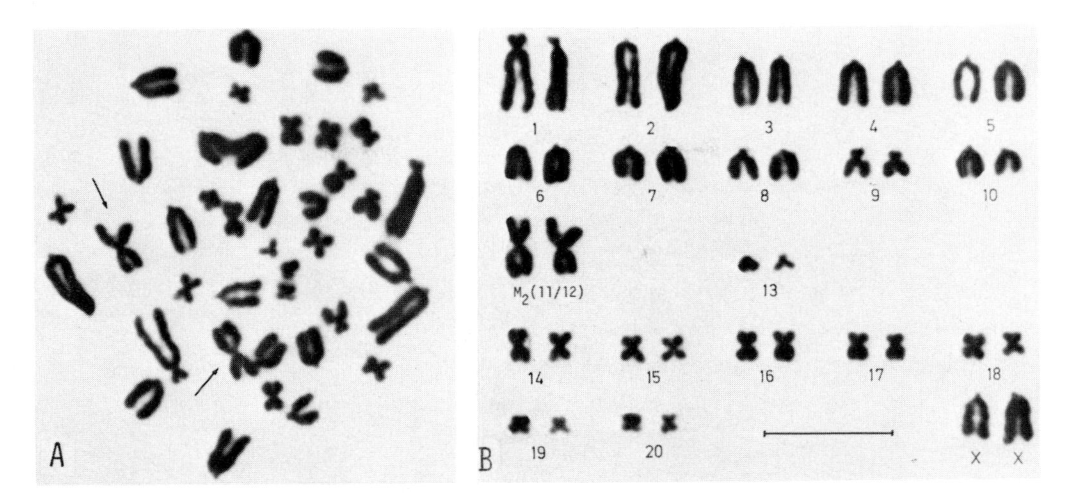

FIGURE 13. Karyotype of Ceylonese-type black rats collected in Kandy, Sri Lanka ($2n = 40$). One large metacentric pair (M_2) (arrows) resulted from Robertsonian fusion of pairs 11 and 12 in the Asian-type black rat. The pairs 4 and 7 are acrocentrics similar to the Asian type. (A) Metaphase, (B) karyotype. (From Yosida, T. H., Kato, H., Tsuchiya, K., Sagai, T., and Moriwaki, K., *Jpn. J. Genet.*, 47, 452, 1972. With permission.)

out the world. The rat with 42 chromosomes may have migrated from the Indo-Malayan area to Europe through Southwest and Central Asia. During this process, fusion might have occurred between chromosome pair no. 4 and 7 and no. 11 and 12 in the Asian-type rat.[8,35] These rats later spread from Europe to the other world with the human immigrants. Chromosome pairs no. 1 and 9 in the Oceanian-type rats were commonly subtelocentrics that probably arose from Asian-type rats with S/S pairs no. 1 and 9 which originated through pericentric inversion. The following sequence was proposed to explain the evolution of the karyotype from the Asian to the Oceanian-type rats:[8]

- Original Asian type ($2n = 42$, A/A pairs no. 1 and 9)
- Derived Asian type ($2n = 42$, S/S pairs no. 1 and 9)
- Intermediate type ($2n = 40$, including one large metacentric pair)
- Oceanian type ($2n = 38$, including two large metacentric pairs)

At the time this scheme was proposed, the existence of a black rat with $2n = 40$, including one large metacentric pair, was only hypothetical, but such rats were suspected to exist somewhere in Southwest or Central Asia as a transient type between the Asian and Oceanian types. The black rats with 40 chromosomes were indeed found in Kandy, Sri Lanka. Their karyotypes indicated that one metacentric pair resulted from the Robertsonian fusion of pairs no. 11 and 12 as predicted.[36,37] From the size, arm ratio, and G-banding analyses, it could be seen that the metacentric pair is the same as the M_2 chromosomes in the Oceanian type (Figure 13). If a second Robertsonian fusion occurred between pairs no. 4 and 7 in the Ceylonese type ($2n = 40$), the Oceanian type would be produced. It was, thus, confirmed that the Ceylonese-type rat is a transient form from the Asian to Oceanian-type black rats. The karyotypes of the three geographical types, Asian ($2n = 42$), Ceylonese ($2n = 40$), and Oceanian ($2n = 38$) thus were assumed to arise in the following sequence: (1) Asian type ($2n = 42$), originally developed in Southeast Asia; (2) it migrated to the Southwest Asia, where the first Robertsonian fusion (M_2) led to the Ceylonese type with 40 chromosomes; (3)

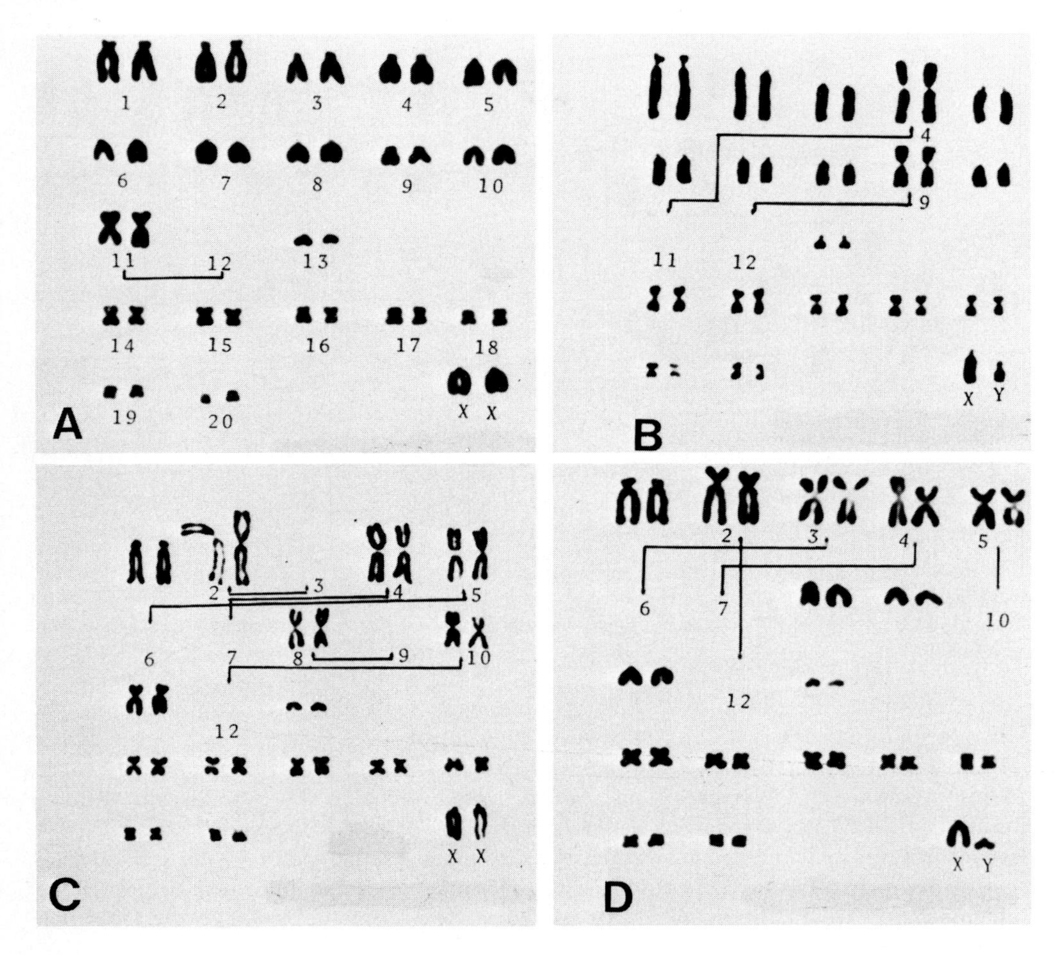

FIGURE 14. Karyotype of 4 *Rattus* species developed by Robertsonian fusion. (A) *R. bowersii* ($2n = 40$) with one large metacentric due to the fusion of pairs 11 and 12 in the basic *Rattus* karyotype; (B) *R. fuscipes* ($2n = 38$), with two large metacentrics caused by fusion of pairs 4 and 11, and 9 and 12; (C) *R. conatus* ($2n = 32$) with fusion of pairs 2 and 3, 4 and 7, 5 and 6, 8 and 9, and 10 and 12; (D) *R. leucopus* ($2n = 34$) with fusion of pairs 3 and 6, 4 and 7, 5 and 10, and 2 and 12. The chromosomes other than the large metacentrics are principally the same as the Asian black rat. (From Yosida, T. H., *Chromosoma (Berlin)*, 40, 285, 1973. With permission.)

the second Robertsonian fusion occurred (M_1 chromosomes) resulting in the Oceanian-type black rat. This type spread widely over southern India; (4) the Ceylonese-type rats were forced out of India by the Oceanian type and pushed onto Sri Lanka; (5) the Oceanian-type rats migrated first to Central Asia and from there to Europe. From Europe they spread widely in the world; (6) the Oceanian-type rats, however, could not migrate to Southeast and East Asia because of the strong barrier posed by the earlier occupiers, the Asian-type black rats. Thus, the three geographical types, Asian, Ceylonese, and Oceanian, which developed by sequential karyotype evolution, later were distributed widely within their territories. Species differentiation should occur from these geographical variants by isolation mechanism and other biological events.

B. Species Differentiation in Genus *Rattus* by Robertsonian Fusion

Robertsonian fusion of chromosomes seems to be another important process in karyotype evolution and species differentiation in *Rattus*. *R. bowersii* in Malaysia had 40 chromosomes including one large metacentric pair. The metacentric pair has been derived from the Robertsonian fusion of pairs no. 11 and 12 in the basic *Rattus* karyotype. Thus, this karyotype is similar to that of the Ceylonese-type black rat (Figure 14A). As the number of acrocentric and subtelocentric chromosomes is different in these two species, the karyotype of *R. bowersii* seems to have developed independently from the Ceylonese-type black rat. The karyotype of *R. fuscipes* with 38 chromosomes, including two large metacentric pairs, is similar to the Oceanian-type black rat, but the constitution of the metacentrics in the former is different from the latter. One of them consists of pairs no. 4 and 11, and the other one of pairs no. 9 and 12 (Figure 14B). *R. leucopus* collected in New Guinea had 34 chromosomes, four pairs being large metacentrics. They arose by Robertsonian fusion of eight acrocentric pairs in the basic karyotype in the *Rattus* group (Figure 14C). Similar fusion was also found in *R. conatus* collected in Australia with 32 chromosomes. Five large metacentric pairs were suggested to have been derived by the Robertsonian fusion of ten acrocentric chromosomes (Figure 14D). Although the variants with similar karyotypes of these *Rattus* species were not found in black rats, they have all the karyological characteristics derived from the basic karyotype in *Rattus* species by Robertsonian fusion.[10,38]

C. Species Differentiation due to Robertsonian Fusion in Rodents other than Genus *Rattus*

Evidence of species differentiation by the Robertsonian fusion was often found in the other rodents. Various chromosome numbers ranging from 36 to 18 have been observed in *Mus (Leggada) minutoides.*[17,18,39,40] The fundamental number (NF) of chromosomes of this species was 36.[17] The karyotype with lower diploid numbers than 36 appeared to result from Robertsonian fusion between acrocentric autosomes of unequal size. Among several species of *Leggada, L. setulosus, L. tenellus, L. indutus,* and *L. bufo* with 36 chromosomes are the most primitive. *L. triton* with $2n = 32$ from the Congo and with $2n = 20$ to 22 from Tanzania are derived from the primitive type ($2n = 38$) by Robertsonian fusion. *L. minutoides* has a complicated karyotype. This species from the Congo had $2n = 38$, but from South Africa had $2n = 18$ to 20 through Robertsonian fusion. A Robertsonian relationship was confirmed by observation of meiotic figures in the hybrids.[39,41] Robertsonian fusion was regarded as the responsible mechanism in karyotype evolution in this group.

The tobacco mouse (*Mus poschiavinus*) with 26 chromosomes had been found in Switzerland. The seven metacentric pairs are proved to have been derived from Robertsonian fusion of acrocentric chromosomes in the original karyotype with $2n = 40$.[42,43] Later on, karyotypes with 38 and 34 chromosomes[44] and 32 and 22 chromosomes[45,46] were reported in this species. They have been attributed to Robertsonian fusion in the basic karyotype of $2n = 40$ found in the feral mouse (*M. musculus*).

The Japanese wood mouse, *Apodemus speciosus*, is common in fields. It is classified into several subspecies by taxonomists, but in respect to chromosome number, only two groups can be identified: one with $2n = 46$ and the other with $2n = 48$.[47] These two types are distributed allopatrically, separated by a boundary line in the central part of Japan. The animals with $2n = 46$ are distributed in the southern part from the boundary line, but those with $2n = 48$ in the northern part. The karyotype difference between these two types is due to Robertsonian fusion.

A European creeping vole, *Microtus*, has two species: *M. Oregoni* with $2n = 17$ and *M. montanus* with $2n = 24$. It has been suggested that the former species has resulted through Robertsonian fusion from the latter.[48] In the gerbil (*Gerbillus phramidium*) living in Central Asia, karyological variants with $2n = 66$, 52, and 40 have been reported.[49] Karyotype evolution has also been explained by Robertsonian fusion.

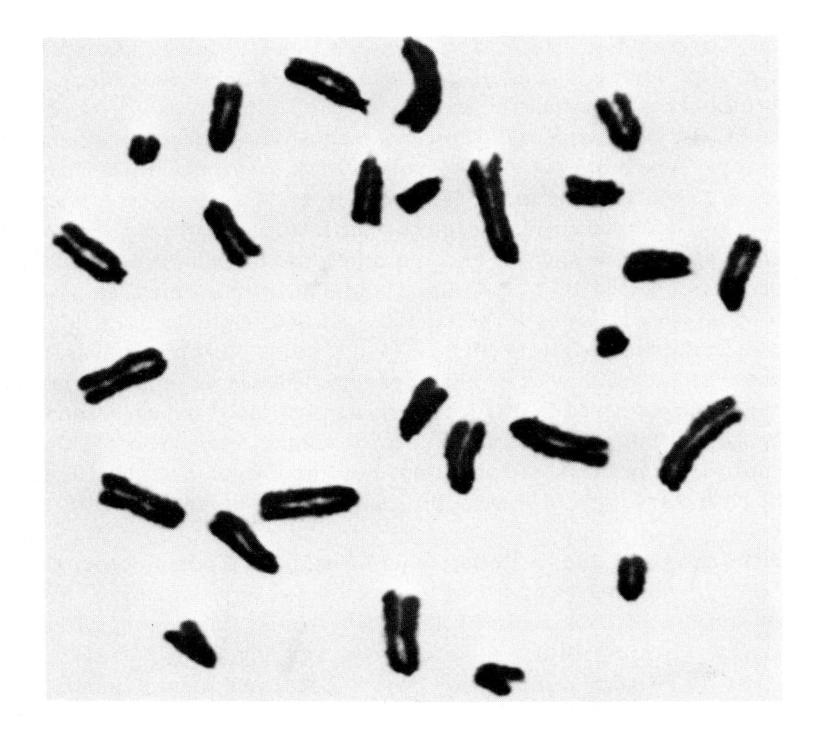

FIGURE 15. Metaphase cell of the Indian spiny mouse (*Mus platythrix*) with 26 acrocentric chromosomes.

IV. RELATIONSHIP BETWEEN TANDEM FUSION OF CHROMOSOMES AND SPECIES DIFFERENTIATION

Tandem fusion of chromosomes is a special type of translocation; it occurs when the main part of two or more chromosomes fuse tandemly after a breakage near telomere in one chromosome and near centromere in the other. Chromosome differentiation by tandem fusion was not observed in *Rattus* species, but it was occasionally found in some mammals.

A. Species Differentiation in the Indian Spiny Mouse by Tandem Chromosome Fusion

A good example of species differentiation by tandem fusion of chromosomes has been demonstrated in the Indian spiny mouse (*Mus platythrix*). The chromosome number of this species is 26, all chromosome pairs being acrocentric (Figure 15).[50-53] On the basis of the measurement of chromosome length as well as of G-band comparison between the spiny mouse and the feral mouse (*M. musculus*), the present author[54,55] has demonstrated that several chromosome pairs in the Indian spiny mouse are the products of tandem fusion of some acrocentrics in the feral mouse (Figure 16). For instance, G-bands on the proximal part of the longest no. 1 chromosome in the spiny mouse are similar to those on pair no. 9 in the feral mouse; and further, the distal half of the spiny chromosome is identical with pair no. 1 of the feral mouse. Pair no. 2 in the spiny mouse appears to be derived from tandem fusion of pairs 5 and 6 in the feral mouse. Similar fusions between two different chromosome pairs 12 and 4, 10 and 13, 16 and 7, and 14 and 17 in the feral mouse have contributed to the formation of pairs 3 to 6 in the spiny mouse.

Some chromosome pairs are longer in the spiny mouse than in the house mouse. The total average length of chromosomes in 30 cells of the feral mouse was 82.6 μm

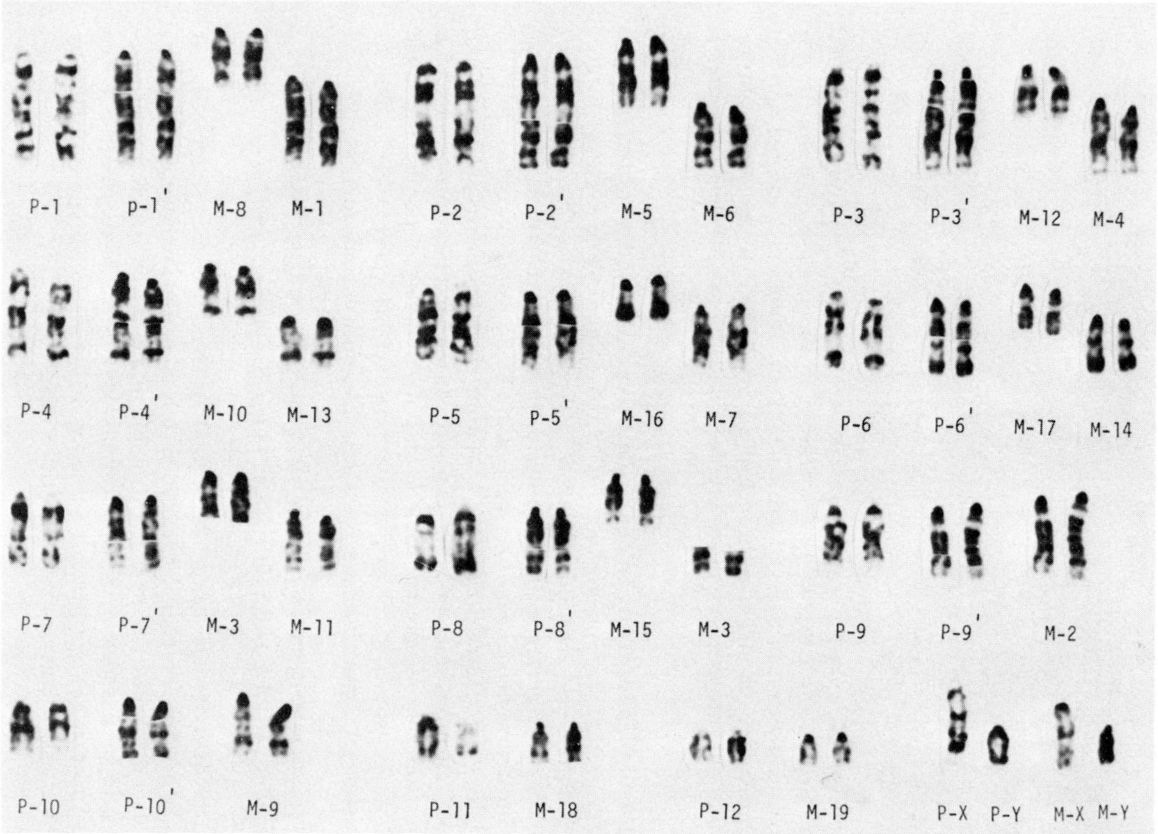

FIGURE 16. Comparison of G-banded chromosomes between Indian spiny *(Mus platythrix)* and house mice (*M. musculus*). In the figure, P-1, P-2, . . . denote the chromosomes of *platythrix,* and M-1, M-2, . . . , those of *musculus.* P-1′, P-2′, . . . show the chromosomes of *platythrix* which are artificially composed from *musculus* chromosomes. (From Yosida, T. H., *Cytologia,* 45, 757, 1980. With permission.)

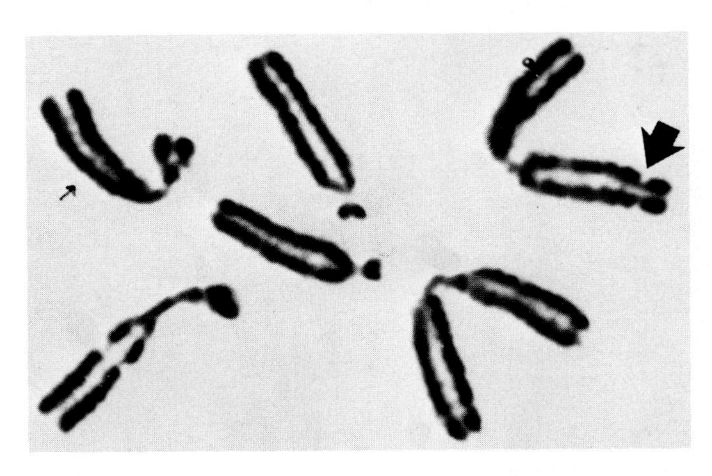

FIGURE 17. Metaphase chromosomes of Indian muntjac *(Muntiacus muntjak)*. $2n = 6$ (♀).

(haploid autosomes + XY); in the spiny mouse it was 72.0 μm. Apparently the total length of all chromosomes was about 12.8% shorter in the spiny mouse than in the feral mouse. This means that in cases of tandem fusion, some chromosome parts of telomeres and centromeres could have been lost during fusion. Karyotype evolution by tandem fusion of chromosomes seems to involve a major loss of the genetic material as described above.

B. Species Evolution by Tandem Fusion of Chromosomes in the Indian Muntjac

Many species of Artiodactyla have 46 to 70 chromosomes. The barking deer or muntjac (*Muntiacus reevesi*), distributed widely in Southeast Asia, has 46 chromosomes. They are all acrocentrics as in the other Artiodactyla mammals. The Indian barking deer or Indian muntjac (*Muntiacus muntjak*) has only six chromosomes (♀), and they are very long acrocentrics (Figure 17). Among them four (two pairs) are long autosomes and XX chromosomes (♀), but the male has seven chromosomes (XXY). These long chromosomes are suggested to have been derived by the sequential tandem fusion of several chromosomes of the original barking deer (*M. reevesi*).[56] The hybrids between *M. reevesi* and *M. muntjak* are reported to have both chromosome sets.[87]

V. RELATIONSHIP BETWEEN ROBERTSONIAN FISSION AND SPECIES DIFFERENTIATION

As compared to Robertsonian fusion, Robertsonian fission, namely, derivation of two acrocentrics from one metacentric by dissociation of its kinetochore, is very rare. Some cytogeneticists doubt the occurrence of the phenomenon, in which the acrocentrics are supposed to be induced by a simple breakage across the centromere of the metacentric chromosome. In the black rat a good evidence of karyotype evolution by simple Robertsonian fission was found in Mauritius Island.

A. Robertsonian Fission in the Mauritius Black Rat

On Mauritius Island in the Indian Ocean black rats with a peculiar karyotype were found (Figure 18). They had a basic Oceanian karyotype because two large metacentric pairs (M_1 and M_2), identical to those of the Oceanian type, were included, but 42 chromosomes were always counted. So far as Asian, Ceylonese, and Oceanian-black rats are concerned, they are always characterized by having seven small metacentric pairs (14 to 20) and one small acrocentric pair no. 13. In contrast, the Mauritius black

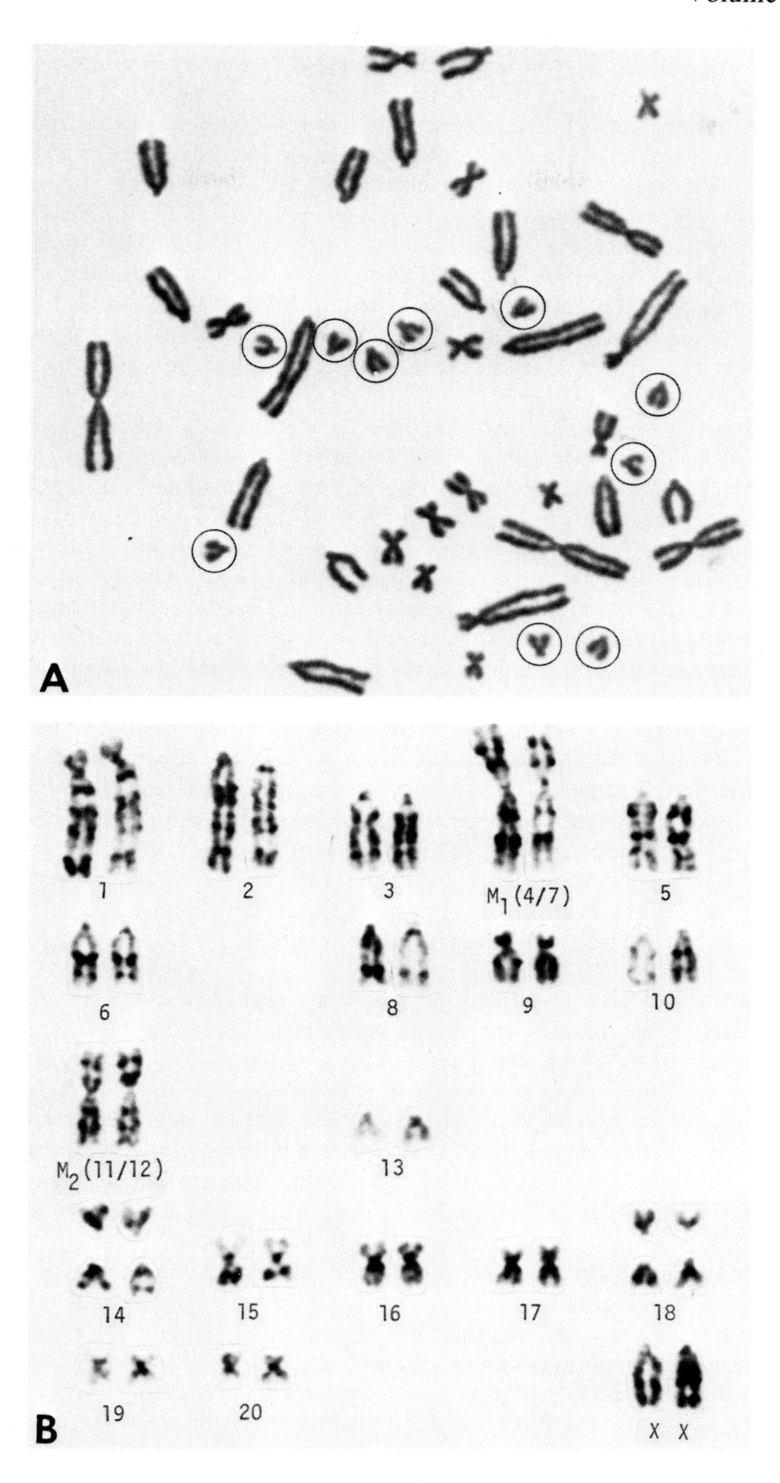

FIGURE 18. Metaphase (A) and G-banded karyotype (B) of the Mauritius black rat with $2n = 42$. In the karyotype two large metacentrics, M_1 and M_2, are included as seen in the Oceanian-type black rat, but eight extra small acrocentrics, which occurred through Robertsonian fission of metacentric pairs no. 14 and 18, are characteristic of this type. Extra small acrocentrics in metaphase are shown by circles. (From Yosida, T. H., Kato, H., Tsuchiya, K., Moriwaki, K., Ochiai, Y., and Monty, J., *Chromosoma (Berlin)*, 75, 51, 1979. With permission.)

rats were distinguished by ten small acrocentrics. Two of them belonged to pair no. 13, but the other eight small acrocentrics were found to be unusual. Accompanying the increase in small acrocentrics, the number of small metacentrics decreased to 10 (5 pairs) from 14 (7 pairs) common in other Asian or Oceanian type rats. Based on the size difference between the small acrocentrics and the metacentrics, it is suggested that eight small acrocentrics correspond to arms of metacentric pairs no. 14 and 18, because these pairs are absent in the Mauritius black rats. G-band analysis showed that four small acrocentrics are homologous to two arms of the small metacentric pair no. 14, and the other four acrocentrics to those of pair no. 18. Thus, it is clearly shown that the extra small eight acrocentrics found in Mauritius black rats are derived from the Robertsonian fission of small metacentrics included in the original Oceanian-type rat.[57,58]

Centric fusions and fissions are included in Robertson's rearrangement system in karyotypes. The former phenomenon is commonly observed in many animal species, including black rats. Robertsonian fission has been considered an important mechanism of karyotype evolution, but adequate evidence has not been obtained so far. Centric fusions are very often reported in cultured animal and human cells and in tumor cells, but centric fission is very rare in them. Clear evidence of centric fission in cultured cells, however, has been reported in the Chinese hamster cell line D-6.[59] In this case, spontaneous fission in the centromeric regions of subtelo- or submetacentric chromosomes, in 2.9% of the cells studied, resulted in the production of functional telocentric chromosomes and the establishment of cell lines with one or two telocentric chromosomes. When telocentric chromosomes arise either spontaneously[60] or following X-irradiation[61] through the fracture of an interstitial centromere, they usually are eliminated or merge together at their broken ends to form a metacentric isochromosome.[62] In black rats, the telocentrics formed by simple centric misdivision are very stable and do not merge in this way. They seem to divide normally at mitosis and meiosis, and their offspring exhibit regular karyological behavior for a number of generations, as occurs on Mauritius Island. This process could be important to karyotype differentiation, although its occurrence is rare in both natural and experimental conditions.

All small acrocentrics developed by fission of metacentrics have clear C-bands in their centromeric regions. The size of the bands is not different from that of the other acro- or metacentrics.[58] This region is considered to consist of highly repetitive satellite DNA.[63,64] According to Hsu and Arrighi,[65] an additional increase in heterochromatin can occur in the course of karyotype evolution. Large, clear C-bands in the small acrocentric chromosomes developed by Robertsonian fission might be due to a repetitive increase in DNA sequences in the heterochromatic region. The acrocentrics developed by the fission might become stable by such mechanism as increase of the C-band heterochromatin.

Female Mauritius black rats ($2n = 42$) were mated to Oceanian-type rats with $2n = 38$, and hybrids were successfully obtained in the laboratory. Chromosome numbers in the hybrids were $2n = 40$, consisting of one haploid set from each parent.[66] In the hybrid metaphase, large metacentric M_1 and M_2 pairs were usually included in the hybrid karyotype, indicating that it is an Oceanian karyotype. In the hybrids, the males were remarkable in having seven small acrocentrics, but the females only six. Among these small acrocentrics in the hybrids, 13 and the Y chromosomes were common to the Oceanian rat. The remaining four small acrocentric autosomes, however, were specific in the hybrids and were, of course, derived from the Mauritius rat. On the other hand, the number of small metacentrics in the hybrids was 12 in both males and females. This number was intermediate between the number of small metacentrics in the Oceanian (14 chromosomes) and the Mauritius rats (10 chromosomes). Analysis of the

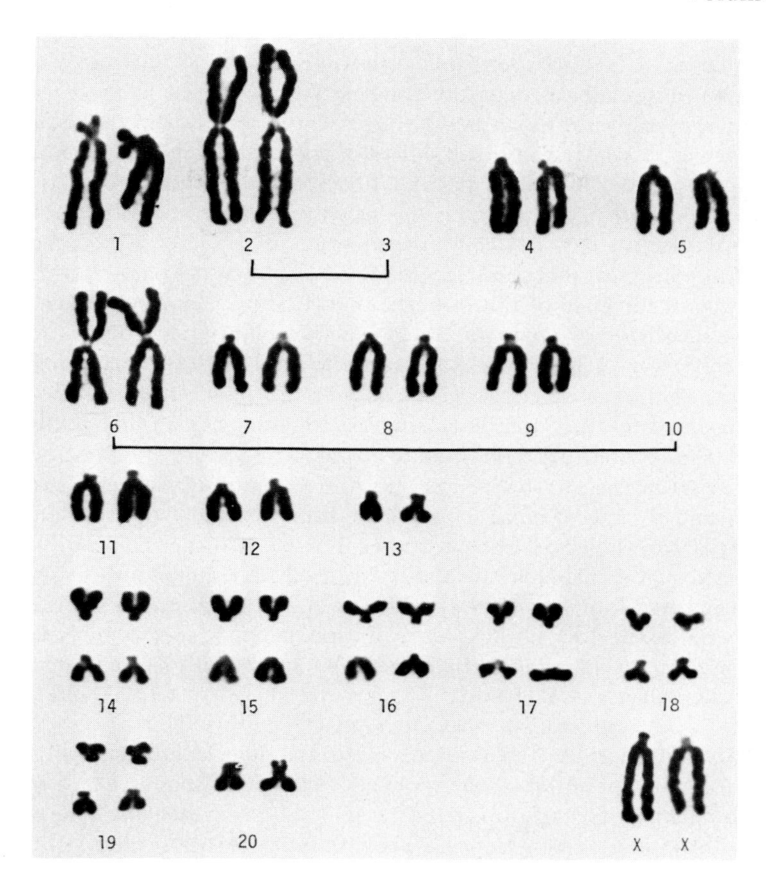

FIGURE 19. G-banded karyotype of *Rattus villosissimus* ($2n = 50$). Two large metacentric pairs developed by Robertsonian fusion of pairs 2 and 3 and 6 and 10 are included, but 24 small acrocentrics (12 pairs) are derived from the Robertsonian fission of metacentric pairs 14 to 19. (From Yosida, T. H., *Proc. Jpn. Acad.*, 55(Ser. B), 497, 1979. With permission.)

G- and C-band karyotypes in the hybrids clearly showed that among four small acrocentrics, two were homologous to the long and short arms of the small metacentric pair no. 14, and the other two to those of the other small metacentric pair no. 18. One partner of each of the small metacentrics of pairs 14 and 18 in the hybrid originated from the Oceanian-type rat, but the acrocentric partners were derived from the Mauritius type. The other chromosome pairs in the F_1 hybrid were all homologous. The F_2 generation offspring are expected to be divided into nine types with respect to the small acrocentrics and metacentrics in pairs 14 and 18. The average litter size of the F_2 offspring was 4.3 ± 1.7, which is almost normal. The segregation ratio of chromosome pairs no. 14 and 18 in the F_2 hybrids was nearly identical with the expected one.[67] From these studies it is concluded that the Mauritius-type black rats are very closely related to the Oceanian-type rats.

B. Species Differentiation in *Rattus* Species by Robertsonian Fission

Although chromosome constitution derived from Robertsonian fission has been found in the black rat, the role of fission to species differentiation has not been fully understood. Australian wild rats, *R. villosissimus (R. sordidus villosissimus),* seem to be a favorite example of karyotype evolution involving Robertsonian fission. The chromosome number of *R. villosissimus* was determined as $2n = 50$ with two large metacentrics.[24] Based on the measurement of the length of two arms of these chromosomes and alignment according to the author's *Rattus* karyotype system,[10] the two metacentrics were regarded to be derived through Robertsonian fusion of acrocentric pairs no. 2 and 3, and 6 and 10. The peculiarity of this species was the presence of 26 small acrocentrics and 2 small submetacentrics in the complement.[24,29] Of the 26 acrocentrics 2 were those of pair no. 13, but the remaining 12 pairs (24 chromosomes) could be arranged in the position of metacentric pairs no. 14 to 19 in the basic *Rattus* karyotype. This suggests the occurrence of the Robertsonian fission. The other two small submetacentrics were located at the pair no. 20. As such, the karyotype of this species corresponds to the basic karyotype of the *Rattus* species (Figure 19). This G-banding pattern of *R. villosissimus* is likewise conservative as seen in the other *Rattus* species. Twelve extra small acrocentric pairs just corresponded in their G-banding patterns to each member of six small metacentric pairs (14 to 19) of the basic *Rattus* karyotype.

As evidenced from the karyotype analysis and G-banding patterns, the Mauritius-type black rat and *R. villosissimus* could have been derived from the same ancestor, but the Robertsonian fission of chromosomes in these two species occurred independently in these two. The chromosome fission in the Mauritius black rat seems to have occurred recently in Mauritius Island, while in *R. villosissimus* in the extreme past. The Mauritius black rat does not appear as an independent species but as a geographical variant. In contrast, *R. villosissimus* seems to have been differentiated as an independent species with the lapse of time.

C. Species Differentiation by Robertsonian Fission in Other Genera

According to Wahrman et al.,[68] the mole rat (*Spalax*), distributed in South Europe, has chromosome number variations of 52, 54, and 60, which may be explained by Robertsonian fission. In the squirrel, a possibility of Robertsonian fission has been reported by Nadler and Harris.[69] According to them, *Cynomys*, the squirrel, is considered to have been evolved from genus *Spermophilus*. The latter species has 46 chromosomes, but the former has 50 chromosomes. They explained that *Spermophilus* developed from *Cynomys* by Robertsonian fission of two metacentric pairs. Chromosome polymorphism of the root vole, *Microtus oeconomus*, has been attributed by Fredga and Bergström[70] to Robertsonian fission.

Importance of fission in karyotype evolution has been argued by Todd,[71] based on studies on Canid phylogeny. According to him, fusions occur sporadically and such events are not abundant. His theory accounts for the wide variation in diploid number among the Mammalia, and is related to the presumed episodes of karyotype fission during known periods of explosive speciation and adaptive radiation.

It is difficult at present to assess the relative importance of Robertsonian fusion or fission in karyotype evolution, at least in the animals. Occurrence of Robertsonian fusion seems to be much higher than fission.[27-29] Todd[71] has suggested that fusion is sporadic in its occurrence, but the present author[29] has noted that occurrence of the fission is rather sporadic for karyotype evolution, at least in mammals.

VI. PARALLEL KARYOTYPE EVOLUTION AND SPECIES DIFFERENTIATION

It is well understood that the karyotype can evolve through mechanisms like pericentric inversion, Robertsonian fusion and fission, tandem fusion, and others. If karyotype differentiation occurred as polymorphism in a certain locality, it would be regarded as chromosome polymorphism. When karyotype evolution occurs at different localities, it is called a geographical variant or type. In black rats, pericentric inversion was found as a polymorphic condition, while Robertsonian fusion and fission were noted as the geographic variants. By comparative studies of karyotypes of these genetical variants, the author realized the clear relationship between the variations of karyotypes as polymorphic or geographic and species evolution. There is, however, the question whether karyotype differentiation can proceed parallel to species differentiation. The answer is in the affirmative in many plants and animals, but not always. In the *Rattus* group, as already described, the Norway rat, the little house rat, the Singapore rat, the swamp rat, and others have 42 chromosomes, and their karyotypes are very similar to the Asian-type black rat with $2n = 42$. They are suggested to have developed from one of the polymorphic karyotypes in the Asian-type black rat. They, however, are independent species isolated from each other. On the other hand, karyotypes of the Ceylonese, Oceanian, and Mauritius-type black rats have deviated greatly from the original Asian-type rats in their chromosome numbers and constitution. Any mating between these geographical variants can easily produce hybrids, although the fertility of the F_1 hybrids is different depending on their combinations. In the independent species, even those with a similar karyotype, the hybrids are never produced in natural matings and also by the artificial insemination procedure as observed in *Rattus* species.[72-74] The above example is good evidence to show that karyotype differentiation does not always proceed parallel to species differentiation. Cytogeneticists have, thus, encountered a great contradiction in explaining the relationship between the karyotype evolution and species differentiation. This problem, however, seems to be explained by the observation of the inner-chromosomal structures such as distribution of heterochromatin (C-bands) and nucleolar organizer regions (NORs) of chromosomes.[29]

VII. ROLE OF DIFFERENTIATION IN PHENOTYPICAL AND INNER-STRUCTURAL KARYOTYPES IN SPECIES EVOLUTION

Karyotype analysis was carried out for a long time by so-called conventional staining by Giemsa, carmine, orcein, and other stains. Chromosome figures stained by such conventional techniques showed a flat structure, and any differential pattern of chromosomes could not be revealed. These figures may be termed phenotypical karyotypes.[29] From such karyotypes, therefore, one can find the mode of differentiation in the superficial morphology but cannot demonstrate the inner-structural differentiation of chromosomes. Recent advances in several differential staining techniques have made it possible to demonstrate the finer details of chromosome structures, such as distribution of the heterochromatin, nucleolar organizer regions, kinetochore, and so on.

A. Differentiation of Constitutive Heterochromatin and Species Evolution

By the conventional staining of chromosomes it is not possible to differentiate the heterochromatin from the euchromatin. But, by C-banding technique the constitutive heterochromatin is manifested as C-bands. The size and shape of the constitutive heterochromatin (C-bands) are different in different species even though they have a similar karyotype. For instance, the black rats, irrespective of their karyotypes, always have large C-bands near centromeric regions of all chromosomes, but the Norway rats, although they have a very similar karyotype to the Asian-type black rat, generally have small C-bands; the bands of some pairs (4, 6 and 10, 12 and 14 to 16) are considerably clear, while those of others (1 to 3, 5, 11, 19, and 20) are small and indistinct (Figure 20). The little house rat (*R. exulans*) and *R. fuscipes* also have small C-bands.[75] These *Rattus* species (*norvegicus, exulans,* and *fuscipes*) have been suggested to be derived from either one of the polymorphic karyotypes in the black rats. If the karyotype

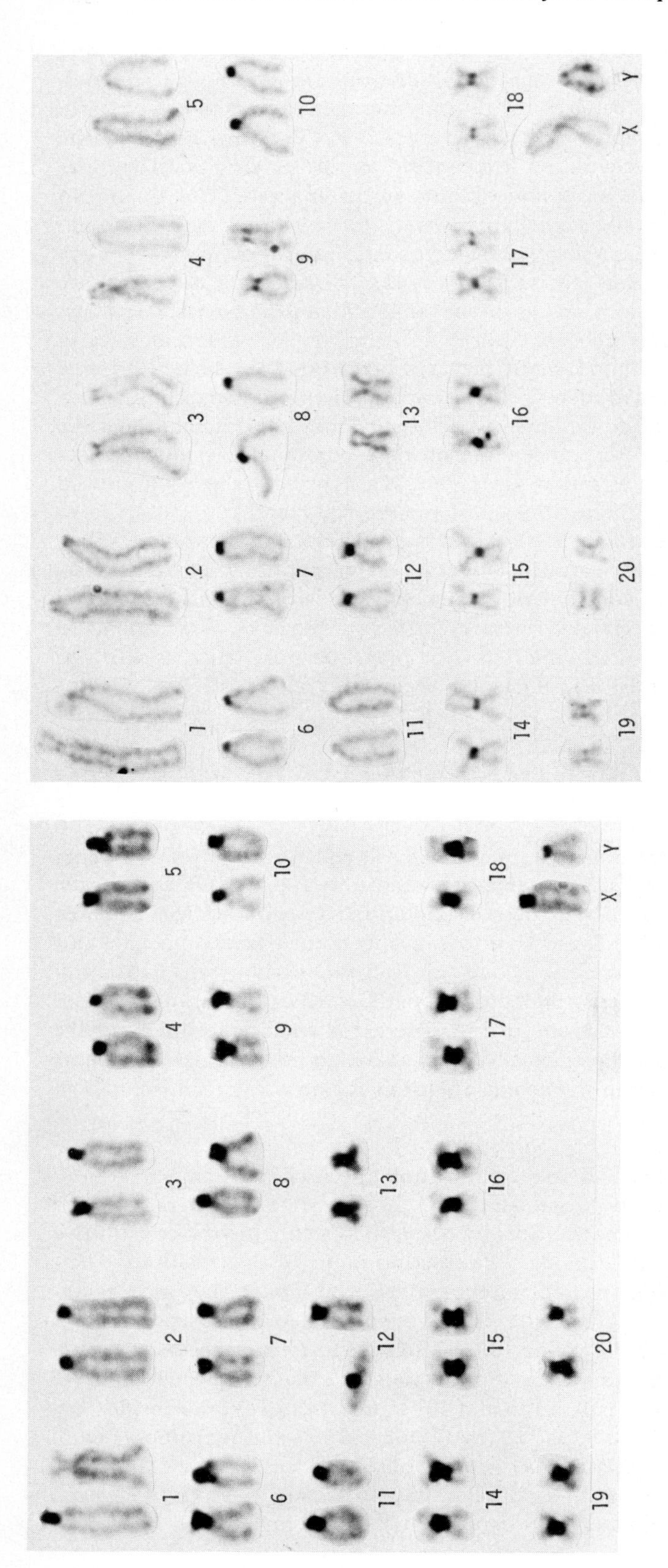

FIGURE 20. Comparison of heterochromatic C-bands in the black (*R. rattus*) and Norway rats (*R. norvegicus*). (A) Black rat, (B) Norway rat. The C-bands in the black rat are markedly larger than in *R. norvegicus*.

evolved in the above direction in genus *Rattus*, it is quite likely that heterochromatic substances can be reduced in the course of species differentiation.

It has been demonstrated by several investigators[63,76,77] that the centromeric regions of chromosomes stained by the C-banding technique are centromeric heterochromatin, consisting of highly repetitive satellite DNA.[63,64,78,79] The significance of the heterochromatin in relation to karyological evolution has been discussed by Hsu and Arrighi,[65] who reported that an additional increase in heterochromatin may occur in one category of karyotype evolution. One the other hand, a diminition or decrease in C-band heterochromatin has been recognized in some subspecies of the black rat as another possible mode of karyotype evolution. *R. sabanus* with a few small metacentrics was suggested to be an older form than the black rat with seven small metacentrics. If the evolutionary tendency of C-bands is always in the direction of reduction, the relation between the black rat and *R. sabanus* would be contradictory insofar as the explanation of C-band evolution is concerned. Thus, heterochromatic C-bands can be seen as having a periodical change, increase and decrease, during the long evolutionary history of the animal.

B. Differentiation of Nucleolus Organizer Regions (Ag-NORs) and Species Evolution

The nucleolus organizer regions (NORs) in black rats by the Ag-AS staining technique are recognized on chromosome pairs 3, 8, and 13, in spite of any differentiation of their karyotypes. With the same Ag-AS staining technique, the NORs in other *Rattus* species, such as *R. norvegicus, R. exulans, R. annandalei, R. losea*, and *R. sabanus*, are found on several chromosomes, which are species specific.[80,81] The Norway rat has a very similar karyotype to the Asian-type black rat, but the location of Ag-NORs differ; in the black rat they are on pairs 3, 8, and 13, but in the Norway rat they are on pairs 3, 12, and 13 (corresponding to pairs 11 and 12 in the nomenclature of "Standard Karyotype of the Rat"; Figure 21).

The karyotype of the little house rat (*Rattus exulans*) from Thailand is similar to the black and Norway rats. The NORs, however, are found on pairs 3, 5, and 13, which is slightly different from the location of those of the black and Norway rats. Clear NORs are observed on pairs 3 and 13 in *R. losea*, although the karyotype of this species is very similar to the Asian-type black rat. In *R. annandalei*, the regions are found on chromosome pairs 8 and 13. *R. sabanus*, which has a karyotype different from the above *Rattus* species, has the NORs on pairs 1, 3, and 5. The regions in these chromosomes are at the ends of the short arms. The other NOR is observed in the telomeric region of one member of acrocentric pair no. 9. *R. neilli* ($2n = 44$) has a karyotype similar to *R. sabanus*, although its chromosome number is different. The NORs in this species are similar to those in *R. sabanus* on pairs 1 and 3 but not in pair no. 5. NORs on the distal ends in pair no. 9 are also observed in this species.[81]

Studies of nucleolar organizers in genus *Rattus* show that most of the species examined here have nucleolus organizers in pairs no. 3 and 13 chromosomes (Table 1). The other NOR in the black rats is found in the centromeric regions of pair 8. Although the region on pair no. 8 is present in *R. annandalei*, it was not found in the other *Rattus* species such as *R. norvegicus, R. losea*, and *R. exulans*. In the Norway rat (*R. norvegicus*) it occurs on pair 12, but in *R. exulans* on pair 5. The karyotypes of the Norway rat and some other *Rattus* species are assumed to be differentiated from that of the black rat.[10] If it is true, the nucleolus organizer located in pair no. 8 of the latter species might have been translocated to pair no. 12 in the Norway rat, and to pair no. 5 in the course of species differentiation. The NOR regions in *R. sabanus* and *R. neilli*, which have different karyotypes, are markedly different from those of the other species.

There is an interesting study on nucleolar organizers in humans and apes.[82] In human

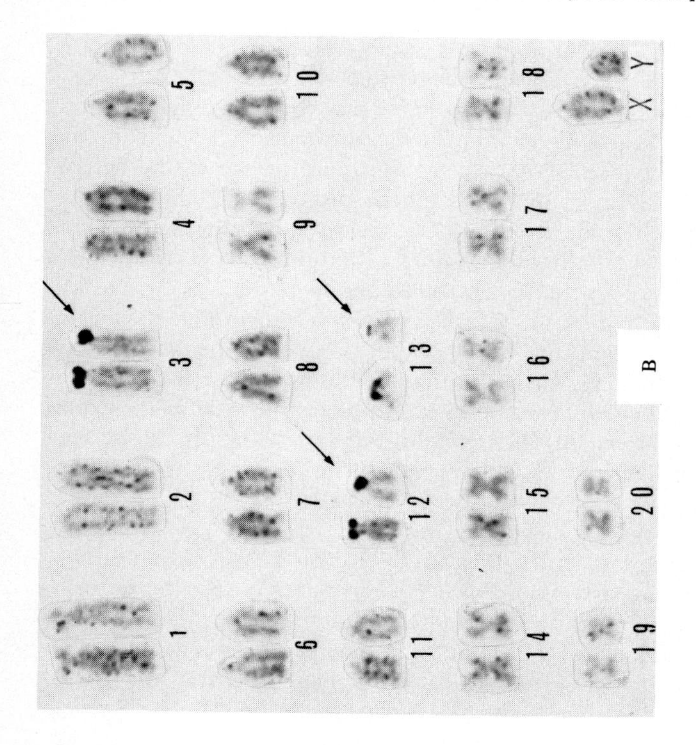

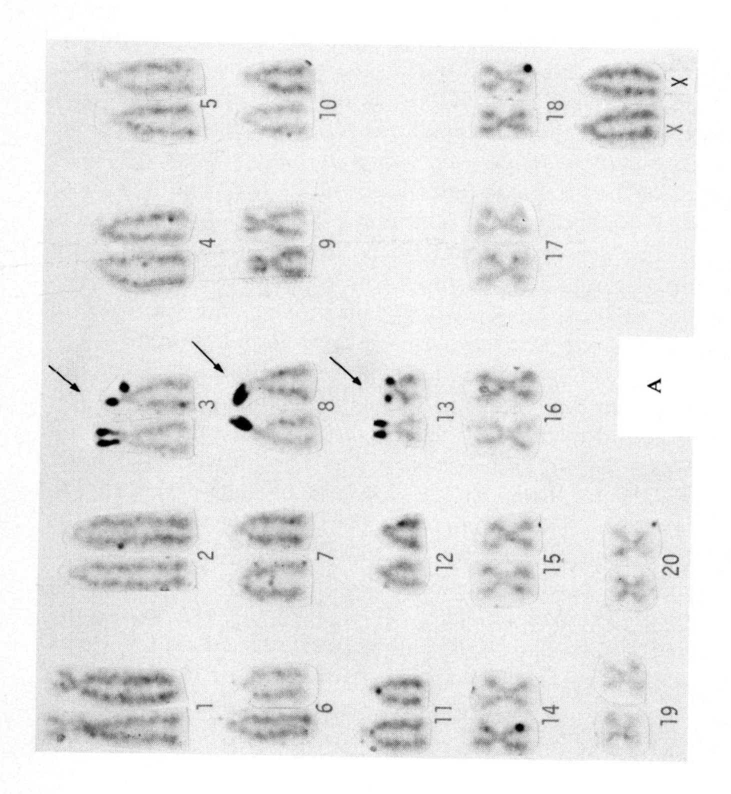

FIGURE 21. Comparison of nucleolar organizer regions (NORs) in the black (Asian type) and Norway rats. (A) Black rat. The NORs were found in pairs 3, 8 and 13. (B) Norway rat. The NORs are recognized in pairs 3, 12, and 13 (corresponding to pairs 11 and 12 in the Standard Nomenclature System).

Table 1
NUCLEOLUS ORGANIZER REGIONS OF CHROMOSOMES IN VARIOUS *RATTUS* SPECIES BY Ag-AS STAININGS

Species	Chromosome pairs 1	3	5	8	9	12	13	Total no. of NORs
R. rattus		+		+			+	3
R. norvegicus		+				+	+	3
R. annandalei				+			+	2
R. losea		+					+	2
R. exluans		+	+				+	3
R. sabanus	+	+	+		+'			4
R. neilli	+	+			+'			3

Note: + = NORs in centromeric regions; +' = NORs in telomeric regions.

From Yosida, T. H., *Proc. Jpn. Acad. (Ser. B)*, 55, 481, 1979. With permission.

chromosomes, the NORs are found in five acrocentric pairs (numbers 13, 14, 15, 21, and 22). The NORs in the chimpanzee also are in five pairs of acrocentric chromosomes, corresponding to human pairs 13, 14, 18, 21, and 22. In the orangutan, the NORs are observed in eight acrocentric pairs, six of which correspond to the human 13, 14, 15, 18, 21, and 22 chromosomes; the other two pairs are specific to the orangutan. This is another evidence of the relationship between the differentiation of nucleolar organizer regions and species differentiation.

By *in situ* hybridization of ^{3}H-labeled ribosomal RNA to the chromosomes of bone marrow cells in the Norway rat, Kano et al.[83] demonstrated that clusters of ribosomal cistrons (rDNA) are located in the secondary constrictions of chromosomes 3 and 12 and near the centromere of chromosome no. 11, both of which are associated with the late DNA-replicating regions (11 and 12 cited by them correspond to chromosomes 12 and 13 of the present author). The sites of nucleolar organizer regions (NORs), revealed by Ag-staining, coincide with the clusters of rDNA cistrons demonstrated by Kano et al.[83] The Ag-NORs are known to be sites of rRNA genes on chromosomes.[84,85] Thus, the differentiation of the NORs, which are an important part of the inner organization of chromosomes, seems to affect gene expression, and may have a dynamic effect on species differentiation as discussed in the studies of *Rattus* species.[29,81]

The relationship between the differentiation of phenotypical karyotypes and species differentiation has been well demonstrated by comparative studies on several *Rattus* species and others as mentioned earlier. The relationship, however, is sometimes unreliable. In this case, if one can compare the inner-structural components, such as heterochromatin and nucleolar organizer regions, in otherwise similar karyotypes, one can understand how the species evolved with the differentiation of the karyotypes. The relationship between karyotype differentiation and species evolution, thus, can well be explained by comparing both karyotypes, phenotypical and inner-structural. At present, it is difficult to describe the relative importance of phenotypical and inner-structural karyotypes in evolution. These two, however, are not inseparable. If there is some problem to solve, one must challenge it from different aspects. By a critical analysis of finer details of chromosomes, the pathways of evolution become comprehensible with concrete evidences based on karyotype differentiation.

REFERENCES

1. **Yosida, T. H., Nakamura, A., and Fukaya, T.,** Chromosomal polymorphism in *Rattus rattus* (L.) collected in Kusudomari and Misima, *Chromosoma (Berlin),* 16, 70, 1965.
2. **Yosida, T. H., Tsuchiya, K., and Moriwaki, K.,** Frequency of chromosome polymorphism of *Rattus rattus* collected in Japan, *Chromosoma (Berlin),* 33, 30, 1971.
3. **Yosida, T. H.,** Frequencies of chromosome polymorphism in pairs no. 1, 9 and 13 in three geographical variants of black rats, *Rattus rattus, Chromosoma (Berlin),* 60, 391, 1977.
4. **Yosida, T. H.,** Segregation on the no. 1 chromosome pair in the black rat (*Rattus rattus*) maintained in a population room, *Proc. Jpn. Acad.,* 52, 130, 1976.
5. **Tate, G. H. H.,** Some Muridae of the Indo-Australian region, *Bull. Am. Mus. Nat. Hist.,* 72, 501, 1963.
6. **Silver, J.,** The house rat, *Wildl. Circ.,* 6, 1, 1941.
7. **Southern, H. N.,** *The Handbook of British Mammals,* Blackwell Scientific, Oxford, 1964.
8. **Yosida, T. H., Tsuchiya, K., and Moriwaki, K.,** Karyotypic differences of black rats, *Rattus rattus,* collected in various localities of East and Southeast Asia and Oceania, *Chromosoma (Berlin),* 33, 252, 1971.
9. **Yosida, T. H.,** Frequencies of chromosome polymorphism (pairs no. 1, 9 and 13) in the black rat of Japan, *Proc. Jpn. Acad.,* 52, 405, 1976.
10. **Yosida, T. H.,** Evolution of karyotypes and differentiation of 13 *Rattus* species, *Chromosoma (Berlin),* 40, 285, 1973.
11. **Yosida, T. H. and Amano, K.,** Autosomal polymorphism in laboratory bred and wild Norway rat, *Rattus norvegicus,* found in Misima, *Chromosoma (Berlin),* 16, 658, 1965.
12. **Yosida, T. H.,** Studies on the karyotype differentiation of the Norway rat. I. Translocation between the pair nos. 1 and 12 chromosomes in the Lewis strain rat, *Proc. Jpn. Acad.,* 56(Ser. B), 268, 1980.
13. **Yosida, T. H.,** Studies on the karyotype differentiation of the Norway rat. II. A mosaic rat carrying the translocation and inversion of pair no. 1 chromosomes with a note on their transmission to offspring, *Proc. Jpn. Acad.,* 56(Ser. B), 322, 1980.
14. **Yosida, T. H.,** Studies on the karyotype differentiation of the Norway rat. III. Segregation of offspring from the 1/12 translocation heterozygotes and fertility of the translocation homozygotes, *Proc. Jpn. Acad.,* 56(Ser. B), 437, 1980.
15. **Yosida, T. H.,** Studies on the karyotype differentiation of the Norway rat. IV. Segregation and fertility of the Norway rats with inversion pair no. 1, *Jpn. J. Genet.,* 55, 397, 1980.
16. **Yosida, T. H.,** Studies on the karyotype differentiation of the Norway rat. V. Hypotrichotic mutant rats appeared in the inversion stock (LEM), *Proc. Jpn. Acad.,* 57(Ser. B), 29, 1981.
17. **Matthey, R.,** Cytologie comparée et polymorphisme chromosomique chez des *Mus africains* appartenant aux groupes *bufo-triton* et *minutoides* (Mammalia Rodentia), *Cytogenetics,* 2, 290, 1963.
18. **Matthey, R.,** Le polymorphisme chromosomique des *Mus africans* du sous genre *Leggada.* Révision général portant sur l'analyse de 213 individus, *R. Suisse Zool.,* 73, 585, 1966.
19. **Nadler, C. F.,** Contributions of chromosomal analysis to the systematics of North American chipmunks, *Am. Midland Nat.,* 72, 298, 1964.
20. **Ohno, S., Weiler, C., Poole, J., Christian, L., and Stenius, C.,** Autosomal polymorphism due to pericentric inversion in the deer mouse (*Peromyscus maniculatus*) and some evidence of somatic segregation, *Chromosoma (Berlin),* 18, 177, 1966.
21. **Hsu, T. C. and Arrighi, F. E.,** Chromosomal evolution in the genus *Peromyscus* (Cricetidae, Rodentia), *Cytogenetics,* 5, 355, 1966.
22. **Baverstock, P. R., Watts, C. H. S., and Hogarth, J. T.,** Polymorphism of the X-chromosome, Y-chromosome and autosomes in the Australian hopping mice, *Notomys alexis, N. cervinus* and *N. fuscus* (Rodentia, Muridae), *Chromosoma (Berlin),* 61, 243, 1977.
23. **Hsu, T. C. and Arrighi, F. E.,** Chromosomes of *Peromyscus* (Rodentia, Cricetidae). I. Evolutionary trends in 20 species, *Cytogenetics,* 34, 417, 1968.
24. **Baverstock, P. R., Watt, C. H. S., Hogarth, J. T., Robinson, A. C., and Robinson, J. F.,** Chromosome evolution in Australian rodents. II. The *Rattus* group, *Chromosoma (Berlin),* 61, 227, 1977.
25. **Yosida, T. H.,** Chromosome polymorphism of the large naked-soled gerbil, *Tatera indica* (Rodentia, Muridae), *Jpn. J. Genet.,* 56, 241, 1981.
26. **Robertson, W. R. B.,** Chromosome studies. I. Taxonomic relationships shown in the chromosomes of Tettigidae and Acrididae. V-shaped chromosomes and their significance in *Acrididae, Locustidae* and *Gryllidae*: chromosomes and variation, *J. Morphol.,* 27, 179, 1916.
27. **Matthey, R.,** The chromosome formulae of eutherian mammals, in *Cytotaxonomy and Vertebrate Evolution,* Chiarelli, A. B. and Capanna, E., Eds., Academic Press, London, 1973, 531.
28. **White, M. J. D.,** *Modes of Speciation,* W. H. Freeman, San Francisco, 1978.

29. **Yosida, T. H.,** *Cytogenetics of the Black Rat — Karyotype Evolution and Species Differentiation,* University Park Press, Baltimore, 1980.
30. **Oguma, K.,** Chromosomes of four wild species of Muridae, *J. Fac. Sci. Hokkaido Univ. Zool.,* 6(4), 35, 1935.
31. **Makino, S.,** Studies on the murine chromosomes. III. A comparative study of chromosomes in five species of *Rattus, J. Fac. Sci. Hokkaido Univ. Zool.,* 6(9), 19, 1943.
32. **Matthey, R.,** La formule chromosomique de quelques Murinae (Muridae-Rodentia-Mammalia), *Arch. Julius Klaus Stift.,* 31, 294, 1956.
33. **Yosida, T. H., Tsuchiya, K., Imai, H. T., and Moriwaki, K.,** New chromosome types of the black rat, *Rattus rattus,* collected in Oceania and hybrids between Japanese and Australian rats, *Jpn. J. Genet.,* 44, 89, 1969.
34. **Yosida, T. H. and Sagai, T.,** Banding pattern analysis of polymorphic karyotypes in the black rat by a new differential staining technique, *Chromosoma (Berlin),* 37, 387, 1972.
35. **Yosida, T. H., Kato, H., Tsuchiya, K., and Moriwaki, K.,** Karyotypes and serum transferrin patterns of hybrids between Asian and Oceanian black rats, *Rattus rattus, Chromosoma (Berlin),* 34, 40, 1971.
36. **Yosida, T. H., Kato, H., Tsuchiya, K., Sagai, T., and Moriwaki, K.,** Ceylon population of black rats with 40 diploid chromosomes, *Jpn. J. Genet.,* 47, 451, 1972.
37. **Yosida, T. H., Kato, H., Tsuchiya, K., Sagai, T., and Moriwaki, K.,** Cytogenetical survey of black rats, *Rattus rattus,* in Southwest and Central Asia, with special regard to the evolutionary relationship between three geographical types, *Chromosoma (Berlin),* 45, 99, 1974.
38. **Yosida, T. H. and Sagai, T.,** Similarity of Giemsa banding patterns of chromosomes in several species of the genus *Rattus, Chromosoma (Berlin),* 41, 93, 1973.
39. **Matthey, R.,** Cytogenetic mechanism and speciations of mammals, *In Vitro,* 1, 1, 1965.
40. **Jotterand, M.,** Le polymorphisme chromosomique des *Mus (Leggada)* africains. Cytogénétique, zoogéographie, évolution, *Rev. Suisse Zool.,* 79, 287, 1972.
41. **Matthey, R.,** Cytogenetique et taxonomie du genre *Acomys, Mammalia,* 32, 621, 1973.
42. **Gropp, A., Tettenborn, U., and von Lehmann, E.,** Chromosomenvariation vom Robertsonischen Typus bei der Tabakmaus, *M. poschiavinus,* und ihren Hybriden mit der Laboratoriumsmaus, *Cytogenetics,* 9, 9, 1970.
43. **Tettenborn, U. and Gropp, A.,** Meiotic nondisjunction in mice and mouse hybrids, *Cytogenetics,* 9, 272, 1970.
44. **Gropp, A., Winking, H., Zech, L., and Müller, H.,** Robertsonian chromosomal variation and identification of metacentric chromosomes in feral mice, *Chromosoma (Berlin),* 39, 265, 1972.
45. **Capanna, E., Cristaldi, M., Perticone, P., and Rizzoni, M.,** Identification of chromosomes involved in the nine Robertsonian fusion of the Apennine mouse with a 22-chromosome karyotype, *Experientia,* 31, 294, 1975.
46. **Capanna, E., Gropp, A., Winking, H., Noack, G., and Civitelli, M. V.,** Robertsonian metacentrics in the mouse, *Chromosoma (Berlin),* 58, 341, 1976.
47. **Tsuchiya, K., Moriwaki, K., and Yosida, T. H.,** Cytogenetical survey in wild population of Japanese wood mouse, *Apodemus speciosus* and its breeding, *Exp. Anim.,* 22(Suppl.), 221, 1973.
48. **Matthey, R.,** La position des genues *Ellobius* Fischer et *Prometheomys satunin* dans la systematique des Microtines, *Rev. Suisse Zool.,* 65, 377, 1958.
49. **Wahrman, J. and Zahavi, A.,** Cytological contributions to the phylogeny and classification of rodent genus *Gerbillus, Nature,* 175, 600, 1955.
50. **Satya Prakash, K. L. and Aswathanarayan, N. V.,** The chromosomes of the field mouse, *Mus platythrix, Mammal. Chrom. Newsl.,* 13, 120, 1972.
51. **Pathak, K.,** The karyotype of *Mus platythrix* Bennett (1932), a favorable mammal for cytogenetic investigation, *Mammal. Chrom. Newsl.,* 11, 105, 1970.
52. **Tsuchiya, K. and Yosida, T. H.,** Indian brown spiny mouse, *Mus platythrix,* having the lowest fundamental chromosome number, *Ann. Rep. Nat. Inst. Genet. Jpn.,* 22, 51, 1972.
53. **Dhanda, V., Mishra, A., Bhat, K. M., and Wagh, U. V.,** Karyological studies on two sibling species of the spiny mice, *Mus saxicola* and *M. platythrix, Nucleus,* 16, 56, 1973.
54. **Yosida, T. H.,** Possible evidence for the karyotype evolution of the Indian spiny mouse due to tandem fusion of the house mouse chromosomes, *Proc. Jpn. Acad.,* 55(Ser. B), 270, 1979.
55. **Yosida, T. H.,** Karyotype of the Indian spiny mouse resulted from tandem fusion of some of the house mouse chromosomes, *Cytologia,* 45, 753, 1980.
56. **Wurster, D. H. and Benirschke, K.,** Indian muntjac, *Muntiacus muntjak:* a deer with a low diploid chromosome number, *Science,* 168, 1364, 1970.
57. **Yosida, T. H., Kato, H., Tsuchiya, K., Moriwaki, K., Ochiai, Y., and Monty, H.,** Black rats from Mauritius with a new karyotype involving the Robertsonian fission (preliminary note), *Proc. Jpn. Acad.,* 55 (Ser. B), 120, 1979.

58. **Yosida, T. H., Kato, H., Tsuchiya, K., Moriwaki, K., Ochiai, Y., and Monty, J.,** Mauritius type black rats with peculiar karyotypes derived from Robertsonian fission of small metacentrics, *Chromosoma (Berlin),* 75, 51, 1979.
59. **Kato, H., Sagai, T., and Yosida, T. H.,** Stable telocentric chromosomes produced by centric fission in Chinese hamster cells in vitro, *Chromosoma (Berlin),* 40, 183, 1973.
60. **Upcott, M. B.,** The external mechanics of the chromosomes. VI. The behaviour of the centromere at meiosis, *Proc. R. Soc.,* B124, 336, 1937.
61. **Rhoades, M. M.,** Cytogenetic study of a chromosome fragment in maize, *Genetics,* 21, 491, 1936.
62. **Darlington, C. D.,** The origin of isochromosomes, *J. Genet.,* 39, 351, 1940.
63. **Pardue, M. L. and Gall, J. G.,** Chromosomal localization of mouse satellite DNA, *Science,* 168, 1356, 1970.
64. **Jones, K. W.,** Chromosomal and nuclear localization of mouse satellite DNA in individual cells, *Nature (London),* 225, 912, 1970.
65. **Hsu, T. C. and Arrighi, F. E.,** Distribution of constitutive heterochromatin in mammalian chromosomes, *Chromosoma (Berlin),* 34, 243, 1971.
66. **Yosida, T. H.,** Karyotype of F_1 hybrids between Mauritius and Oceanian type black rats, *Proc. Jpn. Acad.,* 55(Ser. B), 275, 1979.
67. **Yosida, T. H.,** Segregation of karyotypes in the F_2 generation of the hybrids between Mauritius and Oceanian type black rats with a note on the size of litters, *Proc. Jpn. Acad.,* 55(Ser. B), 557, 1980.
68. **Wahrman, J., Goitein, R., and Nevo, E.,** Mole rat *Spalax:* evolutionary significance of chromosome variation, *Science,* 164, 82, 1969.
69. **Nadler, C. F. and Harris, K. E.,** Chromosomes of the North American prairie dog *Cynomys ludvicianus, Experientia,* 23, 41, 1967.
70. **Fredga, K. and Bergström, U.,** Chromosome polymorphism in the root vole *(Microtus oeconomus), Hereditas,* 66, 145, 1970.
71. **Todd, N. B.,** Karyotypic fissioning and canid phylogeny, *J. Theor. Biol.,* 26, 445, 1970.
72. **Yosida, T. H. and Taya, C.,** Studies on interspecific hybridization in the rodents. I. Artificial insemination between the Norway rat (♀) and black rat (♂) and the resulting karyotypes in the hybrid blastocysts, *Jpn. J. Genet.,* 52, 289, 1977.
73. **Yosida, T. H. and Taya, C.,** Studies on interspecific hybridization in rodents. II. Histological observations on the hybrids developed by artificial insemination between Norway rats and black rats, *Jpn. J. Genet.,* 54, 371, 1979.
74. **Yosida, T. H. and Taya, C.,** Studies on interspecies hybridization in the rodents. III. Artificial insemination between *Rattus norvegicus, R. annandalei* and *R. losea, Proc. Jpn. Acad.,* 5(Ser. B), 141, 1980.
75. **Yosida, T. H.,** Diminution of heterochromatic C-bands in relation to the differentiation of *Rattus* species, *Proc. Jpn. Acad.,* 51, 659, 1975.
76. **Arrighi, F. E. and Hsu, T. C.,** Localization of heterochromatin in human chromosomes, *Cytogenetics,* 10, 81, 1971.
77. **Polani, P. E.,** Centromere localization at meiosis and the position of chiasmata in the male and female mouse, *Chromosoma (Berlin),* 36, 343, 1972.
78. **Chen, T. R. and Ruddle, F. H.,** Karyotype analysis utilizing differentially stained constitutive heterochromatin of human and murine chromosomes, *Chromosoma (Berlin),* 34, 51, 1971.
79. **Saunders, G. F., Hsu, T. C., Getz, M. J., Simes, E. L., and Arrighi, F. E.,** Location of human satellite DNA in human chromosomes, *Nature New Biol.,* 236, 244, 1972.
80. **Yosida, T. H.,** A preliminary note on silver-stained nucleolar organizer regions in the black and Norway rats, *Proc. Jpn. Acad.,* 54(Ser. B), 353, 1978.
81. **Yosida, T. H.,** A comparative study on nucleolus organizer regions (NORs) in 7 *Rattus* species with special emphasis on the organizer differentiation and species evolution, *Proc. Jpn. Acad.,* 55(Ser. B), 481, 1979.
82. **Tantravahi, R., Miller, D. A., Dev, V. G., and Miller, O. J.,** Detection of nucleolus organizer regions in chromosomes of human, chimpanzee, gorilla, orangutan and gibbon, *Chromosoma (Berlin),* 56, 15, 1976.
83. **Kano, Y., Maeda, S., and Sugiyama, T.,** The location of ribosomal cistron (rDNA) in chromosomes of the rat, *Chromosoma (Berlin),* 55, 37, 1976.
84. **Goodpasture, C. and Bloom, S. E.,** Visualization of nucleolar organizer regions in mammalian chromosomes using silver staining, *Chromosoma (Berlin),* 53, 37, 1975.
85. **Miller, D. A., Dev, V. G., Tantravahi, R., and Miller, O. J.,** Suppression of human nucleolus organizer activity in mouse-human somatic hybrid cells, *Exp. Cell Res.,* 101, 235, 1976.
86. **Yosida, T. H.,** The karyotype of *Rattus villosissimus* resulted from Robertsonian fission of small metacentrics, *Proc. Jpn. Acad. (Ser. B),* 55, 497, 1979.
87. **Liming, S. L., Yingying, Y., and Xingsheng, D.,** Comparative cytogenetic studies on the red muntjac, Chinese muntjac, and their F_1 hybrids, *Cytogenet. Cell Genet.,* 26, 22, 1980.

Chapter 4

ALGAL KARYOLOGY AND EVOLUTIONARY TRENDS

Y. S. R. K. Sarma

TABLE OF CONTENTS

I. INTRODUCTION

Studies on algal karyology have gained impetus since the middle of this century, although a few valuable contributions were made earlier. This paper reviews the present status of the subject in the members belonging to the classes Chlorophyceae, Charophyceae, Dinophyceae, and Euglenophyceae and a few others in which limited contributions have been made. The predominantly marine Phaeophyceae and Rhodophyceae are not considered, nor is the Cyanophyceae, the prokaryote group, included.

Earlier literature includes works of Bold,[1] Tischler,[2] and a few others. Godward,[3] for the first time, attempted to collate the scattered information on algal cytology. The present author reviewed the progress of research on various aspects of algal cytology, such as (1) nuclear cytology of the Chlorophyceae,[4] (2) the role of cytology in algal systematics,[5,6] (3) cytology and cytotaxonomy of Indian Charophyta,[7] (4) algal karyology in India,[8] (5) atypical mitosis,[9] (6) chromosomes of algae,[10] (7) algal cytology from Banaras School,[11] and (8) trends in algal cytology and cytogenetics.[12] Ultrastructural aspects of the nucleus and nuclear division of different algal classes have been summarized by Dodge[13] and Pickett-Heaps.[14] The above references provide background information on the present review.

Godward[3] considered the relatively noncommital classification of Fritsch[15] to fit adequately the basic divergence in algal nuclear organization. Keeping this in view, this system of classification[15] has been followed in the present text except for a few departures: (1) the Characeae is considered as a separate class instead of as an order Charales under Chlorophyceae as was done by Fritsch, (2) Prasinophyceae, Haptophyceae, and Eustigmatophyceae are recognized as stable classes, as was done by Dodge,[13] and are considered briefly towards the end.

The algal classes considered from a karyological standpoint in this review, therefore, are the Chlorophyceae, Charophyceae, Bacillariophyceae, Euglenophyceae, Dinophyceae, Cryptophyceae, Chrysophyceae, Chloromonadineae, Prasinophyceae, and Haptophyceae. Most of the important work on algal karyology from 1950 onwards has been covered and only some of the more significant contributions prior to 1950 have been included.

II. CHLOROPHYCEAE

A. Volvocales

The earliest work on the cytology of Volvocales seems to be by Dangeard,[16] who recorded chromosome numbers in three species of *Chlamydomonas*. Several species of *Chlamydomonas* were investigated later, of which *C. reinhardii, C. moewusii,* and *C. eugametos* are the most important.[17] *C. reinhardii* alone was studied cytologically.[18-28] Cave and Pocock[29,30] established the karyological features in the colonial Volvocales. Subsequent additions to these studies are those by Stein,[31,32] Rayns and Godward,[33] Goldstein,[34] Sarma and Shyam,[35-37] and Patel.[38] Except on *Tetraspora lubrica,* a palmelloid member of Volvocales,[39] not much work has been done on the nonmotile group of Volvocales, except for that on *T. apiocystioides,*[40] a new species, and the other on *Tetrasporidium javanicum.*[41]

The chromosomes of the Volvocales are either small or minute[4] and in many cases have no indication of centromeric positions. Chromocenters (heterochromatic bodies) have been reported in the interphase nuclei of some members of colonial Volvocales. Distinct nucleolar organizing chromosomes have been recorded in certain species of *Volvox.*

Mitosis in most colonial Volvocales lasts commonly for about 10 min. During successive nuclear divisions of the reproductive phase, the chromosome size is gradually reduced without any change in chromosome number.

The haploid chromosome numbers reported for *C. reinhardii* by different workers are $n = 8$ or $n = 16$, others being 11 ± 1.[24] Sixteen linkage groups have been reported for this organism.[42,43] Maguire[27] confirmed the haploid chromosome number as $n = 8$ for this species. The discrepancies in chromosome number in the same species of *Chlamydomonas* were attributed to intensive cultural conditions[24] since, under high light intensity, forms with higher chromosome numbers arise due to delay or temporary failure in cytokinesis following mitosis. An alternative explanation could be that different genetic strains could differ by their chromosome numbers, also.

A recent work on *Chlamydomonas* [44] has shown that several Indian taxa exhibit chromosome numbers below $n = 10$. Variability in chromosome number within the same species of *Astrephomene gubernaculifera*[30] (with $n = 4, 6, 7, 8$) and of *Gonium pectorale*[37] (with $n = 16, 17, 18$), 17 being the most common number, has been reported within colonial Volvocales. It may possibly be attributed to nondisjunction of the daughter chromosomes during anaphase, occasionally observed in natural populations. Such variation was also encountered in *Volvulina steinii* ($n = 7, 8$) where it was assumed to account for sex differentiation. Later investigations,[32,45] however, established that both + and − strains of *V. steinii* revealed an identical chromosome number of $n = 7$ only. *V. pringsheimii* also showed $n = 7 \pm 1$.[36] Chromosomal races within the same species are exhibited by several members of Volvocales besides the cases stated above (e.g., *Eudorina elegans, E. illinoisensis*).

A perusal of chromosome numbers of unicellular and colonial Volvocales[16-66] clearly points to the absence of any euploid trends. Aneuploidy seems to have played a considerable role in the evolution of different taxa. A general trend in *Volvox* is that all the species belonging to section Euvolvox, comprising species with more highly evolved morphological organization, have uniformly the lowest chromosome number ($n = 5$) recorded for the genus, while sections Merillosphaera and Janetosphaera, with comparatively less evolved thallus organization, have higher chromosome numbers ($n = 13, 14, 15$).

Ultrastructural studies on nuclear and cell division have been made on *Chlamydomonas reinhardii* by Johnson and Porter[46] and Goodenough,[47] on *C. moewsii* by Treimer and Brown,[48] *Tetraspora* sp. by Pickett-Heaps,[49] and *Volvox tertius*[50] and *V. aureus* by Deason and Darden.[51] Two species of the same genus *Chlamydomonas*, viz., *C. reinhardii* and *moewusii*, differ in their mitotic behavior. In the former, polar fenestrae are evident, in the latter the nuclear envelope is completely intact, the spindle being completely intranuclear. The rootlet tubules of flagellar bases "basal body complex" and the centrioles "centriole complex" are considered as two highly structured and conspicuous forms of "microtubule organizing centers" (MTOCS). These, with minor variations in their structure and function, play an important role in organizing the spindle and later in cell cleavage. Electron microscopic studies show that although the nucleolus disperses by the end of prophase, the nuclear envelope remains intact except for polar fenestrae through which spindle elements enter the nucleus. Deason and Darden[51] point out that kinetochores are not distinct in *Volvox*. Although the "splitters" in the realm of taxonomy have created the Chlamydomonadales and Tetrasporales apart from Volvocales mostly on morphological grounds, nuclear cytology from both light and electron microscopic studies shows a general uniformity barring minor variations throughout the Volvocales as constituted by Fritsch.[15]

Cytological evidence of zygotic meiosis has been obtained for *Chlamydomonas spp.*[21,27,28,52] among unicellular members and for *Gonium*[53] and *Volvox carteri* f. *nagariensis*[54] among colonial members. Treimer and Brown,[52] while presenting details of meiosis in *C. reinhardii* at the ultrastructural level, have reported that by pachytene, typical synaptinemal complexes are formed, with two lateral elements and a central space of 130 nm. A central element may form midway between the lateral elements and be joined to them via transverse elements. Separated synaptinemal complexes appear in diakinesis-diplotene, confirmed by others.[27,28] While Maguire[27] reports eight bivalents in *C. reinhardii,* Storms and Hastings[28] show 16 individual bivalents plus bivalent arms attached to a common mass of condensed chromatin, thereby yielding a haploid chromosome number of 18 to 20. The $n = 8$ and $n = 16$ strains may be haploid and diploid strains of *C. reinhardii* or the 16-chromosome form could have arisen as a result of endoreduplication.[27]

B. Chlorococcales

Since Timberlake's[67] karyological studies in *Hydrodictyon,* valuable contributions have been made on the Chlorococcales by Wanka,[68] Proskauer,[69] Sulek,[70] Sedova,[71,72] Chan,[73,74] Sarma,[75] Chowdary,[76] Patel and Francis,[77] Noor,[78] Mathew,[79] Rai,[80] and Shashikala Jayaraman.[81]

All the taxa investigated so far are uninucleate, except in *Hydrodictyon* where the segments are coenocytic. Interphase nuclei are usually globose, except in *Selenastrum minutum* and *Kirchneriella lunaris* where they are reported to be lens-shaped. They range from 1.2 μm *(S. minutum)* to 4.0 μm in diameter. Exceptionally, those of *Oocystaenium elegans*[79] are the largest within the group, up to 14 μm in diameter. The nuclei present usually a homogeneous appearance with a single nucleolus per nucleus devoid of any chromocenters except in *Hydrodictyon*[75] and *Gloeotaenium.*[79] The nucleolus disappears and the chromosomes organize themselves into a metaphase plate. The chromosomes are small or minute or rarely in the form of short rods, generally not exceeding 1.0 μm in diameter. However, those of *O. elegans*[79] are very large in contrast to the remaining taxa, the chromosomes within the karyotype ranging from 10.0 to 16.4 μm at metaphase, some being probably the longest reported for any alga so far.[79] Medium-sized chromosomes are also observed in *Gloeotaenium*[79] (1.0 to 2.8 μm). The longer chromosomes of *Oocystaenium* and even the medium-sized ones of *Gloeotaenium* showed distinct localized centromeres, the position of which characterize the individual chromosomes as being metacentric, submetacentric, and acrocentric ones. Clear spindle organization was identifiable in unsquashed preparations of many taxa.

Chromosome numbers of all taxa reported by various workers so far[67-106] show a range of $n = 4$ to 80, with the higher numbers $n = 24$ and 32 being characteristic of *G. loitlesbergerianum*[82] and *O. elegans,* respectively, and the highest number of $n = 80$ of *Eremosphaera viridis,*[83] all three taxa belonging to the Oocystaceae. *Kentrosphaera willei*[84] of the Chlorochytriaceae revealed $n = 22$ to 24. Barring these chromosome numbers, all the remaining taxa show a chromosome number of $n = 18$ or below.

During autospore or zoospore formation the nuclear divisions are rapid and successive, the sizes of the chromosomes decreasing progressively. For example, two members of the Chlorococcaceae, *Chlorococcum infusorium* and *Trebouxia humicola,* are characterized by the same chromosome number ($n = 10$), and so are *Dactylococcus infusiorum* and *Nephrochlamys subsolitaris* ($n = 6$), both of the Solenastraceae.[80]

Although there is identity of chromosome numbers between species of the same genus in certain cases (which may be distinguishable by their karyotypes), in certain other genera each species is characterized by its own chromosome number, as in *Scenedesmus.*[85] Chromosomal races in a few taxa occur (e.g., *Ankistrodesmus falcatus:*[79,80] $n = 8$, $n = 10$; and *Kirchneriella lunaris:*[76,79] $n = 12$, $n = 18$).

Taking an overall view, aneuploidy has played a crucial role in the evolution of species and taxonomic differentiation in this group. In view of the distinctive karyology of *Oocystaenium* and its close proximity to that of *Eremosphaera,* while standing apart from the rest of Chlorococcales, the view of creating a family Eremosphaeraceae as suggested by Brunthaler[86] instead of retaining it as subfamily of Oocystaceae has been upheld.[82]

Mainx[87] reported meiosis in the germinating zygotes of *Hydrodictyon;* Palik[88] recorded $n = 2$ and $2n = 4$ for this alga and observed reduction division to take place prior to gametogenesis. Proskauer[69] observed $n = 19 \pm 1$ in three species of *Hydrodictyon.* He disproved the observations of Palik[88] both with respect to the chromosome number and place of reduction division. In view of the conflicting reports, several workers reassessed the karyological features of this alga[76,78,89,90] and confirmed $n = 18$ for *H. reticulatum.* However, Sahay and Sinha[90] renewed the view of Palik in considering the thallus to be diploid and meiosis to take place prior to the gamete formation. But Noor[78] observed 18 bivalents at zygotic meiosis, confirming earlier observations[69] regarding place of meiosis and the fact that the thallus is haploid and not diploid.

Ultrastructural studies were carried out on *Chlorella pyrenoidosa* and *C. fusca,*[14] *Kirchneriella lunaris,*[91] *Ankistrodesmus falcatus,*[92] *Tetraedron bitridens,*[93] some species of *Scenedesmus,*[14] *Hydrodictyon reticulatum,*[94] *Pediastrum,*[95] *Sorastrum,*[96, 97] etc. In *Kirchneriella,* the centrioles, though initially absent, appear as the nucleus enters into prophase. In *Scenedesmus,* the paired interphase centrioles replicate as usual and separate around the nucleus. Replication of centrioles has been observed in *Hydrodictyon* and *Sorastrum,* also. The nuclear envelope is persistent except for polar fenestrae in several taxa investigated.

C. Ulotrichales and Chaetophorales

These orders of Chlorophyceae are considered together since sufficient karyological similarities exist in their nuclear cytology. The karyology of 42 taxa of the Ulotrichales and 45 of the Chaetophorales has been studied by various workers,[107-182] the major contributions emanating from Banaras School. The nuclei of *Microspora*[107] and *Sphaeroplea annulina*[108] contain a variable number of chromocenters; discontinuous staining of early prophase chromosomes seems to be more common in several taxa.[109]

The chromosome numbers range from $n = 3$ to 46 ± 2 in Ulotrichales[107-169] and $n = 4$ to 56 in Chaetophorales.[170-182] Chromosomes in general are extremely small, sometimes even less than 0.25 μm in diameter.[110] However, some chromosomes of the karyotypes of *Ulva lactuca,*[111] *U. fasciata,*[112] *Enteromorpha* spp.,[113] and to some extent in *Ulothrix zonata*[114] and *Draparnaldia plumosa*[115] are longer. Centromeric positions are difficult to ascertain due to the small size of the chromosomes. In such cases where longer chromosomes are discernible, they seem to be telocentric, based upon the orientation of daughter chromosomes at anaphase.

Telocentric, subtelocentric, and metacentric chromosomes have been recognized from the anaphase configurations of daughter chromosomes of some taxa of *Cephaleuros.*[116] The size of the metaphase plate never exceeds the initial diameter of the telophase nucleus. In *Microspora amoena, Sphaeroplea annulina, Uronema terrestre,* and *Ulva lactuca,* nucleolar chromosomes lie in association with the nucleolus during prophase. A unique feature of the karyology of *Microspora*[107] is the rotation of the spindle axis from an initial transverse to a longitudinal orientation either at metaphase, during anaphasic separation, or even after daughter nuclei are organized. This has been substantiated by electron microscopy.[117] Cytological analysis of various taxa of *Stigeoclonium* reveals a continuous series of chromosome numbers from $n = 5$ to 20, only $n = 17$, 18, and 19 having not yet been reported.

Chromosomal races[6] have been found in *Schizomeris leibleinii, Uronema terrestre, U. gigas, Microspora amoena, Ulva lactuca, Enteromorpha compressa, Trentepohlia aurea, Cephaleuros virescens,* and *C. solutus* and various species of *Stigeoclonium.*

Nuclei of different strains of *C. virescens* revealed different chromosome numbers. Thus, the physiological races of this alga also differ cytologically, although morphologically they look alike. More than one cytological race can occur on the same host and the same host plant at different geographical locations need not always have the same cytological race of the alga.[116]

On the basis of karyological investigations, the inclusion of *Sphaeroplea, Microspora, Ulva, Enteromorpha,* and *Cylindrocapsa* under Ulotrichales by Fritsch[15] was found justifiable.[5]

The karyological observations and critical evaluation of all relevant data on *Schizomeris leibleinii*[118] rule out its consideration as an ecophene of *Stigeoclonium tenue.* Even on ultrastructural grounds Birkbeck et al.[120] concluded that *Schizomeris* remains distinct from *Stigeoclonium.*

Karyological studies favor the merger of the genus *Caespitella* with *Stigeoclonium (S. pascheri)*[122] and *Physolinum* with *Trentepohlia.*[123] Nuclear cytology points toward a close affinity between the Ulotrichales and Chaetophorales, justifying even the merger of two orders into one.[5,121] Pickett-Heaps[14] also states that a distinction between Ulotrichales and Chaetophorales seems artificial. On the whole, aneuploidy seems to have played a very important role in the evolution within both the Ulotrichales and Chaetophorales.

Meiotic studies in *Monostroma fuscum,*[124] *Enteromorpha compressa,*[113,125,126] *Ulva lactuca,*[111,127,128] and *U. fasciata*[112] exhibit isomorphic alternation of generations. In an ultrastructural study of *U. mutabilis,* Braten and Nordby[129] reported synaptinemal complexes. Two chromosomal races of *E. compressa*[113] ($2n = 18$ and $2n = 20$) were found on the basis of bivalent analysis.

Ultrastructural studies on the nucleus and nuclear division have been carried out on *Stichococcus chloranthus,*[130] *Klebsormidium flaccidum,*[131] *K. subtilissimum,*[132] *Ulothrix fimbriata,*[133] *Microspora* sp.,[117] *Stigeoclonium helveticum,*[133] *Cylindrocapsa brebissonii,*[134] *Coleochaete scutata,*[135] and *Ulva mutabilis.*[136] Centrioles are present in all except *Stichococcus.* Pickett-Heaps[14] holds the view that their presence is of minor phylogenetic significance and is probably related to the life cycle of the organism. Thus, *Stichococcus,* in which motile stages are not known, does not show centrioles, while *Klebsormidium,* producing motile spores, does exhibit centrioles. The nucleolus disperses in all cases by premetaphase although in *Microspora* nucleolar remnants may remain with the spindles. The persistence or disappearance of a nuclear envelope is variable. In *Stichococcus, Klebsormidium,* and *Coleochaete* it dissolves completely, leading to an open spindle, while in *Ulothrix, Microspora, Stigeoclonium,* and *Ulva* it persists and spindles are closed, polar fenestrae having been reported in *Ulothrix* and *Ulva.* On the basis of variations in mitosis, cytokinesis, and distribution of ultrastructural studies, the Ulotrichales-Chaetophorales complex was further resolved into four orders: Ulotrichales, Microsporales, Chaetophorales, and Coleochaetales.[137]

D. Cladophorales

The members of the Cladophorales have been investigated in considerable detail by Geitler[183] who established euploidy in the group. About 60 taxa belonging to *Cladophora, Chaetomorpha, Rhizoclonium, Urospora,* and *Spongomorpha* have so far been studied cytologically.[183-216]

This group is characterized by multinucleate cells. The resting nuclei exhibit chromocenters and variable numbers of nucleoli. An association between these two structures is particularly noted in *Cladophora glomerata*[184] and in a few other taxa. Sinha[185] correlated the number of nucleoli per nucleus and an increase in size of interphase nuclei with the level of ploidy. Remarkable direct correlation exists between the size of interphase nuclei of haploid and diploid thalli of *C. callicoma.*[186] The chromosomes, in contrast to many algal groups, are appreciably long or medium in size and exhibit discernible localized centromeres.

The chromosome numbers recorded for various taxa range from $2n = 12$ to $2n = 144$.[127,172,183-217] Geitler[183] and Wik-Sjostedt[187] proposed the basic chromosome number for the group as 6. The higher polyploid levels with $2n = 72$, 96, and 144 are achieved only by *C. glomerata*,[183,184,188] while in most of the taxa the common $2n$ numbers are 24, 36, and 48. In such taxa, where the complete life-history has been worked out, both n and $2n$ numbers are recorded.[186,189-191] In other cases, only n or $2n$ chromosome numbers are known. Nucleolar-organizing chromosomes have been recorded in certain taxa of *Cladophora*.[183,185] Chromosome dimensions have been shown to be of considerable cytotaxonomic significance.[191]

It is comparatively easy to study meiosis in this group since meiotic divisions occur in sporangia of diploid sporophyte thalli prior to zoospore formation in species with the haplodiplontic type of life-cycle (with iso- or heteromorphic alternation of generations) or cells of diploid gametophyte thalli prior to gamete formation, as in *C. glomerata* with a diplontic life-cycle. Meiotic configurations, particularly diplotene and diakinesis, largely resemble those of higher plants.[183,186,188,189,191-193] The number of chiasmata varies from one to four in different taxa.

Although euploidy is firmly entrenched within the group, some species at least might have arisen through aneuploidy, e.g., *Cladophora clevigera*[194] and *Rhizoclonium tortuosum*,[195] both exhibiting $n = 11$ and $2n = 22$. Other chromosome numbers, not within the euploid series, are met within *R. hookeri* ($n = 14$), *R. crassipelitum* ($n = 30$), *Pithophora aequalis* ($n = 30$),[196] *Chaetomorpha aerea* ($n = 10$, $2n = 20$),[197] and *Urospora vancouveriana* ($n = 9$).[198]

Intraspecific polyploidy (polyploid races within a species) has been reported in certain species, notable examples being *Cladophora glomerata* ($2n = 24$, 48, 72, 96, ca 144) and *R. heiroglyphicum* ($2n = 24$, 36, 48).

The assumption by Geitler[183] that the basic chromosome number for Cladophorales would be $n = 6$ is vindicated by the observations of Sinha and Ahmad[192] on *C. uberrima* in which they reported $n = 6$ and $2n = 12$. Amitosis has been described in *P. cleveana*.[199]

Ultrastructural studies on the nucleus and nuclear division in members of the Cladophorales are unfortunately very few. Mughal and Godward[200] investigated the kinetochore and microtubules in *C. fracta*, besides other features of mitosis. The nuclear envelope (NE) is complete around the mitotic figures at anaphase. Microtubules mostly reach up to NE although a few pass beyond through its opening in NE. The kinetochore is localized and well defined. It is less differentiated at metaphase than at anaphase, where it becomes stratified with dense and less dense layers. "Externally there is a dense layer into which the bases of the microtubules pass, then less dense and dense layers alternately, giving in total three layers with two less dense layers between them." As many as 30 microtubules might be attached to a kinetochore.

McDonald et al.[201] described the ultrastructure and differentiation in *C. glomerata*. The mitotic spindle was centric, closed, and developed from two half-spindles. The nucleolus was only partially dispersed during mitosis. Structured kinetochores were evident on the chromosomes.

There is a uniformity of cytological features among the genera *Cladophora, Rhizoclonium*, and *Chaetomorpha*, including ploidy. The karyology of Cladophorales is characteristically different from that of the Siphonales, thus justifying the creation of the order Cladophorales.[15]

E. Oedogoniales

Van Wisselingh[218] was probably the first to have studied in detail the karyology of

Oedogonium (O. cyathigerum) and to have reported its chromosome number as n = 19. Tschermak[219] made substantial contributions to its nuclear cytology, followed by others.[172,220-223] Chromosome numbers have been reported in about 60 taxa of Oedogoniales so far, of which only 2 taxa of *Oedocladium* and 2 of *Bulbochaete* are worked out. Thus, there is an urgent need for further karyological investigations on *Bulbochaete* in which a large number of fresh water species are known.

The studies by Srivastava and Sarma[223] in five monoecious, five dioecious macrandrous, four dioecious nannandrous, and two unidentified taxa of *Oedogonium* are very exhaustive and critical. A wide range in the size of the nuclei and the number and size of chromosomes of various taxa of *Oedogonium* has been brought to light. The range of chromosome number is from n = 9 to n = 46. The chromosomes at metaphase are very long (about 12 μm) in some species, the range being from 2 to 12.3 μm. Thus, in comparison to several other algal groups of Chlorophyceae including Cladophorales and Charales, a large number of taxa of *Oedogonium* possess comparatively longer chromosomes. Thus, *Oedogonium* spp. have been chosen as experimental materials in studies for observing karyological effects of radiation and chemicals. Size variations within a karyotype could be wide as is evident in *O. pringsheimii* var. *abbreviatum* (3.0 to 12 μm.)[223] The discernible localized centromeres make karyotypic studies easier, which was not possible in such groups as the Volvocales, Chlorococcales, Ulotrichales, Chaetophorales, etc. One to three nucleolar organizing chromosomes per haploid set have been reported in different taxa.[219-221,223,224] Henningsen[220] observed two types of SAT-chromosomes, one bearing a very short satellite and the second a very long one, with a short secondary constriction. However, such distinction was not evident in other taxa. No correlation between the number of nucleolar organizing chromosomes and chromosome number can be made in the genus *Oedogonium* even in polyploid series. Male and female filaments of heterothallic species are similar in their karyotypic organization,[223] suggesting that sex differentiation in *Oedogonium* is gene-controlled rather than determined by any specific sex chromosome(s). Tschermak[219] generalized that longer-celled species possess longer chromosomes and a higher chromosome number in comparison to smaller-celled ones. Such a generalization does not always hold good in a number of cases[223] and during the evolution of some species, increase in chromosome number did not seem to accompany increase in cell size or total nuclear volume.[223]

A perusal of chromosome numbers of various taxa of *Oedogonium* [218-238] shows that many species possess odd numbers (n = 11, 13, 15, 17, 19, 27, 41), although species with even numbers are also on record, though fewer in number.

Srivastava and Sarma[223] suggest x = 9 as the basic chromosome number, the lowest reported so far in five species of *Oedogonium*. Species with n = 18, 27, 36 seem to be at different ploidy levels. In general, however, aneuploidy seems to have played a more efficient role in the speciation of this genus. The probable role of hybridization in the evolution of aneuploids cannot be ruled out. Further, they can also come into existence as a result of nondisjunction of chromosomes during anaphasic separation. Several species have same chromosome numbers but a comparison of their karyotypes shows that each is distinctive, suggesting a possible role of structural changes in their evolution. In view of the paucity of information on the karyology of *Oedocladium* [240,241] and *Bulbochaete,* [105,231] no worthwhile conclusions could be drawn on their karyology.

The distinctive cytological features, coupled with other unique morphological traits, seem to favor the view of Round[225,226] of the separation of the Oedogoniales from the Chlorophyceae, raising it to the status of an independent class, Oedogoniophyceae.

A time course study of mitosis in *O. cardiacum* showed that it takes about 20 min.[14] Mitosis of two species of *Oedogonium* and *Bulbochaete hiloensis* was also studied ultrastructurally. The nuclear envelope remains essentially intact throughout mitosis except for numerous small gaps.[227,228] The nucleolar material, although persisting for some time during metaphase and anaphase, is eventually eliminated when daughter nuclei are reformed. The kinetochore structure in *O. cardiacum* is conspicuous and seven layered with tufts of microtubules attached to it, the latter becoming organized subsequently into spindle elements. Mitosis proceeds in *Bulbochaete* [229] in the same manner as in *Oedogonium*, kinetochores being complex, spindles being closed, and nucleolar material persisting for a long time but eventually being eliminated at telophase.

F. Conjugales

The Conjugales are distinguished from other orders of Chlorophyceae chiefly by their mode of sexual reproduction through conjugation of aplanogametes. Members are either unicellular as in desmids or filamentous as in *Spirogyra*.

Strasburger[242] and Geitler[243,244] are credited as being the first workers to investigate cell division in *Spirogyra*. Godward[245-251] and others[252] investigated the karyology of several species of *Spirogyra* and established the unique features of conjugalean cytology and also laid the foundation for cytotaxonomic studies. Prasad and Godward made further contributions to the karyology of *Mougeotia* [253] and *Zygnema*.[254] In our own laboratory, extensive work has been carried out on the nuclear cytology and cytotaxonomy of more than 60 taxa of Conjugales,[255-261] covering the filamentous genera of *Spirogyra, Sirogonium, Zygnema,* and *Mougeotia*, and the desmid genera *Cosmarium, Closterium, Micrasterias, Euastrum, Spondylosum,* and *Desmidium*. King,[262] Nizam,[263] Brandham,[264-266] and Kasprik[267] contributed substantially to desmid karyology. A large number of taxa belonging to this fascinating group of algae have so far been analyzed karyologically.[3,174,242-308]

The most significant karyological features of a large majority of the members of the group are (1) complex nucleolus with internal differentiation caused by the presence of nucleolar organizing regions of specific chromosomes, in the interphase nuclei; (2) stainable material derived from the nucleolus designated as "Geitler's nucleolar substance" during metaphase and its association with metaphase and anaphase chromosomes; (3) polycentric chromosomes (or chromosomes with diffuse centromeres) in a large majority of taxa; and (4) parallel disjunction of daughter chromosomes at anaphase in taxa characterized by polycentric chromosomes. Subsequent electron microscopic studies confirmed the above features originally established on the basis of light microscopy.

The interphase nucleus varies in shape: spherical, ovoid, ellipsoid, etc. Several taxa are characterized by chromocenters (as in *S. subechinata*,[250] *S. azygospora*,[255] and *S. paradoxa*[259] which may be rounded *(S. submargaritata)* or long and thread-like, while in some, a few chromocenters are associated with the nucleolus (e.g., *S. submargaritata*[250]). Thus, the presence or absence, number, and shape of chromocenters, if judicially evaluated, could be features of cytotaxonomic interest.[259]

The number of nucleoli per nucleus exhibits wide variation. A correlation seems to exist in some cases between the number of nucleoli and associated chromosomes, as in *S. submargaritata*[250] and *S. crassa*.[246] The nucleolus in several Conjugales, including desmids, is complex in organization, with nucleolar organizer tracts within them through which portions of chromosomes traverse. The extent of complexity varies from species to species. King[268] described four main types of nucleolar structure in desmids: (1) solitary and simple nucleoli, (2) several small but distinct nucleoli, (3) a compound nucleolus presenting a beaded appearance resulting from fusion of several nucleoli, and (4) a complex sausage-shaped or sinuously coiled nucleolus. Kasprik's[267] analysis of the nucleolar structure in 54 strains of 23 species and varieties of *Micrasterias* shows that they can be relegated to different types of nucleoli. Such a variation in the nucleolar structure is seen sometimes even in different clones of a single species (e.g., *Netrium digitus*[264]).

Nucleolar organizing chromosomes of Conjugales are unique structures in that the secondary constrictions cannot be made out although some satellite-like segments may appear at some stages of mitosis. Certain species of *Spirogyra* with their very long nucleolar organizing region and long drawn-out telophase serve as excellent materials to study nucleolar chromosomes.

Ultrastructural studies[269,270] on the nucleolus from telophase to interphase in *S. submargaritata* have produced clear evidence that the chromatin of the nucleolar organizing region traverses the nucleolus and appears as fibrillar material. Chromosomes expand laterally into numerous loops which may be helpful in transcription. It was also suggested that the nucleolar organizing regions of chromosome could be considered as the model of lampbrush chromosomes or puffs of polytene ones.

Geitler,[243] in *S. crassa,* attributed the straight or slightly sinuous chromatids lying at right angles to the long axis of the spindle and later on undergoing parallel disjunction to fluidity of the spindle or rigidity of the chromosomes. Godward[248] interpreted this phenomenon as due to the polycentric nature of the chromosomes. An examination of several species of *Spirogyra* revealed that some are characterized by chromosomes with localized centromeres, while in several others polycentric chromosomes are confirmed.[255,259] *S. submargaritata* is interesting in that it has both types of chromosomes in the same karyotype.[248] The polycentric nature of the chromosomes has been proved through irradiation[249] and ultrastructural studies.[200] Mughal and Godward[200] presented clear evidence for the presence of two kinetochores on the longer chromosomes. The kinetochores do not present a stratified appearance as in *Cladophora* or *Oedogonium.* The majority of desmid taxa are characterized by polycentric chromosomes although *Mesotaenium caldariorum, Cosmarium ehrenberghii, Pleurotaenium* spp., and *Micrasterias* spp. among desmids seem to exhibit localized centromeres.

Chromosome size is highly variable both in filamentous Conjugales and in desmids. Several taxa of *Spirogyra, Mougeotia,* and *Zygnema* are known to be characterized by minute (less than 1 μm in diameter) to small chromosomes (up to 2 μm). On the other hand, there are some taxa like *S. crassa* where large-sized chromosomes having a length up to 12 μm can be seen. King has demonstrated that chromosome size was directly related to cell size, and inversely proportional to chromosome number in three clones of *Cosmarium botrytis.* Kasprik[267] classified all taxa of *Micrasterias* into four types on the basis of chromosome characteristics: (1) species with small chromosomes (mostly 1 μm) showing stickiness; (2) species with well-separated chromosomes about 1 μm long with distinct contours; (3) species with short chromosomes up to 2.5 μm long, which are rather thick and so appear compact; and (4) species characterized by longer chromosomes up to 7.5 μm appearing somewhat joined together. Giant chromosomes have been found by Gerrath[271] in *Triploceros verticillatum* ($n = 15$), up to 20 μm in length at late prophase. Those of *T. gracile* are a little shorter, but longer than any other desmids examined so far.

Chromosome numbers in various taxa of *Spirogyra* are shown to range from $n = 2$ from Japan[272] and from India[259] to $n = 92 \pm 2$.[259] An examination of chromosome numbers reveals that in the majority of taxa they are 24 or below. The base number for *Spirogyra* seems to be $n = 2$. Except for two species of *Spirogonium* with $n = 6$[273] *(S. melanosporum)* and $n = 8$[260] *(S. inflatum),* all other taxa of this genus show n = ca 50.[274] However, one species, *S. pseudofloridanum,* has a high chromosome number of n = ca 100.[274] In *Mougeotia,* the number ranges from $n = 24$ to n = ca 94.[253] Species of *Zygnema* exhibit a range from $n = 14$ to 82. Among desmids examined so far, *Desmidium aptogonum* is characterized by the lowest chromosome number of $n = 8$,[261] the highest number being $n = 592$ in clone M of *Netrium digitus.*[275] Species of *Closterium* are generally characterized by higher chromosome numbers as compared with *Cosmarium*.

There is ample evidence to show that aneuploidy has played a significant role in the evolution of species, together with euploidy in certain taxa of *Spirogyra*.[3] Chromosomal races are prevalent in various species of desmids. It is possible to explain their origin through agmatoploidy.[258, 261]

Polyploid (diploid, triploid, and tetraploid) nuclei have been induced in *Micrasterias* spp. by various means (cold shock, heat treatment, centrifugation, and dinitrophenol treatments).[276] The viability of tetraploids was low. Morphological variation in diploids was more than in haploids. Brandham[265] studied polyploidy in desmid taxa, *C. siliqua, C. botrytis, Staurastrum denticulatum,* and *S. dialatum*. The diploids were morphologically different and, in some cases, can be referred to different species or even genera on the basis of present morphological criteria employed in the classification of desmids. Kasprik[267] explained the origin of aneuploids in *Micrasterias* through agmatoploidy.

The karyological features of the Conjugales are so distinctive as to lend support to the view to upgrade the order to the status of a class Conjugatophyceae or Zygnematophyceae as proposed by Round.[225, 226]

The original finding by Karsten[277] that zygotic meiosis occurs in *Spirogyra* has been substantiated by Godward.[251] She demonstrated and described, in detail, meiosis in 2-day-old zygospores of *S. crassa* having polycentric chromosomes. Two to four chiasmata per bivalent have been reported. The pattern of meiosis follows the same course as in higher plants having chromosomes with diffuse centromeres in that the first meiotic division is equational.[251]

Excellent accounts of meiosis in desmids were given on two varieties of *Cosmarium botrytis*[278] and three species of *Pleurotaenium*.[279] The zygospore nuclei do not fuse until immediately before germination and at the time of germination fusion nucleus may already be at diplotene or diakinesis. The meiosis in inter- and intraspecific hybrids of *Pleurotaenium* is comparable to that of higher plants indicating that the chromosomes have localized centromeres.

Mitotic events have been studied ultrastructurally on *Spirogyra* sp.[280] and *Mougeotia* sp.[281] Transport of cytoplasmic tubules and their incorporation into the prophase spindle has been adequately demonstrated in *Spirogyra* sp.[280]

In *Mougeotia* sp.[281] the prominent nucleolus disperses completely and chromosomes have distinct densely stained kinetochores. On the basis of ultrastructural studies, Mughal and Godward[200] prefer to designate the chromosomes in which more than one kinetochore is detectable along the length of chromosomes as polycentric instead of as possessing diffuse centromeres as in *Luzula,* where a single diffuse kinetochore extends along most of the length of chromosomes.[282]

Electron microscopic studies on *Netrium digitus,*[14] *Closterium littorale,*[283] *Micrasterias rotata,*[284] and *Cosmarium botrytis*[285] show mitosis to be uniform throughout in that the nucleolus disperses and NE disintegrates, spindle being open and centrioles absent. Numerous spindle and interzonal microtubules have been identified towards regions of broad poles. Anaphase separation of daughter chromosomes in the species is reported to be rapid, lasting not more than 5 min.

G. Siphonales

The members are usually marine, *Protosiphon botryoides* and *Dichotomosiphon tuberosus* being nonmarine. They are mainly characterized by siphoneous thallus, and on their karyology they could be considered under two groups: (1) multinucleate forms (e.g., *Bryopsis, Caulerpa*) and (2) uninucleate forms, which are uninucleate for a longer or shorter part of their life span, such as the members of Dasycladaceae.

The small size of the nuclei and the technical difficulties involved in staining, failure to obtain mitotic/meiotic divisions from the material fixed in the field, and, finally, difficulties in maintenance of laboratory cultures for longer periods are some factors which have created hurdles in the detailed karyological investigations in the groups.[309-347] In spite of these handicaps the contributions of Schulze,[309] Werz,[310] Puiseux-Dao,[311] Chowdary,[312] Chowdary and Singh,[313] Chowdary et al.,[314] and Singh and Chowdary[315-319] are noteworthy.

The individual cell of the septate forms or an aseptate segment of the siphoneous forms is multinucleate. The interphase nuclei may be globose or lens shaped, exhibiting generally network-like organization of chromatin. In some, however, the nucleus is more or less homogeneous with fine granular contents, e.g., *Bryopsis, Codium, Dichotomosiphon, Halicystis,* and *Trichosolen.*[320] Chromocenters of various numbers are reported in *Siphonocladus, Anadyomene, Microdictyon, Boergesenia,* and *Codium.*[311]

Members of the Dasycladaceae, however, remain uninucleate throughout their vegetative life cycle except for *Cympolia.* The primary nucleus of *Acetabularia* and *Batophora* becomes greatly enlarged (from 2 to 3 μm to about 100 μm in diameter through endomitosis).[311] It is surmised that the 16 endomitoses are involved. *Cympolia*[310] differs from other members in that the primary nucleus after enlargment "dissociates" into small daughter nuclei which get distributed throughout the thalli. After increasing in size they again divide. This process seems to be repeated a number of times. *Neomeris annulata*[321] resembles *Acetabularia* in that the voluminous primary nucleus persists for a very long time, but may become broken up into small daughter nuclei as in *Cympolia.* An enlargement of the nucleus from 4 to 5 μm up to 30 μm in diameter is also recorded in the gametangium of *Neomeris.*[320] Spring et al.,[322, 323] on the basis of ultrastructural studies of primary nucleus, do not, however, subscribe to this view.

The division of the primary nucleus is probably a sort of nuclear fragmentation as in *Cympolia* and *Neomeris.* Nucleolar emission is often observed from the nuclei. In *Acetabularia* and *Batophora,* the nucleus becomes disaggregated into tubules of varying sizes. A great part of the nuclear substance is left out without being used up in the formation of secondary nuclei. The secondary nuclei divide mitotically and in a regulated manner and a uniform chromosome number is maintained. The precise cytological events as to how the secondary nuclei receive a uniform karyotype from the primary nucleus remain still intriguing. Investigations on the ultrastructure of *Bryopsis hypnoides* by Burr and West[324, 325] have shown that the germling produced from the zygote contains a very large primary nucleus comparable to that in *Acetabularia.* The giant nucleus eventually is dissolved, giving place to two nuclei which subsequently undergo mitoses.

The mitotic divisions of the nuclei in the multinucleate forms and of secondary nuclei in the uninucleate forms are identical and largely asynchronous. While in a majority of forms the nucleolus disappears during prophase, in certain taxa of *Bryopsis,*[314,326] *Neomeris,*[320] and *Codium fragile*[327] it persists, undergoing changes in shape and size, sometimes covering the premetaphase chromosomes like a cap or a rounded mass.

Prophase chromosomes appear discontinuously stained. Mitosis was considered to be intranuclear in *Codium tomentosum,*[328] *Bryopsis plumosa,*[326] *B. hypnoides,* and *Anadyomene wrightii.*[329] This has been refuted by Singh[320] who observed that NE disappears by late prophase except in *A. stellata,* where the NE partially remains beyond premetaphase stages. Ultrastructure of *Acetabularia wettsteinii* at anaphase clearly indicates the persistence of NE as shown by Godward et al.[330] Borden and Stein[327] opined that the nucleolus-like body not only persists but also divides at anaphase and is included in daughter nuclei at telophase. Metaphase plates are normal. However, centromeric positions are not very clear in certain taxa where the chromosomes are minute or small and dot-like (such as *Bryopsis, Trichosolen,* and *Dichotomosiphon*). In some taxa such as *Valonia, Boergesenia forbesii,* and *Chaemodoris,* centromeric position could be more or less made out. The longer chromosomes in the karyotypes of a few others show distinct centromeres[320] *(Acetabularia calyculus, Valoniopsis, Microdictyon,* and *Codium,* for example). Evidence has been obtained on the basis of ultrastructural study for the presence of centromeres (kinetochores) with attached microtubules in *A. wettsteinii.*[330] However, they are relatively simple in organization and lack stratification, unlike the stratified centromeres of *Oedogonium,*[228] *Cladophora,*[200] and *Chlamydomonas.*[331] Spindles have been observed in many cases, which may be short or large and squat.

Among the eusiphoneous members, two distinct lines of chromosomal characters are found: (1) dot-like chromosomes (or short rods) with no distinguishable chromatids and centromeres *(Bryopsis, Trichosolen,* and *Dichotomosiphon)* and (2) longer chromosomes with clear chromatids and centromeric positions. Most septate members have karyotypes with some long chromosomes, but *Valonia* stands out distinctly in that chromatids are identifiable in all chromosomes which also have distinct centromeres.

The plants are diploid (except in *Protosiphon* and *Dichotomosiphon*) and meiosis before gametogenesis is assumed in many cases.[15,311] In the sporophyte of *Derbasia,* one nucleus from each series of meiotic divisions proceeds towards spore formation, the other three nuclei aborting.[332,333] Meiotic division occurs before cyst formation in *Acetabularia.*[334] Koop,[335] however, is of the opinion that meiosis occurs in *Acetabularia* during the division of the primary nucleus (although no crossing over or synaptinemal complexes have been observed) on the basis of chromosome numbers, ultrastructure, and genetics.

Examination of chromosome numbers reveals a limited euploidy in genera such as *Acetabularia, Boodlea,* and *Cladophoropsis.* Occurrence of polyploid races within same species and different races occurring at different geographical regions *(Borgesenia forbesii and Valoniopsis pachynema)* has been recorded.[320] On karyological grounds, inclusion of *Anadyomene* under Cladophorales by some taxonomists does not seem to be sound.

The classification of the Siphonales has suffered many vicissitudes.[225,226] The earlier order of Siphonocladales was dismembered by Fritsch,[15] raising Cladophoraceae to the order Cladophorales, transferring Sphaeropleaceae to Ulotrichales, and combining the remaining families of Siphonocladales with the order Siphonales. Round[226] has created a separate class Bryopsidophyceae, amalgamating and rearranging the members of Cladophorales and Siphonales of Fritsch,[15] while bringing *Sphaeroplea* into this orbit. On karyological grounds, Cladophorales constitutes a well-defined group with established polyploidy with a basic number of $n = 6$. No such trend is seen in the truly siphoneous members. Further, *Sphaeroplea*[108] has been shown to be very close to Ulotrichales rather than to Cladophorales and probably not even to truly siphoneous members. Thus, karyological evidence does not favor the creation of Bryopsidophyceae.[226] However, there exists a case for the separation of the members of Dasycladaceae from Siphonales and raising it to the status of an order Dasycladales.

Ultrastructural studies have revealed that the giant nucleus of *Acetabularia* and *Bryopsis* is not only surrounded by a true nuclear envelope but also by another porous, more inflated cisterna continuous with the vacuolar spaces of cytoplasm. A perinuclear zone lies sandwiched between true and secondary envelope.[336]

Lamp-brush-type chromosomes with defined axes, chromomeres, and lateral loops in the primary nucleus of the green alga, *Acetabularia,* were studied. Since bivalent structure has not been found in the alga as in the lamp-brush chromosomes of typical oocytes, such chromosomes are regarded as not restricted to meiosis of oocytes of certain amphibians alone.[322,323] The majority of loops, however, are smaller, measuring 5 to 10 μm but some reaching even up to 15 to 20 μm. Loops are covered by shorter transcriptional units[337] (0.3 to 10 μm with a modal peak of 4 μm, corresponding to sizes of primary transcripts between 900 and 30,000 nucleotides) with apparently non-transcribed spacer regions on either side of the transcribed portion. This type of chromosomal structure naturally makes the nuclei less stainable. Endomitosis may not at all be involved in the enlargement of primary nucleus, which seems due to only loosening and expansion of the chromatin which may be a temporary phase for specialized transcriptional purposes. When once the functional requirements are completed, the chromosomes assume their normal size leading to the restoration of the karyotype which divides to give rise to secondary nuclei.

Liddle et al.[338] brought forth evidence on the basis of ultrastructural studies that the primary nucleus of *Batophora* is organized in a similar manner as in *Acetabularia* in that it shows lampbrush type of organization. Meiosis occurs in the primary nucleus. Thus, according to these workers also, meiosis is completed before the secondary nuclei are formed, as summarized by Koop.[335] On the basis of these new findings the primary nuclei of *Cympolia* and *Neomeris* need investigation at the ultrastructural level.

III. CHAROPHYCEAE

Most of the chromosome numbers of this large cosmopolitan group have been determined largely from the dividing nuclei of spermatogenous filaments from the antheridium and only a few from vegetative thalli. Literature on the subject reveals that more than 150 taxa belonging largely to the genera *Chara, Nitella,* and *Tolypella* and a few to *Nitellopsis, Lychnothamnus,* and *Lamprothamnium* seem to have been analyzed karyologically.[172,348-474]

The resting nuclei are globular, generally with a single nucleolus. They exhibit chromocenters of variable number and size. The early prophase chromosomes are always discontinuously stained. The chromosomes are appreciably large in many of the taxa, the overall range being 1 to 12 μm long at metaphase. Among the different genera, some taxa of *Nitella* with low chromosome numbers, such as *N. mirabilis* (n = 6), have strikingly long chromosomes[348] (6.0 to 12 μm). Within *Chara,* the longest chromosomes do not generally exceed 6.00 μm in length at metaphase.[349] The maximum variation in length of chromosomes is met within the karyotype of *Chara corallina* [350] (n = 42, 0.3 to 3.9 μm long). The width of chromosomes in different taxa seems to range from 0.7 to 1.4 μm.[350]

The chromosomes are characterized by localized centromeres. A notable correlation between chromosome number and chromosome size has been observed both in *Chara* and *Nitella.* Taxa having low chromosome numbers are characterized by longer chromosomes *(N. mirabilis, C. bharadwajae),* while those with higher numbers have shorter chromosomes[7] (e.g., *N. furcata* subsp. *flagellifera* and *C. globularis* var. *virgata*).

The absence of any observable nucleolar organizing chromosomes with secondary constrictions, even in those taxa with appreciably long chromosomes,[351,352] is unusual; a few taxa bear definite secondary constrictions.[253,254] In a number of taxa one or two chromocenter-like deeply stained bodies could be made out in association with the nucleolus at the interphase nucleus.[351] Comparison of the karyotype of the spermatogenous filaments of male plants with that of the apical cell (vegetative) of the female plant of *Nitella inokasiraensis* shows that in both the chromosome number is 6 with identity in all chromosomes between the two sexes, except in one case (3M in male and 3F in female) which varies to some extent.[354,355] On that basis 3M and 3F chromosomes may be heterotypic sex chromosomes. Further, Kanahori[356] found differences in chromosome numbers of male (n = 14) and female (n = 15) clones of *Nitellopsis obtusa.*

It is essential that comparisons should be made between the karyotypes obtained from mitotically dividing nuclei of the apical cells of both male and female plants of a large number of dioecious taxa to uncover any sex chromosome mechanism determining sex differentiation.

Khan and Sarma[348-350] and Ramjee and Sarma[357] described the detailed cytology and discussed affinities of several Indian charophytes from different regions of the country. Kanahori,[354] however, adopted their data on arm ratios of individual chromosomes for karyotypic analysis.

A survey of chromosome numbers from various countries of the world reveals distinct euploidy in different charophycean genera:[358] *Chara*, $n = 7, 14, 28, 35, 42, 49, 56, 70$; *Lychnothamnus*, $n = 14, 28$; *Lamprothamnium*, $n = 14, 28, 42, 56, 70$; *Nitella*, $n = 6, 9, 12, 15, 18, 21, 24, 27, 36, 48$; and *Tolypella*, $n = 9, 12, 15, 33, 42$. The basic chromosome numbers which are acceptable as valid can be considered as "7" for the tribe Charae and "3" for *Nitella* of the tribe Nitellae.[359-361] These euploid series could be the result of active ploidy coupled with chromosomal rearrangements or of hybridization as surmised in *N. flexilis*.[362,363] Occurrence of aneuploid numbers has been found to be a common feature in European taxa in comparison to Indian taxa,[359] e.g., *Chara* ($n = 8, 9, 12, 16, 18, 20, 24, 26, 27, 30, 32, 33, 37, 38, 40$); *Lamprothamnium* ($n = 24, 25, 50$); *Nitellopsis* ($n = 13, 16, 18, 20$); *Nitella* ($n = 14, 29$); *Tolypella* ($n = 10, 20, 25, 50$) may be the result of nondisjunction which led to increased or decreased chromosome numbers at different ploid levels. In world distribution, the taxa with chromosome number $n = 14, 28$ in *Chara* and $n = 18$ in *Nitella* with maximum percentage of frequency appear to have more adaptable genomes in comparison to others.[358] Sarma and Ramjee[359] and Khan and Sarma[358] reviewed the world distribution of euploids and aneuploids of charophyte taxa and polyploidy in relation to various factors.

A review of the conflicting views expressed by various workers regarding the relative primitiveness or advanced state of the tribe Charae and Nitellae indicates that (1) Charae are more advanced than Nitellae, (2) within the genus *Chara*, corticated forms are more evolved than ecorticated taxa, and (3) within the genus *Nitella*, arthrodactylous taxa are more advanced than anarthrodactylous. An evolutionary sequence in Charophyta based on cytological findings was presented by Sarma et al.[364] The theory proposed by Desikachary and Sunderalingam[365] that Charophyta could have evolved from a chaetophoralean genus such as *Draparnaldiopsis* was refuted by Sarma, Khan, and Ramjee,[364] and Sarma[8] on the basis of comparative assessment of nuclear cytology of Chaetophorales and Charales. Two opposing views are present with respect to the place of the reduction division in the life cycle of charophytes: (1) before formation of gametes[366,367] and (2) zygotic meiosis.[368] In spite of very extensive karyological studies, meiosis has not been observed so far by any worker in this group. However, the identity of chromosome numbers in the vegetative thallus and spermatogenous filaments of the same plant in certain taxa[350] (*Chara globularis* var. *virgata*, *C. fibrosa* var. *hydropytis*, *C. zeylanica* var. *diaphana* f. *verstediana*, and a few other taxa) rules out the fact that meiosis takes place before gamete formation and offers indirect evidence for zygotic meiosis, pointing out to the fact that charophytes are haplonts.

Moutschen[369] studied the structure of the nuclear modifications in *C. vulgaris* during phylloid and axis morphogenesis and reported two types of nuclei, one of which (micronucleus) is capable of resuming its mitotic activity. The other giant nucleus (macronucleus) may be ribbon shaped or polylobated.

In a revision of the Characeae, Wood and Imahori[370] did not consider monoecious and dioecious conditions as important criteria in speciation, as in their view they reflect minor genetic variation and dioecious taxa represent genetic strains of monoecious taxa with half the usual complement characteristic of the latter. They have either completely merged several dioecious species with morphologically similar monoecious species or reduced them to lower intraspecific ranks. This view was shown to be incorrect on cytological grounds by Sarma and Khan[371] and Proctor.[372] Thus, monoecious and dioecious conditions are important criteria in specific delimitation of Characeae. Polyploidy has played a significant role in cytological diversification both in monoecious and dioecious taxa independently.[359]

Cytological analysis of a large number of charophyte taxa shows that merger/combination of certain valid taxa by Wood and Imahori[370] to erect new species is not justified on cytological grounds. A large number of specific cases, too many to be included here, have been discussed.[349, 350, 353, 357, 373]

Intraspecific polyploidy has been established in a number of species. Frequency of polyploidy in Indian taxa has been estimated to be about 41%.

In a cytogeographic survey of world Charophyta with particular reference to India, the higher frequency of distribution and chromosome numbers in Indian Charophyta indicates the suitability of Indian habitats for charophytes, with Bihar and Uttar Pradesh as primary centers of origin of world charophycean flora, since several taxa with low chromosome numbers are prevalent in these regions.[358]

Fritsch[15] treated charophytes under the order Charales and placed it under Chlorophyceae. However, the morphological complexity of the thallus, developmental history, and nature of reproductive organs coupled with the karyological studies warrants the separation of charophytes from Chlorophyceae to treat them under a separate class Charophyceae or a division Charophyta.

Ultrastructural studies on *Chara* and *Nitella* have brought to light several interesting features.[14] The dense nucleolar material is reported to disperse somewhat during prophase coating the chromosomes as in some species of *Spirogyra*[247] and is carried along with the daughter chromosomes, to be incorporated into newly formed nucleoli in daughter nuclei. This persistent behavior of the nucleolus is considered to be relatively primitive feature of mitotic systems.[374] Cells of older threads exhibit two centrioles and at mitosis two are present at each pole of the spindle, acting as the foci of spindle microtubules.

IV. XANTHOPHYCEAE

In contrast to the Chlorophyceae, only a few members *(Vaucheria, Tribonema,* and *Botrydium)* of this class have been investigated karyologically. Gross[475] studied the mitosis in vegetative and antheridial filaments of *V. geminata* and pointed out that divisions in both are essentially similar. Karyokinesis was shown to be completely intranuclear. Hanatschek[476] and Mundie[477] earlier reported chromosome numbers in *V. geminata* (n = 7 to 10) and *V. sessilis* (n = 5), respectively.

Ott and Brown,[478] from ultrastructure of mitosis, confirmed intranuclear mitosis as reported on the basis of light microscopy. The nuclear envelope does not show any polar fenestrae or opening during mitosis. The intranuclear spindle consists of both continuous and chromosomal fiber. The centriole is outside the nucleus and after prophase one may be seen at each pole. The microtubules produced from the centrosome seem to be essentially involved in the movement of nuclei within the coenocyte. The telophase nuclei are separated by invagination of inner nuclear membrane. Individuality of chromosomes at metaphase or anaphase has not been observed and kinetochores were not evident even where individual chromosomes were encountered.

Abbas and Godward[479] reported the existence of two cytologically distinct strains with n = 15 and n = 17 in *T. utriculosum*. Chowdary recorded a chromosome number of 16 to 20 in *B. lateralis*.[174]

V. BACILLARIOPHYCEAE

Existence of cytological peculiarities are known, since the time of researches of Lauterborn[480] and Klebahn,[481] reports of chromosome counts in Bacillariophyceae are rare. Geitler[482-487] and Cholnoky[488-491] studied the karyology of some diatoms, but no definite chromosome counts seem to be on record.

Iyengar and Subrahmanyan[492] described reduction division of the centric diatom, *Cyclotella meneghiniana*. Subrahmanyan[493] described somatic and meiotic divisions in *Navicula halophila*, a pennate diatom, the first record of actual reduction division in the genus *Navicula*. He concluded that sex determination occurs phenotypically. A sex chromosome mechanism has not been observed in diatoms. While greater attention has been paid to the life-cycle pattern and place of reduction division in the life cycle, mitotic counts are rare (Table 1). The few counts recorded so far are based on late prophase stages. The crowding of the chromosomes at metaphase and their minute size seem to be the probable reasons for any lack of extensive karyological work on the group with LM. Further, the mitotic events are so fast[494] that the mitosis was completed in *Terpsione musica* and *Biddulphia mobiliensis* within 30 sec to 3½ min.

The essential features of mitosis based on the observations in *N. halophila* are (1) the interphase nucleus has a lightly stained reticulum and a single nucleolus which at times appears vacuolated; (2) short and rod-shaped chromosomes occur in a ring-like fashion around a central spindle; (3) the centrosome probably plays a role in the organization of the spindle. It is not clear from the descriptions whether the nuclear membrane remains intact or disperses. Chromosomes of *Synedra ulna*[501] are minute (0.7 μm in diameter).

Manton et al.[495-498] described the intricacies of the origin, structure, and functioning of spindle in *Lithodesmium*, both at mitosis and meiosis. The ultrastructural studies reveal that a spindle precursor consisting of a rectangular body made up of a series of parallel plates is largely connected with the formation of spindle microtubules. The nuclear envelope breaks down, allowing the spindle to elongate. Microtubules have been shown to radiate from a single spherical structure of 500 nm in diameter in *Surirella ovalis* and *Melosira varians*.[499] A number of unusual features during mitosis have been observed by Tippit and Pickett-Heaps [500] in the pennate diatom, *S. ovalis*, in which a large spherical organelle named as microtubule center (MC) is associated with the formation of the extranuclear central spindle. During anaphasic separation of chromosomes at least three different mechanisms occur. Microtubules have been observed to associate laterally to the chromatin and no distinct kinetochore tubules have been observed. Tippit et al.[501] have also studied cell division in the centric diatom *Melosira*. Pickett-Heaps et al.[502] described cell division in the pennate diatom *Diatoma vulgare* where "persistent polar complexes" (PPC) are concerned with the formation of cytoplasmic microtubules. The central core of the doubled PPC alters its structure from a flat plate at prophase to a rod-shaped structure by metaphase. The tubules of the central spindle terminate in a layer described as "spindle insertion" (S.I.) close to PPCs. Some microtubules penetrating the chromatin have been considered to be true kinetochore tubules. The nuclear envelope ruptures during the course of division.

VI. EUGLENOPHYCEAE

Blochmann[503] and Keuten[504] reported for the first time certain peculiarities of nuclear structure in euglenoid flagellates. A number of papers followed with controversial reports, the survey of which is not contemplated here since Leedale[505-508] described very elegantly nuclear cytology of this interesting group, setting at rest several controversies of the past and clarifying the exact position. His initial studies are based upon

25 euglenoid species belonging to 12 genera. Nuclear structure and mitosis were described in detail for *Euglena gracilis* and other members of Euglenophyceae.[509-511] Despite some differences the nuclear cytology presents a common pattern as in *E. gracilis,* as summarized below.

Interphase nuclei show a range of sizes, shapes, and positions within the group. The nucleolus in these organisms was known as "endosome" since it was seen to persist and divide during mitosis, which is established even at LM level. The numbers of nucleoli per nucleus varies within species, sometimes of the same genus (*E. gracilis* — one nucleolus; *E. acus* — more than five nucleoli). Apparently these nucleoli are not associated with nucleolar organizing chromosomes. Before the commencement of division all the nucleoli in a nucleus fuse into one. The nucleolus elongates during metaphase stage at right angles to the long axis of the cell and parallel to the plane of division of the nucleus and assumes a dumbbell-shaped appearance at anaphase, and becomes constricted into two daughter nucleoli during telophase. Thus, this is one very distinctive feature of euglenoid mitosis. Interphase nuclei contain discrete condensed chromosomes which separate out in squash preparations of telophase nuclei. The extent of condensation seems to be variable but not to such an extent as in the Dinophyceae.

The pairs of chromatids are never seen to be attached at any one point, since they are free from each other from one end to another. Their separation also is not parallel at anaphase. Consequently, the evidences indicate that the chromosomes do not possess centromeres nor are they polycentric or with diffuse centromere. Thus, they are distinctly atypical in nature.

Division of chromosomes (termed as reduplication by Singh[512]) is linear resulting into chromatids, but there is considerable diversity of opinion as to when it takes place. Baker[513] concluded that it occurs in prophase in *E. agilis* and Hollande[514] that it occurs at metaphase in *E. granulata.* However, Pickett-Heaps[515] reports the presence of kinetochores with attached microtubules in *E. gracilis,* showing that the mechanisms involved in chromosomal movement are not unusual as thought by Leedale.[509,516] A critical examination of many more species of *Euglena* is warranted from this angle.

The arrangement of chromosomes at metaphase is variable.[509] They may be arranged (1) in circlet of single chromatids by which time the separation of chromatids had already taken place, (2) in a circlet of chromosomes which subsequently duplicate into chromatids, or (3) to show orientation of pairs of chromatids into a metaphase circlet. In the last case, duplication into chromatids must have already taken place in the preceding telophase. These variations are met with even in different species of *Euglena: E. gracilis, E. acus,* and *E. viridis.* The arrangement may even be irregular as in *Eutreptia viridis.* The metaphase figure in most cases consists of chromatids lying along the division axis parallel to one another and to the elongating endosome. There is no true equatorial plate. Anaphase is marked by the movement of individual chromatids towards the poles. The migration is staggered and consequently anaphase is prolonged. During telophase, the chromosomes assume their interphase state.

The nuclear envelope persists throughout mitosis. Spindle elements are not seen at any time during mitosis. The movement of the daughter chromosome is thought to be autonomous, although electron microscopy suggests that spindle elements or microtubules of some sort are present in the nucleus. Although spindle elements are from pole to pole there is no evidence that these establish contact with the chromosomes. Further, there is no apparent relation between nucleoplasmic microtubules and the flagellar basal bodies. That the spindle is not normal is shown by the fact that all attempts by employing 4% colchicine remained unsuccessful to disrupt it. Even 30,000-rad X-rays did not do any damage to chromosomes, although this high dose proved lethal to the alga. The duration for various division phases in *Euglena gracilis* was 17 min for anaphase, 4 min for telophase, and 133 min for cell cleavage (longest duration). Several other euglenoids also have been studied from this viewpoint.[508] Thus, the unique features of euglenoid mitosis are (1) chromosomes lying parallel at middivision, (2) absence of centromeres, (3) intact nuclear membrane, and (4) nucleoplasmic microtubules forming a primitive spindle. The chromosomes are in a condensed state in interphase nuclei. However, ''chromosomes of Euglenophyceae consist of DNA, RNA, and proteins and are the repositories of the genetic information of the cell'' as is characteristic of eukaryotes, while those of Dinophyceae contain only DNA without association of histones characteristic of DNA-histone complexes characteristic of eukaryotes. Similarity is obvious between Euglenophyceae and Dinophyceae in their karyology in spite of differences in the chromatin structure. Thus, nuclear division in euglenoid flagellates is a ''mitotic process, albeit a peculiar one with many departures from the conventional mitosis.

As in Dinophyceae, it is difficult to obtain exact counts of chromosome numbers due to improper spreading and consequent overlapping.[505-522] In six taxa the actual counts ranged from 42 to 92, [509] while in others only estimates were given ranging from ca 15 to 177. Shashikala and Sarma[517] reported chromosome numbers of three Indian taxa of euglenoid flagellates. The counts of $n = 14$, $n = 45$, and $n = 95$ in *E. gracilis* [509,513,517] indicate the possibility of existence of polyploid strains. A low chromosome number of $n = 16$ is recorded in *E. limnophila.*[515] On the basis of microdensitometer studies employing feulgen staining, DNA replication is seen to terminate by the end of prophase.

A process of nuclear fragmentation erroneously interpreted as ''amitosis'' by some workers was shown by Leedale[508] to result in the formation of binucleate cells in *E. acus* and *E. spirogyra.* These half-products are capable of independent division resulting in two binucleate cells or occasionally into one uninucleate and one trinucleate cells. The fact that cell lineages containing reduced chromosome numbers (up to half) are maintained shows that the original nuclei should be high polyploids.

Meiosis occurs in *Phacus pyrum* [519] and *Hylophacus occlatus* [520] since the chromosomes present a bivalent-like configuration in certain nuclei. Such meiosis, a rare phenomenon, seems to follow autogamy. It may also occur probably in response to poor growth conditions. Leedale concludes, thus, ''In a taxonomically isolated and caryologically peculiar group such as euglenoid flagellates, especially where most species may be asexual clones, the possible combination of primitively high chromosome numbers, endopolyploidy, ancestral polyploidy with preserved anomalies of varying types, and the sudden loss or gain of half the total chromosome complement, would seem to have fair chance of defying cytological analysis for some time to come.''

VII. DINOPHYCEAE

The members of the Dinophyceae, commonly referred to as dinoflagellates, constitute a diverse and heterogenous group of microorganisms, included under Protozoa by zoologists and Algae by botanists. They stand at the crossroads where prokaryotes and eukaryotes part, which makes their study highly exciting.

Earlier studies from 1880 onwards described the structure and nuclear division of *Ceratium, Oxyrrhis,* but the events could not be properly interpreted.[523] With advancement in culture techniques, several marine dinoflagellates and subsequently fresh-water ones have been analyzed karyologically and their chromosome numbers determined.[525-574]

The karyology of Dinophyceae is marked by several unusual nuclear and chromosomal characteristics and drastic departures from the normal mitotic pattern (''ideosyncratic mitosis,'' to use the expression of Taylor[539]). While prokaryotes do not have conventional nuclei and chromosomes, the eukaryotes, in general, possess them. The dinophycean nucleus or the dinokaryon is unique in combining characteristics of both and, in addition, exhibiting its own. Such an extraordinary combination of nuclear characters renders the group of special evolutionary interest, serving to some extent to connect the gap between two levels of genetic organization. Dodge[529] coined the term "Mesokaryota" to include members of this group. However, the organization of the cell as a whole, barring the nucleus, is basically eukaryotic. Rizzo and Nooden[540-542] and Taylor[539] do not support Dodge's view for "Mesokaryota", considering the dinoflagellates as primitive eukaryotes. Loeblich[537] summarized the prokaryotic and eukaryotic features of dinoflagellates and is inclined to consider that the Dinophyceae occupies a primitive position in the eukaryotic line. The distinctive karyological features of Dinophyceae are (1) condensed state of chromosomes throughout the cell cycle including interphase; (2) variation in shapes of interphase nuclei: spherical, subspherical, ovoid, ellipsoid, triangular, crescent shaped, deeply U shaped, tetragonal, etc., coupled with variation in their situation in the cell; (3) nucleolus distinct (in *Gonyaulax tamarensis*,[527] large nucleolus-like bodies situated between the arms of the nucleus and external to the latter) or indistinct; (4) absence of visible centromeres in the chromosomes at any stage; (5) considerable departure from metaphase and anaphase arrangements of chromosomes during cell division as compared with higher eukaryotes; (6) absence of conventional spindle and staggered movement of chromatids; and (7) persistence of nucleolus (not in all) and nuclear envelope during mitosis.

Considerable variation exists in the size of the chromosomes. Some are characterized by very long and rod-shaped chromosomes, longest being *Prorocentrum micans* (15 μm)[529] followed by some chromosomes of *Gymnodinium dodgei* (10 μm[536]). On an average, the range of chromosome length in a majority of taxa is 1.5 to 5.5 μm. Within a karyotype most of the chromosomes appear remarkably uniform in length. The width range is 0.5 to 1.0 μm. However, no correlation is found between the chromosome length and chromosome number nor between initial size of interphase nucleus and the chromosome number.[545]

It is extremely difficult to arrive at exact chromosome numbers of various taxa. All the chromosome numbers reported give a range, the difference in lowest and highest estimates ranging from 2 to 20 (e.g., *Amphidinium* sp., 24 to 26, *Ceratium hirundinella*, 264 to 284,[529] *Gyrodinium pusillum*, 94 to 96, and *Gymnodinium dodgei*, 190 to 200[536]). While chromosome numbers are low in *Prorocentrum triestum* (20 to 27) and *Amphidinium* sp. (24 to 26), others such as *Woloszynskia hiemale* (216 to 224) and *Ceratium hirundinella* (264 to 284) are characterized by high chromosome numbers. Some parasitic dinoflagellates, on the other hand, bear very low chromosome numbers, e.g., *Syndinium* and relatives (Syndiniophyceae grouped along with Dinophyceae by Loeblich[537] under division Pyrrophyta has four to ten). Comparatively low numbers (18 to 69) are exhibited by desmokonts, also.[539] Some of the numbers appear to be double or triple the lower counts for the lineage, showing that ploidy changes may have occurred.[536]

Two taxa designated as *Gymnodinium* OBG 70[534] and *Gymnodinium* NBG 76[546] revealed an unusual organization. Chromosomes appear as small dots, an estimate of ca 600 and 514 ± 10 having been made for the two taxa, respectively. The nuclei are dinokaryotic in nature, on the basis of which *Gymnodinium* OBG 70 and NBG 76 have the smallest chromosome size with the highest chromosome numbers (ca 600 in the former and 514 ± 10 in the latter) known so far among the dinoflagellates barring *Endodinium chattonii* Hovassi (Zooxanthlales) in which 500 to 1000 was reported.[537]

Loeblich et al.[547] proposed to utilize nuclear organization in the classification of dinoflagellates. Shyam and Sarma[536] used nuclear and chromosomal features to provide additional characteristics to resolve interspecific differences within a genus. Physicochemical characterization of DNA and determination of molecular weights of chromosomal DNA were initiated by Allen et al.[548] and Roberts et al.,[549] respectively.

Zingmark[550] established two types of nuclei in Dinophyceae: dinokaryotic with condensed chromosomes at interphase and noctikaryotic, in which chromosomes seem to be relaxed, at least in interphase. While the first type is found in the vast majority of Dinophyceae, the latter seems to be characteristic of vegetative cells of *Noctiluca*. Ultrastructurally, the nuclear envelope is similar to the eukaryotic one, while in the noctikaryotic type, it is composed of annulated vesicles associated with a two-membraned outer envelope. Although the vegetative cells of *Noctiluca* exhibit noctikaryotic nucleus, the nuclei of gametes were dinokaryotic.[550] However, in *Gymnodinium fuscum*, the nuclei possess dinokaryotic chromosomes and a noctikaryotic nuclear envelope.[554] Thus, such a distinction between dinokaryotic and noctikaryotic nuclei seems to be not warranted. Binucleate cells were seen in *Glenodinium foliaceum*[552] and *Peridinium balticum*.[553] Of the two nuclei, one was dinokaryotic but the other was found to be eukaryotic. Tomas et al.[553] suggested that *P. balticum* obtained its eukaryotic nucleus after the invasion of an endosymbiont.

A unique feature of the dinokaryon is the lack of conventional histones in the chromosomes,[526,529,554] making the chromatin of Dinophyceae comparable to the genophores of bacteria.[555] There is no dispute that DNA is the major component of the dinophycean chromosomes.[548,549] Immunofluorescent techniques showed the presence of DNA-histone complex, though the protein content in association with DNA was very low and their exact nature was not known.[541]

Biochemical analysis of the chromatin of *Gymnodinium cohnii* and *P. trochoideum* shows that the chromatin is composed of DNA, RNA, and acid-soluble and acid-insoluble proteins as in higher plants and animals, but the amount of acid-soluble protein relative to DNA (0.02 to 0.08) is much lower than that of typical eukaryotes (about 1).[540, 541] Electrophoretic studies indicated that acid-soluble chromosomal proteins gave a banding pattern quite different than that of typical histones, the mobility being less than that of histone IV from corn.[592] It is a basic protein but differs from most histones. Further, it is present in log phase cells but absent in stationary phase cells. For these reasons, the major acid-soluble protein is probably not a histone. Biochemical studies employing gel electrophoresis show that the proteins of the chromatin from *Olisthodiscus luteus*, an eukaryotic flagellate, and the eukaryotic nuclei of *P. balticum* revealed four major bands which comigrated with H_2a, H_2b, H_3, and H_4 of calf thymus and a minor band which comigrated with H_1.[556,557] Thus, the dinokaryon does not contain histones, while the eukaryons of algal flagellates do contain histones identical with those of higher plants and animals. Loeblich[537] stated that the log phase cells have, on the average, twice as much DNA per cell as stationary phase cells, and that dinoflagellates have basic proteins although these proteins differ quantitatively and qualitatively from those of higher eukaryotes. Dodge[529] expressed the view that in dinoflagellates replication of DNA is perhaps continuous. But Loeblich,[537] on the basis of available experimental evidence, stated that there is "discrete" phase of DNA synthesis in the cell cycle.

Comparative ultrastructural study of the chromatin fibers from isolated lysed nuclei of *O. luteus* (Chrysophyta or chloromonad?), *Crypthecodinium cohnii* (uninucleate dinoflagellate as in most of the Dinophyceae), and *P. balticum* (an exceptional binucleate dinoflagellate, one nucleus being dinokaryotic and the other eukaryotic) was made.[558] Chromatin from nuclei of *Olisthodiscus* and eukaryotic nuclei of *P. balticum* presented repeating subunits similar to the size and morphology of mouse nucleosomes. Thus, the association of DNA with histones is obvious. On the other hand, the chromatin fibers from *Crypthecodinium* and dinokaryotic nuclei of *P. balticum* appeared as smooth threads without any indication of presence of nucleosome subunits, thereby showing that histones are not associated with DNA. Nuclear preparations containing mixtures of dino- and eukaryotic nuclei of *P. balticum* contained both types of fibers, smooth ones emanating from dinokaryotic and those with subunit structures (nucleosomes) from eukaryotic nuclei. This conclusively proves that DNA of dinokaryons is not associated with the conventional histones as in eukaryons.[537]

The opinion that the large differences in complexity between prokaryotes and eukaryotes may represent the extremes of an evolutionary continuum and the dinoflagellates may represent the transition forms between the two extremes with respect to evolution of nuclear organization in plants seems to be reasonable. On the basis of nuclear cytology and chromosome organization, Loeblich[537] considers three possibilities for the origin of dinoflagellates: (1) they might have diverged from higher eukaryotic lineage before evolution of eukaryotic chromatin but after the evolution of repeated DNA; (2) they diverged early and the repeated DNA arose separately in different groups; or (3) they represent degenerate eukaryotes that secondarily lost eukaryotic chromatin. He is more inclined to accept the first one to be more plausible. This highly interesting group of algal flagellates needs further intensive probe for a clearer understanding of evolutionary trends between prokaryotes and eukaryotes.

Ultrastructural studies of mitotic division have revealed[559,560] that it begins with the chromosomes splitting into pairs of chromatids. The Y- and V-shaped configurations suggest that splitting commences at one end and proceeds towards the other end. While the nucleus is enlarging, irregular invagination of the nuclear envelope penetrates the nucleus. Microtubules run through in these tunnels into the cytoplasm. This seems to be the "metaphase" condition. During "anaphase" the chromatids move towards opposite poles. As the division progresses the envelope remains intact although the tunnels are "drawn out". Telophase is recognized by the formation of two groups of daughter chromosomes being enveloped, as a result of constriction of the original nuclear envelope of the mother nucleus. Electron microscopy of *Gleonodinium*[529] revealed attachment of ends of chromosomes to the nuclear envelope and probably with the tunnel microtubules in *Amphidinium*.[561] Such terminal ends of chromosomes may represent the early beginnings of kinetochores. This association between ends of chromosomes and the nuclear envelope is comparable to the situation in prokaryotes where genophores are attached to the membrane and separation of daughter genophores is membrane mediated.

Nucleolus in *Oxyrrhis* is present throughout mitosis and divides like the "endosome" of Euglenophyceae during mitosis. Association between some chromosomes and nucleolus has been observed in *Gonyaulax* like the nucleolar-organizing chromosomes.

Ultrastructure of the chromosomes shows fibrillar nature of the chromosomes in all studies, diameter of fibrils, however, varying from 40 to 100 μm in different reports. [524,529] The variations in appearance (fibrillar, banded, condensed, or expanded) are attributed to variations in fixation procedure and to cultural conditions.[529]

Dinoflagellates *(Ceratium, Gymnodinium,* and *Woloszynskia)* are haploid in nature, meiosis occurring in the diploid zygote.[562,563] In marine forms, the zygote has a protracted period of growth (planozygote), while in fresh-water forms, a resting cyst is often formed (hypnozygote) and meiosis takes place in the hypnozygote. The occurrence of a nuclear cyclosis has been reported by von Stosch[564,565] to mark the beginning of meiosis, nuclear contents having been observed to rotate in the zygotic nucleus after the latter enlarging to maximum volume. This process seems to be related to synapsis. Pfiester[566] also recorded zygotic meiosis in *P. cinctum* f. *ovoplanum*. Zingmark,[550] however, considers the nucleus of *Noctiluca* as diploid, which, following meiosis, gives rise to gametes.

VIII. OTHER GROUPS OF ALGAL FLAGELLATES

Very little information is available on the algal flagellate groups such as the Cryptophyceae, Chrysophyceae, Chloromonadineae, and Haptophyceae.

A. Cryptophyceae

Thakur,[575] in 15 taxa of Cryptophyceae, 10 from fresh water and 5 from marine situations, recorded chromosome numbers ranging from ca 24 to ca 216, except in *Cyanophora paradoxa* in which a low chromosome number of 10 was recorded. His work contradicts the view of Hollande[576,577] that the cryptomonads possess a single long chromosome identified as metaphase later by Godward.[578] Ultrastructure of mitosis in the cryptophycean flagellate, *Chroomonas salina,* shows microtubules outside the nucleus before commencement of division.[579] By metaphase it is not possible to identify individual chromosomes which aggregate to constitute the dense plate. Spindle tubules pass through the plate or some may be attached to it. Two dense clumps of chromosomes separate at anaphase. Mitosis and cell division in *Cryptomonas* are more or less similar to that of *Chroomonas.*[580]

B. Chloromonadophyceae

In *Gonyostomum* chromatin of the interphase nuclei appeared as fine strands and partially condensed, a condition not very common for eukaryotic nuclei.[581, 582] Discontinuous staining of early chromosomes was evident. Metaphase chromosomes which were estimated to be 65 to 75 were spherical or rod shaped with no evident indication of centromeres. The orientation of anaphase chromosomes, however, showed localized centromeres. While Howasse[581] reported persistence of the nuclear envelope, Heywood[582] reported that it disappears during mitosis, but that it may have been destroyed by the fixative employed. Ultrastructural studies on some Chloromonadineae[583] and on *Vacuolaria* [584] also present contradictory data. While Mignot[583] records disappearance of the nuclear envelope, Heywood and Godward[584] state that "the nuclear envelope is probably entire throughout mitosis" in *Vacuolaria.* This feature is not constant in all members of the class. Chromosomes of *Vacuolaria* have simple localized kinetochores (0.25 to 0.34 μm thick) to which up to six tubules of intranuclear spindle are attached.[584]

C. Chrysophyceae

Chromosome numbers are not available for any member of Chrysophyceae. Slankis and Gibbs[585] investigated mitosis of *Ochromonas danica* at electron microscope level. Earlier to prophase, the basal bodies of two flagella replicate. The rhizoplast is attached to the basal body of long flagellum. During preprophase a daughter rhizoplast appears. During preprophase, two pairs of basal bodies, each with its accompanying rhizoplast and Golgi body, begin to separate. Extranuclear spindle microtubules proliferate outside the nucleus. Gaps appear in the nuclear envelope through which microtubules enter the nucleus and form the spindle. Spindle tubules are also seen directly into the rhizoplasts. Some tubules run from pole to pole and some are attached to the chromosomes. Kinetochores have not been observed. The nuclear envelope finally breaks down.

D. Haptophyceae

Haptophyceae was established[586] by separation from Chrysophyceae chiefly on the basis of "Haptonema". Chromosome counts are not available for any member of the class. At the ultrastructural level, mitosis in *Prymnesium parvum* showed that flagellar basal bodies act as centrioles at poles of the microtubular spindle which originate outside the nuclear envelope at prophase.[587] Mitosis is open, the nuclear envelope disappearing before metaphase.

E. Prasinophyceae

On the basis of ultrastructural studies, members of Chlorophyceae having tiny scales on the flagella have been separated to constitute a separate Class Prasinophyceae.[586] No chromosome counts are available. Mitosis of two flagellates, *Platymonas subcordiformis* and *Pyramimonas parkeae,* have been studied by Stewart et al.[588] and Pearson and Norris,[589] respectively. The mitotic behavior is different in both. While the nuclear envelope persists at anaphase in *Platymonas,* open mitosis is observed in *Pyramimonas.* The flagellar bases probably act as centrioles. In *Platymonas,* the extranuclear spindle also is seen arising from a granular region formed as a result of dissolution or dispersion of the rhizoplast. It was surmised that a large part of the rhizoplast is consumed in the formation of the spindle and that basal bodies do not participate in spindle formation. On the basis of a comparison of mitoses in *Platymonas* and *Chlamydomonas,* Stewart et al.[588] do not favor creation of a separate Class Prasinophyceae. Pickett-Heaps and Ott[590] investigated the ultrastructure of the nuclear cytology of *Pedinomonas minor,* another member of the Prasinophyceae. The nuclear envelope was found to be completely intact without any polar fenestrae, the spindle being a completely closed one. Basal bodies and flagellum remain at each pole throughout mitosis. Cytokinesis was by cleavage between widely separated daughter nuclei. Spindle was seen to persist with no trace of phycoplast. These cytological features indicate that the pediomonad cell type is indeed a primitive progenitor of other green algae and its tentative position is at the base of the phyogenetic tree.[594]

Ultrastructural details of the three Prasinophyceae considered above clearly show that this group at present represents a heterogenous assemblage of genera, some of very ancient stock *(Pediomonas),* which may well have been ancestral to all other green algae, even to the chlamydomonad cell type.

F. Eustigmatophyceae

Hibberd and Leedale[591] created a new class of algal flagellates separating genera from Xanthophyceae on the basis of the ultrastructure of cells. No information is available about their karyology.

IX. OVERALL VIEW AND CONCLUSIONS

The living organisms have been classified in various ways,[592] but the most recent attempt to classify them is based on the nature of genetic organization. According to this system, all living organisms have been grouped into Prokaryota and Eukaryota, besides others,[593] showing clear-cut basic differences in genetic organization. Blue-green algae have been shown to lack a nuclear organization, the DNA not being associated with histones. Thus, these algae are of prokaryotic nature like bacteria and are being referred to as Cyanobacteria;[594] in certain systems of classification blue-green algae are grouped with the bacteria.[595] The gap between the two levels of genetic organization (prokaryotic and eukaryotic) is astounding and attempts have been made to bridge it in recent years. The karyological studies on the Dinophyceae proved rewarding in this direction. The dinokaryon has a well-developed nucleus and chromosomes and undergoes mitosis. However, the DNA of the chromosomes is not complexed with the conventional histones of the eukaryotes, and the mitosis is an aberrant one not conforming to the pattern of eukaryotes. Further, the dinokaryon is unique in that the interphase nucleus contains chromosomes in highly condensed state. Dodge was tempted to treat Dinophyceae as a separate kingdom "Mesokaryota" to bridge the gulf between Pro- and Eukaryota. However, several workers, while not favoring such a treatment, would like to consider them as the most primitive eukaryotes.

The differences in the manner through which genome separation is achieved in the two kingdoms are striking. The general view regarding bacteria is that the genome is attached to the cell membrane which, in some mysterious way, is connected with genome separation. Information concerning this phenomenon in blue-green algae — the bacterial allies — is still not available. In Dinophyceae the nuclear membrane seems to be involved in some way in chromosome separation. In contrast to prokaryotes, the genome separation in eukaryotes is secured through the well-defined mechanism of mitosis. When and at what stage in the evolutionary history mitosis came into being is still a baffling question.

The above review on algal karyology of different groups clearly brings to the forefront the very wide range of variation that exists in the shape, size, and structure of the interphase nuclei. In Dinophyceae itself one comes across three types of nuclei: dinokaryotic, noctikaryotic, and eukaryotic. In *Peridinium balticum* the dinokaryon and eukaryon coexist within the same cell. While in *Noctiluca* the nucleus of the vegetative cell is noctikaryotic, that of the gamete is dinokaryotic.

The algal nuclei may be homogeneous showing fine granular appearance or thread-like network. They may or may not exhibit chromocenters. The nucleolus also presents a variety of structure both in number per nucleus and in complexity. The largest extent of variation in nucleolar structure is exhibited by the desmids. The members of Conjugales have revealed intimate relationship of nucleolus-organizing chromosomes with the nucleolus of interphase nucleus, while in some algae such relationship is only inferred. The smallest nuclei are met within some of the Chlorococcales, although interestingly enough, some of the members of Oocystaceae belonging to this order (e.g., *Oocystaenium*) have unusually large nuclei.

A wide range of chromosome structure has been established among algae. On the one hand, the chromosomes of the Charales, Oedogoniales, and Cladophorales are appreciably larger in size and possess localized centromeres. Although the chromosomes of Dinophyceae and Euglenophyceae are also larger in general, they do not exhibit any centromeres and are, thus, atypical. Again the appreciably longer chromosomes within certain taxa of filamentous Conjugales and desmids are polycentric in nature. Chromosomes of a larger majority of the Volvocales, Chlorococcales, Ulotrichales, Chaetophorales, Siphonales, and Bacillariophyceae are small, presenting a dot-shaped or rod-shaped appearance, the smallest being less than 0.25 μm in diameter. However, in each of the above orders, taxa of some genera are characterized by long chromosomes constituting their respective karyotypes. Asymmetric karyotypes with long and short/minute chromosomes are depicted by some species within each of the above orders.

Chromosome numbers so far recorded in algae present a wide spectrum ranging from $n = 2$ in certain taxa of *Spirogyra* to $n =$ ca 600 in a species of *Gymnodinium* and $n = 592$ in a clone of *Netrium digitus*. Among Volvocales, several Indian species of *Chlamydomonas* are reported to have chromosome numbers less than $n = 10$, as also some species of section Euvolvox. Although the Chlorococcales in general are characterized by low numbers, the Oocystaceae exhibit high numbers of chromosomes of very large size. No taxon has as yet been reported with more than $n = 46$ in the Ulotrichales and $n = 56$ in Chaetophorales. The Oedogoniales are interesting in that a majority of taxa of *Oedogonium* are characterized by odd numbers. The Cladophorales ($x = 6$) and Charophyceae ($x = 3, 7$) exhibit mostly euploid chromosome numbers, aneuploidy being exceptional in the former and not so common in the latter. Among Siphonales, the chromosome numbers seem to range from $n = 6$ to $n = 20$. While chromosome numbers in Dinophyceae are usually on the high side, only certain taxa of Euglenophyceae show high numbers. Information on chromosome numbers is fragmentary in Bacillariophyceae and Cryptophyceae, while no information is on record on members of Prasinophyceae, Chlamydomonadophyceae (except one report), Chrysophyceae, and Eustigmatophyceae.

Chromosome numbers within a particular order give an insight to probable lines of evolutionary diversification. Coupled with chromosome numbers, the other karyological features are also taken into consideration while making taxonomic evaluation and analyzing evolutionary trends. Cytotaxonomic studies achieved in algae have been briefly reviewed by the present author earlier.[5,6] It was shown therein that cytological data have proved of immense use in clarifying the disputed systematic position of some genera and species, besides being helpful in resolving taxonomic and phylogenetic problems at higher levels. Adequate evidence was presented to show the relative roles of euploidy and aneuploidy in the evolution of algal groups. A single species may be constituted by different cytological races which largely account for the prevalence of polymorphism in algae. Besides polyploidy, structural changes, gene mutations, and hybridization have also contributed to the evolution of a number of algal taxa.

Ultrastructural studies have contributed quite a lot in our understanding of interphase nuclei and mitosis, although much less of the chromosomes themselves except for the recent studies by Rizzo and Burgardt.[558] A wide range of variation exists in the presence or absence of the nuclear envelope, behavior of the nucleolus, variety of microtubule organizing centers, types of spindle microtubules, their origin and function, and finally the structure of the centromere. On the basis of the information available, certain evolutionary trends have been established.[515]

Studies on the interphase nuclei on the primary nucleus of *Acetabularia* and *Batophora* are the most outstanding in that they are shown to present the lamp-brush type of chromosome structure, on the basis of which an opinion has been gaining ground that meiosis takes place in the primary nucleus. Even a synaptenemal complex has been demonstrated in *Batophora.* Another noteworthy study is on the ultrastructure of the nucleolus of *Spirogyra britannica* and *S. ellipsospora* wherein an intimate relationship between the nucleolus and nucleolar chromosome was critically examined. There is one kind of 150-Å fibril inherent in the composition of denser nucleolar material and the nucleolus shows no evidence of having separate ribosomal particles, either free or connected to the fibrils. The less dense regions occur as large and smaller lacunae in the interphase nucleolus. The nucleolar chromosome occupies such lacunae. This structure of the nucleolus in *Spirogyra* is rather unusual as compared with vertebrate oocytes and *Vicia.* On the other hand, in *Crypthecodinium,* the nucleolus has been shown to possess an extensive inner fibrillar component and a peripheral component consisting of ca 25 μm granules which could be dissolved by ribonuclease, thus, conforming to the nucleolar structure of higher eukaryotes.

The idea of classical mitosis in eukaryotes has now been exploded in view of the ultrastructural studies made on representative algae belonging to different orders/ classes (though their number is still very few). These studies have shown that a variety of mitotic patterns of varying complexity exist with reference to the behavior of nuclear membrane, nucleolus, origin and orientation of microtubules, and kinetochore microtuble interactions. On the basis of extraordinary variation in the pattern of mitosis it has been shown by Dodge and Leedale that Dinophyceae and Euglenophyceae stand out distinctly from the rest of the algal groups. Within Chlorophyceae itself Pickett-Heaps[14] has drawn certain general conclusions on the basis of variation in mitosis and cytokinesis: (1) centrioles per se are not necessary for mitosis, although the absence of the centriole from vegetative cells does not mean that these cells have lost the ability to form these organelles; at the same time absence of centrioles in certain cases may be a sign of advancement; (2) closed spindles are primitive; (3) closed centric spindles (intranuclear mitosis) are more primitive; (4) furrowing is a primitive mechanism of cytokinesis; (5) cell plate formation is a comparatively advanced form of cytokinesis; (6) the phycoplast is a fairly primitive cytokinetic structure; and (7) the phragmoplast is an advanced cytokinetic system. However, these evolutionary lines do not go hand in hand with morphological complexity. Floyd et al.[596] remark that assessment of phylogenetic events based on mitotic events must be approached with great caution. In different strands of *Chlorella pyrenoidosa* the centrioles are reported to be present or absent. Difference has been observed with respect to the behavior of the nuclear envelope in different species of *Chlamydomonas.* In *C. moewusii* it persists fully, mitosis being completely intranuclear, while in *C. reinhardii* polar fenestrae are clearly observed at the poles. Similarly, different situations are reported in three taxa of Prasionophyceae. *Pedinomonas minor* exhibits a completely intranuclear division, nuclear envelope being fully intact without fenestrae. In *Platymonas subcordiformis* it shows polar fenestrae while in *Pyramimonas parkeae* mitosis is open. These three cases provide evolutionary steps in the evolution of open mitosis from closed mitosis. *Vaucheria* of Xanthophyceae and possibly *Vacuolaria* of Chloromonadophyceae are other examples of algae exhibiting closed mitosis. *Klebsormidium, Coleochaete,* some members of Conjugales and Charophyceae, and unicellular flagellated haptophycean alga *Prymnesium parvum,* to quote a few, are clear examples of open mitosis. Between the two extremes of closed and open mitosis lie a number of intermediary types in which the nuclear envelope progressively breaks down.[597]

The behavior of the nucleolus during mitosis likewise is highly variable in different algae. For example, members of Euglenophyceae are unique in that the nucleolus (endosome) is persistent throughout mitosis. On the other hand, in several algae the nucleolus undergoes complete dissolution by the end of prophase. Between these two extremities are cases where the nucleolus gets dispersed during mitosis but is associated with the dividing chromosomes as ''nucleolar substance''. Attachment of microtubules to nucleolar material (in which chromosomes are embedded) instead of being attached to the kinetochores directly has been observed to occur extensively in *Spirogyra.*[200] The nucleolus in the case of *Oedogonium,* although dispersed, remains amid the intranuclear spindle throughout the division as a loosely knit skein of granular material. Persistent nucleolar material seems to be present in *Chara,* also. Such a situation seems to prevail in a number of other algae than was originally thought of. In *Acetabularia wettsteinii* the nucleoli are extruded from the nucleus at anaphase.

There is a wide range of variation also in the microtubule organizing centers as revealed by electron microscopy. Centrioles seem to be such centers for the origin of spindle microtubules in several algae. In some cases centrioles may be present but they do not play such a role. In others, centrioles are totally absent. The origin of microtubules was also found to be associated with the rhizoplast *(Ochromonas, Platymonas),* flagellar basal bodies, spindle precursors, etc.

Microtubule-nuclear envelope associations have been frequently observed in some algae. In some Dinophyceae, cytoplasmic microtubules push into the ''tunnels'' of the nucleus. The nuclear envelope and microtubule association in *Oedogonium* is striking and probably helps in the nuclear-shaping process. Perinuclear microtubules in the secondary nuclei of *Acetabularia*[596] seem to serve as connections between the nuclear and plasma membranes. In several studies both continuous and chromosomal spindlemicrotubules have been found in algae which, however, are not met with in Dinophyceae and Euglenophyceae. In Cryptophyceae, while mitosis is of an open type, spindle microtubules run across the metaphase chromatin mass dividing the latter into several parts, although a few remain attached with the chromatin masses. Mitotic mechanisms in flagellates were recognized to have five major variations.[597]

Centromeres (kinetochores) exhibit wide variation in their presence or absence and, if present, in the extent of complexity. While presence of a localized centromere is the characteristic feature of most of the algal chromosomes (Oedogoniales, Cladophorales, Charales), polycentric chromosomes have been described in *Spirogyra* and some desmids. Electron microscopy shows that the longer chromosomes of *S. majuscula* possess two kinetochores each. Thus, Mughal and Godward consider the chromosomes of *Spirogyra* as polycentric rather than as with diffuse centromeres. Kinetochores have not been observed at all in Euglenophyceae and Dinophyceae nor was any association of microtubules directly with the chromosomes demonstrated. In most of the groups where the chromosomes are very short, the presence of centromeres could not be made out with light microscopy.

Ultrastructural studies, however, revealed a wide range in the organization of centromere from its indistinctness from the rest of the chromosomes to a very complex structure. For example, chromosomes of *Ulothrix, Stigeoclonium, Ulva,* and *Coleochaete* have not revealed any definite centromeric organization except that chromosomal microtubules establish contact with certain point of the chromosome or with the chromatin mass in general. On the other hand, in some algae such as *Oedogonium, Bulbochaete, Chlamydomonas,* and *Cladophora* the centromeres are stratified and complex with dark and light layers alternating with each other. The Conjugales possess less well-defined centromeres as in *Spirogyra majascula* and *Mougeotia,* besides being polycentric at least in the chromosomes of the former. The localized centromeres of *Acetabularia wettstenii* are also comparatively less complex. That the morphological complexity of the whole organism does not go hand in hand with complexity of internal structures is again reflected by the fact that while in the simple unicellular *Chlamydomonas,* complex stratified kinetochores are recorded, they are not distinct in the more highly evolved *Volvox.* The origin of the centromere itself is still intriguing. In *Syndinium,* a Dinophyceae alga, a special structure considered to be an early stage of the evolution of kinetochore[538] attaches the chromosome to the nuclear membrane. The terminal ends of chromosomes which are attached to the nuclear envelope in the dinoflagellates *Glenodinium* and *Amphidinium* may likewise represent primitive kinetochores.

On the basis of ultrastructural features revealed during mitosis and cytokinesis, Pickett-Heaps[14] believes that the pedinomonad cell type (as revealed by *Pedinomonas* of Prasinophyceae) is the architypical green cell rather the chlamydomonad type (as represented by *Chlamydomonas* of Chlorophyceae) as is generally believed, since *Pedinomonas* reveals more primitive characteristics of nuclear division than *Chlamydomonas,* the most important being the completely closed spindles and the persistance of the spindle between widely separated daughter nuclei, and cytokinesis taking place through furrowing without any trace of phycoplast. During evolutionary history, evolution of the phycoplast has taken place as could be seen in chlamydomonad cell type. Finally, phycoplast has given place to phragmoplast as the most highly evolved structure. It was commonly accepted that *Fritschiella* might have been the progenitor of land plants. But, on the basis of recent findings, it does not seem possible, as it contains a phycoplast instead of a phragmoplast. Discussing the similarities of the cytokinetic system of *Fritschiella* [598,599] with higher plants, Pickett-Heaps was of the opinion that in *Fritschiella* cytokinesis takes place by phycoplast and, thus, differs from *Coleochaete* and higher plants which are characterized by phragmoplast. At the most *Fritschiella* could be one of the most advanced members of the phycoplast-containing ulotrichalean algae. On the other hand, forms like *Coleochaete* (centric open spindles in mitosis and cleavage and phragmoplast in cytokinesis) and Charales (acentric open spindles at mitosis and cell plate formation and phragmoplast formation at cytokinesis) are favored as the closest links to bryophytes and higher land plants, as they contain a phycoplast instead of a phragmoplast. But to propose a classification of green algae based exclusively on the basis of phycoplast/phragmoplast, in addition to some minor characters,[601] would be carrying the classification too far.[602] Parallelism in evolution has been an accepted phenomenon in bringing about morphological organization in different classes of algae. Likewise, it is possible to comprehend that different patterns of mitosis, spindle organization, behavior of the nuclear envelope, and nucleolus during mitosis might also have evolved independently, not only in different classes but also within different species of the same genus or genera of the same class. It is perhaps an oversimplified view, but a larger number of representative genera will have to be examined before evaluating the variations in mitosis as of sufficient value in propounding classifications involving changes at higher levels.

While ultrastructural studies, no doubt, have contributed to better understanding of the interphase nucleus and nuclear division, they are not expected to say much about chromosome numbers or about the karyotype. The latter information will have to be obtained only with the help of the light microscope based on refined squash techniques. In order to comprehend fully the karyology of any alga, coordinated studies will have to be carried on at both levels simultaneously.

The last three decades have witnessed a spurt in studies on algal karyology, although it must be admitted that this field of investigation has as yet not attracted as many workers as it should have. All the knowledge accumulated only touches the fringe. Some of the algal groups have not been studied karyologically, even at the light microscope level, and only a single or at best a few species in any family or order have been studied at the ultrastructural level. The banding techniques extensively used for studying animal chromosomes and a few higher plants have not as yet been successfully applied to study the chromosome structure in algae. A breakthrough in this direction may prove profitable in defining interrelationships and evolution of algal species, at least in such cases where chromosomes are of appreciable size. There are several areas in algal karyology which need further intensive study, the more so because the limited studies have brought to the forefront the existence of considerable variability in behavior and structure of different mitotic and cytokinetic components of closely related genera and species.

ACKNOWLEDGMENTS

I wish to express my gratitude to Dr. Y. B. K. Chowdary, Dr. M. Khan, Dr. B. R. Chaudhary, Dr. S. B. Agrawal, Dr. S. C. Agrawal, and Dr. (Mrs.) P. Abhayavardhani for their sincere help and cooperation in the preparation of the manuscript. Thanks are also due to the University Grants Commission and Council of Scientific and Industrial Research, New Delhi, for financial support and to the authorities of Banaras Hindu University for facilities to carry out certain aspects of the work presented in this chapter, from our laboratory.

Finally, I have pleasure in dedicating this article to my teacher, Prof. M. B. E. Godward, M.Sc., Ph.D., F.L.S., F.R.M.S., Emeritus Professor, Department of Plant Biology and Microbiology, Queen Mary College, University of London, London (U.K.).

POSTSCRIPT

The chromosome numbers of algae belonging to various classes discussed in this paper have been published by the author in *The Nucleus* (25, 66—108, 1982), and provide supplementary reading.

REFERENCES

1. **Bold, H. C.**, Cytology of algae, in *Manual of Phycology,* Chronica Botanica, 1951.
2. **Tischler, G.**, *Handbuch der Pflanzenorotomie 2. Allgemeine Pflanzenkaryologie,* Berlin, 1951.
3. **Godward, M. B. E.**, *The Chromosomes of the Algae,* Godward, M. B. E., Ed., Edward Arnold, London, 1966.
4. **Sarma, Y. S. R. K.**, Some recent advances in the nuclear cytology of chlorophyceae, in *Proc. Symp. on Algology,* UNESCO-ICAR, New Delhi, 1959, 46.
5. **Sarma, Y. S. R. K.**, Cytology in relation to systematics of algae with particular reference to Chlorophyceae, *Nucleus (Paris),* 7, 127, 1964.
6. **Sarma, Y. S. R. K.**, Cytotaxonomy in algae, in *Glimpses in Plant Research, Modern Trends in Plant Taxonomy,* Vol. 5, Nair, P. K. K., Ed., Vikas Publishing House, New Delhi, 1980, 19.
7. **Sarma, Y. S. R. K.**, Cytology and cytotaxonomy of Indian Charophyta—a resume, *Nucleus (Paris),* 11, 128, 1968.
8. **Sarma, Y. S. R. K.**, Algal karyology in India, in *Advancing Frontiers in Cytogenetics, Hindustan Publishing, Delhi,* 1973, 266.
9. **Sarma, Y. S. R. K.**, On certain aspects of atypical mitosis in algae and their significance in the evolution of conventional mitosis, *J. Cytol. Genet.,* 9, 85, 1975.
10. **Sarma, Y. S. R. K.**, Chromosomes of algae, in *Recent Trends and Contacts between Cytogenetics, Embryology and Morphology,* UGC Sponsored Seminar Volume, Nagpur, 1977, 155.
11. **Sarma, Y. S. R. K.**, Two decades of algal cytology from Banaras School (Presidential address at the Annual Conference of the Society for Advancement of Botany, India), *Acta Botanica Indica,* 7, 1, 1979.
12. **Sarma, Y. S. R. K. and Chaudhary, B. R.**, Recent trends in algal cytology and cytogenetics, in *Current Trends in Botanical Research,* 139, Nagaraj, M. and Malik, C. P., Eds., Kalyani Publishers, New Delhi, 1980.
13. **Dodge, J. D.**, *The Fine Structure of Algal Cells,* Academic Press, New York, 1973.
14. **Pickett-Heaps, J. D.**, *Green Algae,* Sinaner Associates, Sunderland, Mass., 1975.
15. **Fritsch, F. E.**, *The Structure and Reproduction of the Algae,* Vol. 1, Cambridge University Press, New York, 1935.
16. **Dangeard, P. A.**, Mémoire sur les Chlamydomonadinées ou I histoire d'une cellule, *Botaniste,* 6, 65, 1899.
17. **Lewin, R. A.**, Ed., The genetics of algae, in *Botanical Monographs,* Vol. 12, Blackwell Scientific, Oxford, 1976.
18. **Schaechter, M. and De Lamater E.**, Studies on mitosis and meiosis in *Chlamydomonas, Trans. N.Y. Acad. Sci.,* 16, 371, 1954.
19. **Schaechter, M. and De Lamater, E.**, Mitosis in *Chlamydomonas, Am. J. Bot.,* 42, 417, 1955.
20. **Wetherell, D. F. and Krauss, D. W.**, Colchicine induced polyploidy in *Chlamydomonas, Science,* 124, 25, 1956.
21. **Buffaloe, N. P.**, A comparative cytological study of four species of *Chlamydomonas, Bull. Torrey Bot. Club,* 85, 157, 1958.
22. **Levine, R. P. and Folsome, C. E.**, The nuclear cycle of *Chlamydomonas reinhardii, Zeit. Verebungsl.,* 98, 192, 1959.
23. **Bischoff, H. W.**, Some observation on *Chlamydomonas microhalophila* sp. nov., *Biol. Bull.,* 117, 54, 1959.
24. **Abbas, A. and Godward, M. B. E.**, Chromosome numbers in some members of Chaetophorales, *Labdev. J. Sci. Technol.,* 3, 269, 1965.
25. **McVittie, A. and Davies, D. R.**, The location of the Mendelian linkage groups in *Chlamydomonas reinhardii, Mol. Gen. Genet.* , 112, 225, 1971.
26. **Loppes, R., Matague, R., and Strukert, P. J.**, Complementation at the Axg-7 locus in *Chlamydomonas reinhardii, Heredity,* 28, 239, 1972.
27. **Maguire, M.**, Mitotic and meiotic behaviour of chromosomes of the octet strain of *Chlamydomonas, Genetika,* 46, 479, 1976.
28. **Storms, R. and Hastings, P. J.**, A fine structure analysis of meiotic pairing in *Chlamydomonas reinhardii, Exp. Cell Res.,* 104, 39, 1977.
29. **Cave, M. S. and Pocock, M. A.**, Karyological studies in Volvocaceae, *Am. J. Bot.,* 38, 800, 1951.
30. **Cave, M. S. and Pocock, M. A.**, The variable chromosome number in *Astrephomene gubernaculifera, Am. J. Bot.,* 43, 122, 1956.
31. **Stein, J.**, A morphologic and genetic study of *Gonium pectorale, Am. J. Bot.,* 45, 664, 1958.
32. **Stein, J. R.**, A morphological study of *Astrephomene gubernaculifera* and *Volvulina steinii, Am. J. Bot.,* 45, 388, 1958.

33. **Rayns, D. G. and Godward, M. B. E.**, Cytology of colonial Volvocales of south-eastern England (abstr.), *Br. Phycol. Bull.*, 2, 102, 1961.
34. **Goldstein, M.**, Speciation and mating behaviour in *Eudorina, J. Protozool.*, 6, 249, 1964.
35. **Sarma, Y. S. R. K. and Shyam, R.**, On certain aspects of mitotic division in *Eudorina elegans* Ehrn., *Nucleus*, 14, 93, 1973.
36. **Sarma, Y. S. R. K. and Shyam, R.**, Studies on North Indian Volvocales. II. *Volvulina pringsheimii* Starr (Abstr.), *Int. Symp. Taxonomy Algae, University of Madras*, 36, 1974.
37. **Sarma, Y. S. R. K. and Shyam, R.**, On certain aspects of mitotic division in *Gonium pectorale, Nucleus (Paris)*, 18, 129, 1975.
38. **Patel, R. J.**, Morphology, reproduction and cytology of *Volvox prolificus* Iyengar from Gujarat-India, *J. Ind. Bot. Soc.*, 57, 28, 1978.
39. **McAllister, F.**, Nuclear division in *Tetraspora lubrica, Ann. Bot.*, 27, 681, 1913.
40. **Chowdary, Y. B. K., Suryanarayana, G., and Sarma, Y. S. R. K.**, A new species of *Tetraspora* (*T. apiocystioides* sp. nov.) and its cytology, *Hydrobiologia*, 30, 572, 1967.
41. **Sarma, Y. S. R. K. and Suryanarayana, G.**, Observations on morphology reproduction and cytology of *Tetrasporidium javanicum* Moebius from North India, *Phycologia*, 8, 171, 1969.
42. **Hastings, P. J., Levine, E. E., Cosbey, E., Huddock, M. O., Gillham, N. W., Suzycki, S. J., Loppes, R., and Levine, R. P.**, The linkage groups of *Chlamydomonas reinhardii, Microbiol. Gen. Bull.*, 23, 17, 1965.
43. **Levine, R. P. and Ebersold, W. T.**, The genetics and cytology of *Chlamydomonas, Ann. Rev. Microbiol.*, 14, 197, 1960.
44. **Singh, A. K.**, Cytological and Physiological Studies on Some Isolates of *Chlamydomonas*, Ph.D. thesis, Banaras Hindu University, Varanasi, India, 1981.
45. **Carefoot, J. R.**, Sexual reproduction and intercrossing in *Volvulina steinii, J. Phycol.*, 2, 150, 1966.
46. **Johnson, U. G. and Porter, K. R.**, Fine structure of cell division in *Chlamydomonas reinhardii, J. Cell Biol.*, 38, 403, 1968.
47. **Goodenough, U. W.**, Chloroplast division and pyrenoid formation in *Chlamydomonas reinhardii, J. Phycol.*, 6, 1, 1970.
48. **Treimer, R. E. and Brown, R. M., Jr.**, Cell division in *Chlamydomonas moewusii, J. Phycol.*, 10, 419, 1974.
49. **Pickett-Heaps, J. D.**, Cell division in *Tetraspora, Ann. Bot.*, 37, 1017, 1973.
50. **Pickett-Heaps, J. D.**, Some ultrastructural features of *Volvox* with particular reference to the phenomenon of inversion, *Planta*, 90, 174, 1970.
51. **Deason, T. R. and Darden, W. H., Jr.**, The male initial and mitosis in *Volvox*, in *Contributions in Phycology*, Parker, B. C. and Brown, R. H., Eds., Allen Press, Lawrence, Kan., 1971, 67.
52. **Treimer, R. E. and Brown, R. M., Jr.**, Ultrastructure of meiosis in *Chlamydomonas reinhardii, Br. Phycol. J.*, 12, 23, 1976.
53. **Starr, R. C.**, Sexuality in *Gonium sociale* (Dujardin) Warming, *J. Tenn. Acad. Sci.*, 30, 90, 1955.
54. **Starr, R. C.**, Meiosis in *Volvox carteri* f. *nagariensis, Arch. Protistenk.*, 117, 187, 1975.
55. **Belar, K.**, *Der Formwechsel der Protistenkeine, Gustav Fischer, Jena*, 1926.
56. **Kater, J. M.**, Morphology and division of *Chlamydomonas* with reference to the phylogeny of the flagellate neuromotor system, *Univ. Calif. Publ. Zool.*, 33, 125, 1929.
57. **Zimmermann, W.**, Zur Entiwicklungsgeschichte und Zytologie von *Volvox, Jahrb. Wiss. Bot.*, 60, 256, 1921.
58. **Chaudhary B. R.**, Studies on a new cytotype of *Eudorina elegans* Ehrn., *Cell Chromosome Newsl.*, 2, 2, 1979.
59. **Chaudhary, B. R. and Agrawal, S. B.**, Some observations on a new cytotype of *Eudorina elegans* Ehrenberg (Volvocales), *Microb. Lett.*, 13, 75, 1980.
60. **Hartmann, M.**, Die dauernd agame Zucht von *Eudorina elegans* Experimentelle Beitrage zum Befruchtung und Todproblem, *Arch. Protistenk.*, 43, 223, 1921.
61. **Doraiswami, S.**, On the morphology and the cytology of *Eudorina indica, J. Ind. Bot. Soc.*, 19, 113, 1940.
62. **Hovasse, R.**, Queleques donnees cytologique nouvelles sur *Eudorina illinoisensis* (Kofoid). Contribution a l'etude des Volvocales, *Bull. Biol. Fr. Belg.*, 71, 220, 1937.
63. **Merton, M.**, Über den bau und die Fortpflanzung von *Pleodorina illinoisensis* Kofoid, *Zeitschr. Wiss. Zool.*, 90, 445, 1908.
64. **Akins, V.**, A cytological study of *Carteria crucifera, Bull. Torrey Bot. Club*, 68, 429, 1941.
64a. **Bold, H. C. and Starr, R. C.**, A new member of the Phacotaceae, *Bull. Torrey Bot. Club*, 80, 178, 1953.
64b. **Shyam, R.**, Studies on North Indian Volvocales. VI. On the life cycle and cytology of a new member of Phacotaceae, *Dysmorphococcus sarmaii* sp. nov., *Can. J. Bot.*, 59, 726, 1981.
65. **Dangeard, P. A.**, Observations sur le development du *Pandorina morum, Botaniste*, 7, 193, 1900.

66. **Coleman, A. W.**, Sexual isolation in *Pandorina morum, J. Protozool.*, 6, 249, 1959.
67. **Timberlake, H. G.**, Development and structure of the swarmspores of *Hydrodictyon, Trans. Wis. Acad. Sci.*, 13, 486, 1902.
68. **Wanka, F.**, Ultrastructural changes during normal and colchicine inhibited cell division of *Chlorella, Protoplasma*, 66, 105, 1968.
69. **Proskauer, J.**, On the nuclear cytology of *Hydrodictyon, Hydrobiologia*, 4, 399, 1952.
70. **Sulek, J.**, Nuclear division in *Scenedesmus quadricauda*, in *Ann. Rep. Algolog. Lab. for 1968*, Trebon, Necas, J. and Lhotsky, O., Eds., Czeckoslovakian Academy of Science, 1969, 37.
71. **Sedova, T. B.**, A comparative cytological investigation of unicellular green algae. I. Certain specific features of mitosis in *Oocystis, Botanicheskii Zh.*, 54, 1997, 1969.
72. **Sedova, T. B.**, Certain results and prospects of comparative cytological investigation of unicellular green algae, *Botanischeskii Zh.*, 55, 947, 1970.
73. **Chan, K. Y.**, Nuclear cytology of the green algae (Chlorophyceae), *New Asia Coll. Acad. Annu.*, 15, 29, 1973.
74. **Chan, K. Y.**, Comparative nuclear cytology of *Coelastrum, Can. J. Bot.*, 52, 2365, 1974.
75. **Sarma, Y. S. R. K.**, Some observations on the karyology of *Hydrodictyon reticulatum* and the effects of colchicine on the alga, *Caryologia*, 15, 131, 1962.
76. **Chowdary, Y. B. K.**, Cytological observations on some chlorococcoid green algae, *Caryologia*, 20, 233, 1967.
77. **Patel, R. J. and Francis, M. A.**, Observation on morphology and cytology of *Characiosiphon rivularis* Iyengar, *Phykos*, 6, 91, 1967.
78. **Noor, M. N.**, Cytotaxonomic studies on *Hydrodictyon reticulatum* (L.) Lagerheim, *Res. J. Ranchi Univ.*, 9, 155, 1973.
79. **Mathew, T.**, Cytological and Physiological Observations on Some Chlorococcoid Green Algae, Ph.D. thesis, Banaras Hindu University, Varanasi, 1976.
80. **Rai, U. N.**, Cytological and Physiological Studies on Some Common Indian Chlorococcales, Ph.D. thesis, Banaras Hindu University, Varanasi, 1980.
81. **Sasikala, K.**, Cytological and Physiological Studies on Some Chlorococcoid Green Algae, Ph.D. thesis, Banaras Hindu University, Varanasi, 1980.
82. **Chowdary, Y. B. K. and Mathew, T.**, Cytological observations on some members of Oocystaceae, *Phykos*, 21, 1, 1982.
83. **Mainx, F.**, Untersuchungen über Ernahrung und Zellteilung bei *Eremosphaera viridis* De Bary, *Arch. Protistenk.*, 57, 1, 1927.
84. **Reichardt, A.**, Beitrage sur cytologie der Protisten, *Arch. Protistenk.*, 59, 301, 1927.
85. **Mathew, T. and Chowdary, Y. B. K.**, Comparative nuclear cytology of *Scenedesmus, Phykos*, 21, 19, 1982.
86. **Brunthaler, J.**, Protococcales, Die *Süsswasserflora Deutschlands, Oesterreishes und der Schweiz.*, Heft-5, Chlorophyceae, Pascher, A., Ed., 2, 52, 1915.
87. **Mainx, F.**, Gametencopulation und Zygotenkeimung bei *Hydrodictyon, Arch. Protistenk.*, 75, 502, 1931.
88. **Palik, P.**, Kernteilung bei *Hydrodictyon utriculatum* Roth, *Index Horti. Bot. Univ. Budapestensis*, 7, 150, 1949.
89. **Sinha, J. P. and Noor, M. N.**, Chromosome numbers in some members of Chlorophyceae of Chotanagpur (India), *Phykos*, 6, 106, 1967.
90. **Sahay, B. N. and Sinha, A. K.**, Cytological investigations in *Hydrodictyon reticulatum* Lageh, *Proc. 58th Ind. Sci. Congr.*, (Bot. Abstr.), 1971.
91. **Pickett-Heaps, J. D.**, Mitosis and autospore formation in the green alga *Kirchneriella lunaris, Protoplasma*, 70, 325, 1970.
92. **Pickett-Heaps, J. D.**, Variation in mitosis and cytokinesis in plant cells. Its significance in the phylogeny and evolution of ultrastructural systems, *Cytobios*, 5, 59, 1972.
93. **Pickett-Heaps, J. D.**, Cell division in *Tetraedon, Ann. Bot.*, 36, 693, 1972.
94. **Merchant, H. J. and Pickett-Heaps, J. D.**, Ultrastructure and differentiation of *Hydrodictyon reticulatum*. I. Mitosis of the Coenobium, *Aust. J. Biol. Sci.*, 23, 1173, 1970.
95. **Merchant, H. J.**, Mitosis, cytokinesis and colony formation in *Pediastrum boryanum, Ann. Bot.*, 38, 883, 1974.
96. **Merchant, H. J.**, Mitosis, cytokinesis and colony formation in the green alga *Sorastrum, J. Phycol.*, 10, 107, 1974.
97. **Atkinson, A. W., Jr., Gunning, A. E. S., John, P. C. L., and McCullough, W.**, Centrioles and microtubules in *Chlorella, Nature (London)*, 234, 24, 1971.
98. **Mathew, T. and Chowdary, Y. B. K.**, On the karyology of some chlorococcoid green algae, *Phykos*, 21, 10, 1982.
99. **Tschermark, E.**, Über Vierteilung und succedane Autosporenbildung als gesetzmassigen Vorgang, dargestellt on *Oocystis, Planta*, 32, 5, 1942.

100. **Bristol, B. M.,** On the life history and cytology of *Chlorochytrium grande* sp. nov., *Ann. Bot.*, 31, 107, 1917.
101. **Taümer, L.,** Morphologic, cytologic und Fortpflanzung von *Rhopalocystis oleifera* Schüssnig, *Arch. Protistenk.*, 104, 2, 1959.
102. **Smith, G. M.,** Cell structure and autospore formation in *Tetraedron minimum* (A. Br.) Hansg., *Ann. Bot.*, 32, 459, 1918.
103. **Yamanouchi, S.,** *Hydrodictyon africaunum*, a new species, *Bot. Gaz.*, 55, 74, 1913.
104. **Smith, G. M.,** Cytological studies in the Protococcales. I. Zoospore formation in *Characium seiboldii* A. Braun, *Ann. Bot.*, 30, 459, 1916.
105. **Rai, U. N. and Chowdary, Y. B. K.,** Some observations on the cytology of *Dimorphococcus lunatus* A. Braun (Chlorococcales), *Cell Chromosome Newsl.*, 3, 13, 1980.
106. **Peshkov, M. A. and Rodionous, G. V.,** Karyologicheskaya Kharakteristika razmuozheniya *Chlorella vulgaris, Dokl. Akad. Nauk. U.S.S.R.*, 154, 967, 1964.
107. **Sarma, Y. S. R. K.,** Contributions to the karyology of the Ulotrichales. III. *Microspora* Thuret., *Nucleus (Paris)*, 6, 49, 1963.
108. **Sarma, Y. S. R. K.,** Nuclear cytology of *Sphaeroplea annulina* (Roth) Ag. and its bearing on the systematic position of *Sphaeroplea, Cytologia*, 27, 72, 1962.
109. **Sarma, Y. S. R. K.,** Cytological and Cultural Studies of Some Members of Ulotrichales and Other Chlorophyceae, Ph.D. thesis, London University, London, 1958.
110. **Sarma, Y. S. R. K.,** Contributions to the karyology of the Ulotrichales. II. *Uronema* Lagh. and *Hormidium* Klebs, *Caryologia*, 16, 515, 1963.
111. **Sarma, Y. S. R. K.,** Contributions to the karyology of the Ulotrichales. IV. *Ulva* L., *Phykos*, 3, 11, 1964.
112. **Sarma, Y. S. R. K. and Chaudhary, B. R.,** An investigation on the cytology of *Ulva fasciata* Delile, *Bot. Mar.*, 18, 179, 1975.
113. **Sarma, Y. S. R. K.,** Contributions to the karyology of the Ulotrichales. V. *Enteromorpha* Link, *Phykos*, 9, 29, 1970.
114. **Sarma Y. S. R. K.,** Contributions to the karyology of the Ulotrichales. I. *Ulothrix, Phycologia*, 2, 173, 1963.
115. **Sarma, Y. S. R. K.,** Some observations on the morphology and cytology of *Draparnaldia plumosa, Rev. Algol.*, 7, 123, 1964.
116. **Jose, G. and Chowdary, Y. B. K.,** Karyological studies on *Cephaleuros* Kunze, *Acta Botanica Indica*, 5, 114, 1977.
117. **Pickett-Heaps, J. D.,** Cell division and wall structure in *Microspora, New Phytol.*, 72, 347, 1973.
118. **Sarma, Y. S. R. K. and Chaudhary, B. R.,** On a new cytological race of *Schizomeris leibleinii* Kütz., *Hydrobiologia*, 47, 181, 1975.
119. **Campbell, E. O. and Sarafis, V.,** *Schizomeris* — a growth form of *Stigeoclonium tenue, J. Phycol.*, 8, 276, 1972.
120. **Birkbeck, T. E., Stewart, K. D., and Mattox, K. R.,** The cytology and classification of *Schizomeris leibleinii* (Chlorophyceae). II. The structure of quadriflagellate zoospores, *Phycologia*, 13, 71, 1974.
121. **Abbas, A. and Godward, M. B. E.,** Cytology in relation to taxonomy of Chaetophorales, *J. Linn. Soc.*, 58, 499, 1964.
122. **Shyam, R. and Sarma, Y. S. R. K.,** Observations on the morphology, reproduction and cytology of *Stigeoclonium pascheri* (Vischer) Cox et Bold (Chaetophorales — Chlorophyceae) and their bearing on the validity of the genus *Caespitella* Vischer, *Hydrobiologia*, 70, 83, 1980.
123. **Chowdary, Y. B. K.,** On the cytology and systematic position of *Physolinum monilia* Printz., *Nucleus*, 6, 43, 1963.
124. **Dube, M. A.,** On the life history of *Monostroma fuscum* (Postels et Ruprecht) Wittrock, *J. Phycol.*, 3, 64, 1967.
125. **Ramanathan, K. R.,** On the cytological evidence for an alternation of generation in *Enteromorpha* (Pr. N.), *J. Ind. Bot. Soc.*, 15, 255, 1936.
126. **Ramanathan, K. R.,** The morphology, cytology and alternation of generations in *Enteromorpha compressa* (L.) Grev. var. *lingulata* (J. Agr.) Hank, *Ann. Bot. (London)*, 3, 375, 1939.
127. **Föyn, B.,** Untersuchugen über die Sexualitat und Entwicklung von Algen. IV. Mitteilung über die Sexualitat und den Generationswechsel von *Cladophora* und *Ulva, Ber. Dtsch. Bot. Ges.*, 47, 495, 1929.
128. **Föyn, B.,** Lebenzyklus und Sexualitat der Chlorophyceae *Ulva lactuca* L., *Arch. Protistenk.*, 83, 154, 1934.
129. **Braten, T. and Nordby, Ø.,** Ultrastructure of meiosis and centriole behaviour in *Ulva mutabilis* Föyn, *J. Cell Sci.*, 13, 69, 1973.
130. **Pickett-Heaps, J. D.,** Cell division in *Stichococcus, Br. Phycol. J.*, 9, 63, 1974.
131. **Floyd, G. L., Stewart, K. D. and Mattox, K. R.,** Cellular organisation, mitosis and cytokinesis in the ulotrichalean alga *Klebsormidium, J. Phycol.*, 8, 176, 1972.

132. **Pickett-Heaps, J. D.**, Cell division in *Klebsormidium subtilissimum* (formerly *Ulothrix subtilissima*) and its possible phylogenetic significance, *Cytobios*, 6, 167, 1972.
133. **Floyd, G. L., Stewart, K. D., and Mattox, K. R.**, Comparative cytology of *Ulothrix* and *Stigeoclonium*, *J. Phycol.*, 8, 68, 1972.
134. **Pickett-Heaps, J. D. and McDonald, K. L.**, *Cylindrocapsa:* cell division and phylogenetic affinities, *New Phytol.*, 74, 235, 1975.
135. **Marchant, H. J. and Pickett-Heaps, J. D.**, Mitosis and cytokinesis in *Coleochaete scutata*, *J. Phycol.*, 9, 461, 1973.
136. **Lovlie, A. and Braten, T.**, On mitosis in the multicellular alga *Ulva mutabilis* Foyn, *J. Cell Sci.*, 6, 109, 1970.
137. **Stewart, K. D., Mattox, K. R., and Floyd, G. L.**, Mitosis, cytokinesis, the distribution of plasmodesmata, and other cytological characters of Ulotrichales, Ulvales and Chaetophorales. Phylogenetic and taxonomic considerations, *J. Phycol.*, 9, 128, 1973.
138. **Fujiyama, T.**, On the life history of *Prasiola japonica* Yatbe, *J. Fac. Fisheries Anim. Husb.*, 1, 15, 1955.
139. **Friedmann, I.**, Structure, life-history and sex determination of *Prasiola stipitata* Suhr., *Ann. Bot. (London)*, 23, 571, 1959.
140. **Akhaury, K. D. N.**, Cultural, Cytotaxonomic Studies and Effect of Mutagenic Chemicals on Some Freshwater Green Algae (Chlorophyceae) of Ranchi (Bihar), Ph.D. thesis, Ranchi University, Ranchi, India, 1971.
141. **Schüssnig, B.**, Die mitotische Kernteilung bei *Ulothrix zonata* Kuetzing, *Z. Zellforsch.*, 10, 642, 1930.
142. **Shyam, R. and Saxena, P. N.**, Morphological and cytological investigations of *Ulothrix zonata* and taxonomy of the related species, *Pl. Syst. Evol.*, 135, 151, 1980.
143. **Lind, E. M.**, A contribution to the life-history and cytology of two species of *Ulothrix*, *Ann. Bot.*, 46, 711, 1932.
144. **Cholnoky, B. V.**, Planogonidien- und Gametenbildung bei *Ulothrix variabilis* Kg., *Beih. Bot. Centralbl.*, 49, 221, 1932.
145. **Cholnoky, B. V.**, Beitrage zur Kenntnis det Karyology von *Microspora stagnorum*, *Z. Zell Forsch.*, 16, 707, 1932.
146. **Neuenstein, H.**, Über den Bau Zellkernsbee den Algen und seine Bedentung fur ihre systematik, *Arch. F. Zell Forsch.*, 13, 1, 1914.
147. **Gremling, G.**, Sur la division cellulaire chez. *Microspora amoena* (Kütz.), *Bull. Soc. R. Botan. Belgique*, 72, 49, 1939.
148. **Das, R. N.**, Cytological and Cultural Studies of Some Members of Chaetophorales and Other Green Algae of Ranchi (Bihar), Ph.D. thesis, Ranchi University, Ranchi, India, 1968.
149. **Kostrun, G.** Entwicklung der Keinlunge der Polaritatsverhalten bei Chlorophycees, *Weiner Botanischen Z.*, 93, 172, 1944.
150. **Sarma, Y. S. R. K. and Suryanarayana, G.**, On the occurrence of *Enteromorpha intermedia* Bliding in India and its cytology, *Phykos*, 6, 100, 1967.
151. **Chaudhary, B. R.**, Cytological and Cytotaxonomic Investigations of Green Algae with Particular Reference to Ulotrichales, Ph.D. thesis, Banaras Hindu University, 1975, 152.
152. **Chaudhary, B. R.**, Nuclear cytology of *Enteromorpha compressa* (L.) Grev. from Indian water, *Bot. Mar.*, 22, 229, 1979.
153. **Carter, N.**, An investigation into the cytology and biology of the Ulvaceae, *Ann. Bot. (London)*, 40, 665, 1926.
154. **Yabu, H. and Tokida, J.**, Nuclear and cell dimensions in zoospore formation of *Ulva pertusa* Kjellman, *Bot. Mag. Tokyo*, 73, 182, 1960.
155. **Levan, A. and Levring, T.**, Some experiments on c-mitosis reactions with Chlorophyceae and Phaeophyceae, *Hereditas*, 24, 471, 1942.
156. **Linskens, H. F. and Vennegoor, C. J. G. M.**, Mitose und Meiose in Sporophyt von *Ulva mutabilis*, *Port. Acta Biol. Ser. A*, 10, 89, 1968.
157. **Nordby, Ø.**, Light microscopy of meiotic zoosporogenesis and meiotic gametogenesis in *Ulva mutabilis* Föyn., *J. Cell Sci.*, 15, 443, 1974.
158. **Chowdary, Y. B. K.**, A cytological race of *Uronema terrestre* Mitra, *J. Ind. Bot. Soc.*, 43, 249, 1964.
159. **Chaudhary, B. R.**, Some observations on the morphology, reproduction and cytology of the genus *Uronema* Lagh. (Ulotrichales, Chlorophyceae), *Phycologia*, 18, 299, 1979.
160. **Prasad, B. N. and Srivastava, P. N.**, Some observations on *Uronema gigas* Vischer, *J. Ind. Bot. Soc.*, 43, 113, 1964.
161. **Sarma, Y. S. R. K.**, Observations on the akinete formation and karyology of *Cylindrocapsa involuta* Reinsch., *Hydrobiologia*, 20, 373, 1962.
162. **Srivastava, S. and Sarma, Y. S. R. K.**, Cultural and cytological observations on *Cylindrocapsa geminella* Wolle, *Phykos*, 20, 49, 1981.

163. **Chowdary, Y. B. K.**, Observations on the cytology of *Cylindrocapsa scytonemoides* Randhawa, *Cytologia*, 28, 360, 1963.
164. **Chowdary, Y. B. K.**, A cytological study of *Schizomeris leibleinii* Kütz., *Rev. Algol.*, 8, 302, 1967.
165. **Prasad, B. N. and Srivastava, P. N.**, Observations on the morphology cytology and asexual reproduction of *Schizomeris leibleinii* Kütz., *Phycologia*, 2, 148, 1963.
166. **Patel, R. J.**, On morphology and cytology of *Schizomeris leibleinii* Kützing, *Phykos*, 6, 87, 1967.
167. **Chowdary Y. B. K.**, Further observations on cytological races in *Schizomeris leibleinii* Kutz., *Phykos*, 12, 5, 1973.
168. **Chaudhary, B. R.**, Karyology of an Indian strain of *Sphaeroplea annulina* (Roth) Ag., *Nova Hedwigia*, in press.
169. **Chaudhary, B. R. and Sarma, Y. S. R. K.**, Observation on certain aspects of reproduction and karyology of *Hormidium rivulare* Kützing and *Klebsormidium flaccidum* (Kütz.) Silva et al., *J. Ind. Bot. Soc.*, 58, 185, 1978.
170. **Singh, R. N.**, Nuclear phases and alternation of generations in *Draparnaldiopsis indica* Bharadwaja, *New Phytol.*, 44, 118, 1945.
171. **Noor, M. N.**, Cytotaxonomic and Cultural Studies of Common Members of Chlorophyceae of Chotanagpur, Ph.D. thesis, Ranchi University, Ranchi, India, 1965.
172. **Sinha, J. P. and Noor, M. N.**, Chromosome numbers in some members of Chlorophyceae of Chotanagpur plateau (India), *Phykos*, 6, 106, 1967.
173. **Singh, R. N.**, *Fritschiella tuberosa* Iyengar, *Ann. Bot. Lond.*, 11, 159, 1947.
174. **Chowdary, Y. B. K.**, Studies on Some Aspects of Cytology and Physiology of Common Indian Algae, Ph.D. thesis, Banaras Hindu University, Varanasi, 1962.
175. **Suematu, S.**, The somatic nuclear division in *Trentipohlia aurea*, the aerial alga, *Bull. Liberal Art. Coll. Wakayama Um. Nat. Sci.*, No. 10, 111, 1960.
176. **Jose, G. and Chowdary, Y. B. K.**, On the cytology of some *Trentepohlias* from India, *Acta Botanica Indica*, 6, 159, 1978.
177. **Sarma, Y. S. R. K. and Shashikala Jayaraman**, Karyological studies on certain taxa of *Stigeoclonium* and *Chaetophora* (Chaetophorales, Chlorophyceae), *Phycologia*, 19, 253, 1980.
178. **Abbas, A. and Godward, M. B. E.**, Chromosome numbers in Chaetophorales, *Labdev. J. Sci. Technol.*, 3, 269, 1965.
179. **Chowdary, Y. B. K.**, The chromosome numbers of some species of the genus *Stigeoclonium* Kuetz., *Cytologia*, 32, 174, 1967.
180. **Sinha, J. P. and Das, R. N.**, Cytological study of two species of *Chaetophora* Shrank, *Phykos*, 4, 74, 1965.
181. **Chowdary, Y. B. K.**, Cytology of *Trentepohlia* and *Cephaleuros*, in *Proc. Symp. Algol.*, UNESCO-ICAR, New Delhi, 1959, 65.
182. **Selby, C. H.**, Ph.D. thesis, London University, London, 1956.
183. **Geitler, L.**, Vergleichende Untersuchungen über den feineren Kern und Chromosomenbau der Cladophoraceen, *Planta*, 35, 530, 1936.
184. **Sinha, J. P.**, Cytotaxonomical studies on *Cladophora glomerata*, four freshwater forms, *Cytologia*, 32, 507, 1968.
185. **Sinha, J. P.**, Cytological and Cultural Studies of Some Members of Cladophorales and Oedogoniales, Ph.D. thesis, London University, London, 1958, 186.
186. **Shyam, R.**, On the life cycle, cytology and taxonomy of *Cladophora callicoma* from India, *Am. J. Bot.*, 67, 619, 1980.
187. **Wik-Sjostedt, A.**, Cytogenetic investigations in *Cladophora*, *Hereditas*, 66, 233, 1970.
188. **Schussnig, B.**, Gonidiogenese, Gametogenes und Meiose bei *Cladophora glomerata* (L.) Kuetz., *Arch. Protistenk.*, 100, 287, 1954.
189. **Sinha, J. P.**, Cytotaxonomical studies on *Cladophora flexuosa* (Griff.) Harv. a marine sp., *Cytologia*, 28, 1, 1963.
190. **Sinha, J. P.**, Cytotaxonomical studies on *Cladophora Hutchinsiae* Harv. and *C. refracta* Aresch., two marine spp., *Cytologia*, 30, 375, 1965.
191. **Patel, R. J.**, Ph.D. thesis, London University, London, 1961.
192. **Sinha, J. P. and Ahmad, S.**, A study on the cytology and life-history of *Cladophora uberrima* Lambert. from Ranchi, India, *Cytologia*, 38, 99, 1973.
193. **Sinha, J. P. and Ahmad, S.**, Cytological observations in three species of *Rhizoclonium* Kuetz., *Adv. Frontiers Cytogenet.*, p. 254, 1973.
194. **Balakrishnan, M. S.**, A cytotaxonomical investigation of Cladophoraceae, *J. Ind. Bot. Soc.* (Agarkhar Comm. Vol.), 40, 34, 1961.
195. **Patel, R. J.**, Cytotaxonomical studies on *Rhizoclonium* — *R. implexum* and *R. tortuosum*, *Sea Weed Res. Util.*, 1, 8, 1971.
196. **Prasad, B. N. and Vijaya Kumari**, On the cytology of three Indian species of *Pithophora*, 19, 186, 1980.

197. **Hartmann, M.**, Uber die sexualitat und Generationswechsel von *Chaetomorpha* und *Enteromorpha, Ber. Dtsch. Bot. Ges.*, 47, 485, 1929.
198. **Hanic, L. A.**, Life-History Studies on *Urospora* and *Codiolum* from Southern British Columbia, Ph.D. thesis, University of British Columbia, Vancouver, 1965.
199. **Sinha, J. P. and Verma, B. N.**, On the occurrence of amitotic nuclear division in *Pithophora cleveana* Wittr., *Phykos*, 7, 139, 1968.
200. **Mughal, S. and Godward, M. B. E.**, Kinetochore and microtubules in two members of Chlorophyceae, *Cladophora fracta* and *Spirogyra majuscula, Chromosoma*, 44, 213, 1973.
201. **McDonald, K. L. and Pickett-Heaps, J. D.**, Ultrastructure and differentiation in *Cladophora glomerata*. I. Cell division, *Am. J. Bot.*, 63, 592, 1976.
202. **Gajaria, S. C. and Patel, R. J.**, Taxonomical and karyological studies of freshwater Cladophorales of Gujarat., *J. Ind. Bot. Soc.*, 59 (Suppl.), 5, 1980.
203. **Verma, B. N.**,Chromosome numbers in four Indian species of *Cladophora* Kuetz., *J. Ranchi Univ.*, 4—5, 61, 1967—68.
204. **Chaudhary, B. R.**, Meiotic chromosome number in *Cladophora callicoma* Kuetzing, *Isr. J. Bot.*, 27, 212, 1978.
205. **Masaki, K.**, Cytological observations on *Cladophora speciosa* Saraki, *Bull. Fac. Fish. Hokkaido Univ.*, 29, 322, 1978.
206. **Soderstrom, J.**, Remarks on some species of *Cladophora* in the sense of van den Hoek and of Soderstrom, *Bot. Mar.*, 8, 169, 1965.
207. **Kornmann, P.**, Das Wachstum einer *Chaetomorpha* — Art von List/Sylt, *Helgoland. Wissen. Meeresunter*, 18, 194, 1968.
208. **Noor, M. N.**, Effects of colchicine on *Rhizoclonium implexum, J. Ranchi Univ.*, 3, 9, 1966.
209. **Venkataraman, G. S. and Natarajan, K. V.**, Propionocarmine squash technique and chromosome spreading in algae, *Stain Technol.*, 34, 233, 1959.
210. **Noor, M. N.**, A contribution to the cytology of three species of *Rhizoclonium* Kuetz. of Chotanagpur Plateau, *J. Ranchi Univ.*, 2—3, 61, 1963—64.
211. **Sinha, J. P. and Verma, B. N.**, Artificial induction of polyploidy in algae *Rhizoclonium hieroglyphicum* (Ag.) Kuetz. var. macromeres, *Ind. J. Sci. Ind. Res.*, 2, 99, 1968.
212. **Sinha, J. P. and Akhaury, K. D. N.**, Artificial induction of polyploidy in *Rhizoclonium hieroglyphicum, 57th Ind. Sci. Congr. Assoc.*, (Bot. Abstr.), 1970, 274.
213. **Prasad, B. N., Dutta, S., and Jain, V.**, Cytological studies in the genus *Rhizoclonium, Phykos*, 12 (1—2), 106, 1973.
214. **Verma, B. N.**, Cytological studies in four species of *Pithophora* Wittr., *Cytologia*, 44, 29, 1979.
215. **Noor, M. N.**, A new record on the cytology of *Pithophora oedogonia* Wittrock from India, *Phykos*, 7, 58, 1968.
216. **Sarma, Y. S. R. K. and Agrawal, S. B.**, Effects of gamma rays on the karyology of *Rhizoclonium hieroglyphicum* (Ag.) Kuetz., *Phykos*, 19, 190, 1980.
217. **Verma, B. N.**, Cytological and Cultural Studies on Some Members of the Order Cladophorales, Ph.D. thesis, Ranchi University, Ranchi, India, 1969.
218. **Van Wisselingh, C.**, Über Karyokinese bei *Oedogonium, Bot. Centralbl.*, 23, 157, 1908.
219. **Tschermak, E.**, Vergleichende und experimentelle cytologische Untersuchungen an der Gattung *Oedogonium, Chromosoma*, 2, 493, 1943.
220. **Henningsen, K.**, Chromosome numbers in five species of *Oedogonium, Phycologia*, 3, 29, 1963.
221. **Hoffman, L. R.**, Chromosome numbers in twelve species of *Oedogonium, Am. J. Bot.*, 54, 271, 1967.
222. **Sinha, J. P.**, Cytological studies on *Oedogonium cardiacum* Wittrock and one another *Oedogonium* sp., *Cytologia*, 28, 191, 1963.
223. **Srivastava, S. and Sarma, Y. S. R. K.**, Karyological studies on the genus *Oedogonium* Link (Oedogoniales, Chlorophyceae), *Phycologia*, 18, 228, 1979.
224. **Hesitschka-Janschka, G.**, Vergleichende Untersuchungen an haploiden und durch Colchicine in wirkung diploid geowordenen Stammen von *Oedogonium cardiacum, Österr. Bot. Zeitschr.*, 107, 194, 1960.
225. **Round, F. E.**, Taxonomy of Chlorophyta, *Br. Phycol. Bull.*, 2, 224, 1963.
226. **Round, F. E.**, The taxonomy of Chlorophyta. II, *Br. Phycol. Bull.*, 6, 235, 1971.
227. **Pickett-Heaps, J. D. and Fowke, L. C.**, Cell division in *Oedogonium*. I. Mitosis, cytokinesis and cell elongation, *Aust. J. Biol. Sci.*, 22, 857, 1969.
228. **Pickett-Heaps, J. D. and Fowke, L. C.**, Cell division in *Oedogonium*. II. Nuclear division in *O. cardiacum, Aust. J. Biol. Sci.*, 23, 71, 1970.
229. **Pickett-Heaps, J. D.**, Cell division in *Bulbochaete*, II. Hair Cell formation, *J. Phycol.*, 10, 148, 1974.
230. **Van Wisselingh, C.**, Zehnter Beitrag zur Kenntnis der Karyokinese. II. Die Methoden für die Untersuchung von Kernen und Kernteilungsfiguren, *Beih. Bot. Centralbl.*, 38, 286, 1921.

231. Srivastava, S. and Sarma, Y. S. R. K., Chromosome numbers in three taxa of Oedogoniales, *Cell Chromosome Newsl.*, 3, 32, 1980.
232. Agrawal, S. B., Effects of Radiations and Chemicals on Some Green Algae with Particular Reference to Their Cytology, Ph.D. thesis, Banaras Hindu University, Varanasi, 1980.
233. Ohashi, H., Cytological study of *Oedogonium, Bot. Gaz.*, 90, 177, 1930.
234. Das, R. N., The cytological studies on three species of *Oedogonium* Linn., from Ranch (Bihar), *J. Ind. Bot. Soc.*, 50, 175, 1971.
235. Kretschmer, H., Beitrage zur cytologie von *Oedogonium, Arch. Protistenk.*, 71, 101, 1930.
236. Chowdary, Y. B. K., Cytology of *Oedogonium terrestris* Randhawa, *J. Ind. Bot. Soc.*, 43, 253, 1964.
237. Sarma, Y. S. R. K. and Tripathi, S. N., Some observations on the effects of colchicine and maleic hydrazide on the karyology of a green alga *Oedogonium acmandrium* Elfving, *Phykos*, 12, 28, 1973.
238. Sarma, Y. S. R. K. and Singh, S. B., The induction of chromosome aberrations in *Oedogonium acmandrium* Elfving by ultraviolet radiation, *Acta Botanica Indica*, 2, 61, 1974.
239. Howard, A. and Horsley, R. H., Filamentous green algae for radiobiological study, *Int. J. Rad. Biol.*, 2, 319, 1960.
240. Upadhyaya, S. N., Cytological observations on *Oedocladium operculatum* Tiff., *Proc. 46th Ind. Sci. Congr.*, (Bot. Abstr.) 1959.
241. Chowdary, Y. B. K., Some cytological observations on *Oedocladium himalayense* Randhawa, *Nucleus*, 8, 59, 1965.
242. Strasburger, E., *Über Zellbidung und Zellteidung*, 2nd ed., Jena, 1875.
243. Geitler, L., Über die Kernteilung von *Spirogyra, Arch. Protistenk.*, 71, 10, 1930.
244. Geitler, L., Neue Untersuchungen über die Mitose von *Spirogyra, Arch. Protistenk.*, 85, 10, 1935.
245. Godward, M. B. E., Somatic chromosomes of Conjugales, *Nature*, 165, 653, 1950.
246. Godward, M. B. E., On the nucleolus and nucleolar organizing chromosomes of *Spirogyra, Ann. Bot. Lond. N.S.*, 14, 39, 1950.
247. Godward, M. B. E., Geitler's nucleolar substance in *Spirogyra, Ann. Bot. Lond. N.S.*, 17, 403, 1953.
248. Godward, M. B. E., The diffuse centromere or the polycentric chromosomes in *Spirogyra, Ann. Bot. Lond. N.S.*, 18, 143, 1954.
249. Godward, M. B. E., Irradiation of *Spirogyra* chromosomes, *Heredity*, 8, 293, 1954.
250. Godward, M. B. E., Cytotaxonomy of *Spirogyra*. I. *S. submargaritata, S. subechinata* and *S. britannica*, spp. *Novae, J. Linn. Soc. (Bot.)*, 55, 532, 1956.
251. Godward, M. B. E., Meiosis in *Spirogyra crassa, Heredity*, 16, 53, 1961.
252. Godward, M. B. E. and Newnham, R. E., Cytotaxonomy of *Spirogyra*. II. *S. neglecta* (Hass.) Kütz., *S. punctulata* Jao, *S. majuscula* (Kütz.) Czurda, *S. ellipsospora* Transeau, *S. porticalis* (Muller) Cleve, *J. Linn. Soc.*, 59, 99, 1965.
253. Prasad, B. N. and Godward, M. B. E., Cytological studies in the genus *Mougeotia, Br. Phycol. Bull.*, 2, 111, 1962.
254. Prasad, B. N. and Godward, M. B. E., Cytological studies in the genus *Zygnema, Cytologia*, 31, 375, 1966.
255. Vedajanani, K. and Sarma, Y. S. R. K., Karyological studies on Indian Conjugales. I. *Spirogyra* Link, *Phykos*, 17, 1, 1978.
256. Vedajanani, K. and Sarma, Y. S. R. K., Karyological studies on Indian Conjugales. II. *Zygnema* Agardh, *Phykos*, 17, 17, 1978.
257. Vedajanani, K. and Sarma, Y. S. R. K., Karyological studies on Indian Conjugales. III. *Mougeotia* Agardh, *Phykos*, 17, 27, 1978.
258. Vedajanani, K. and Sarma, Y. S. R. K., Karyological studies on Indian Conjugales. V. Desmids, *Phykos*, 17, 63, 1978.
259. Abhayavardhani, P. and Sarma, Y. S. R. K., Cytological and cytotaxonomic studies on the genus *Spirogyra* Link, *Cytologia*, in press.
260. Abhayavardhani, P. and Sarma, Y. S. R. K., Karyological studies on two taxa of *Sirogonium* Kützing (Conjugales, Chlorophyceae), *Caryologia*, 34, 351, 1981.
261. Abhayavardhani, P. and Sarma, Y. S. R. K., Karyological studies on some desmid taxa (Conjugales, Chlorophyceae), *Phycologia*, 21(2), 1981.
262. King, G. C., The cytology of desmids: the chromosomes, *New Phytol.*, 59, 65, 1960.
263. Nizam, J., Ph.D. thesis, London University, London, 1960.
264. Brandham, P. E., Ph.D. thesis, London University, London, 1964.
265. Brandham, P. E., Polyploidy in desmids, *Can. J. Bot.*, 43, 405, 1965.
266. Brandham, P. E., Some new chromosomes counts in the desmids, *Br. Phyc. Bull.*, 2, 451, 1965.
267. Kasprik, W., Beiträge zur Karyologie der Desmidiaceen-gattung *Micrasterias* Ag., *Beih. Nowa Hedwigia*, 42, 115, 1973.
268. King, G. C., The nucleoli and related structures in the desmids, *New Phytol.*, 58, 22, 1959.
269. Ashraf, M. and Godward, M. B. E., The nucleolus in telophase, interphase and prophase, *J. Cell. Sci.*, 41, 321, 1980.

270. **Godward, M. B. E. and Jordan, E. G.**, Electron microscopy of the nucleolus of *Spirogyra britannica* and *Spirogyra ellipsospora, J. R. Microsc. Soc.*, 84, 347, 1965.
271. **Gerrath, J. F.**, Giant chromosomes in *Triploceros, Br. Phycol. Bull.*, 3, 154, 1966.
272. **Tatuno, S. and Iiyama, I.**, Cytological studies on *Spirogyra*. I, *Cytologia*, 36, 86, 1971.
273. **Hoshaw, R. W. and Waer, R. D.**, Polycentric chromosomes in *Sirogonium melanosporum, Can. J. Bot.*, 45, 1167, 1967.
274. **Wells, C. V. and Hoshaw, R. W.**, The nuclear cytology of *Sirogonium, J. Phycol.*, 7, 279, 1971.
275. **King, G. C.**, Chromosome numbers in the desmids, *Nature*, 172, 592, 1953.
276. **Waris, H. and Kallio, P.**, Morphogenesis in *Micrasterias, Adv. Morphogenes*, 4, 45, 1964.
277. **Karsten, G.**, Die Entwicklung der Zygoten von *Spirogyra jugalis* Kuetz., *Flora*, 99, 1, 1908.
278. **Brandham, P. E. and Godward, M. B. E.**, Meiosis in *Cosmarium botrytis, Can. J. Bot.*, 43, 1379, 1965.
279. **Ling, H. U. and Tyler, P. A.**, Meiosis, polyploidy and taxonomy of the *Pleurotaenium mamillatum* complex (Desmidiaceae), *Br. Phycol. J.*, 11, 315, 1976.
280. **Fowke, L. C. and Pickett-Heaps, J. D.**, Cell division in *Spirogyra*. I. Mitosis, *J. Phycol.*, 5, 240, 1969.
281. **Bech-Hansen, C. W. and Fowke, L. C.**, Mitosis in *Mougeotia* sp., *Can. J. Bot.*, 50, 1811, 1972.
282. **Bokhari, F. S. and Godward, M. B. E.**, The ultrastructure of the diffuse kinetochore in *Luzula nivea, Chromosoma (Berlin)*, 79, 125, 1980.
283. **Pickett-Heaps, J. D. and Fowke, L. C.**, Mitosis, cytokinesis and cell elongation in the desmid, *Closterium littorale, J. Phycol.*, 6, 189, 1970.
284. **Pickett-Heaps, J. D.**, Scanning electron microscopy of some cultured desmids, *Trans. Am. Microsc. Soc.*, 93, 1, 1974.
285. **Pickett-Heaps, J. D.**, Cell division in *Cosmarium, J. Phycol.*, 8, 343, 1972.
286. **Merriman, M. L.**, Nuclear division of *Spirogyra*. II, *Bot. Gaz.*, 61, 311, 1916.
287. **Harada, A. and Yamagishi, T.**, A review of the recent studies on the Zygnemataceae, *Gen. Ed. Rev. Coll. Agric. Vet. Med. (Nihon Univ.)*, 8, 75, 1972.
288. **Van Wisselingh, C.**, Uber keruteilung bei *Spirogyra, Flora*, 87, 355, 1900.
289. **Doraiswami, S.**, Nuclear division in *Spirogyra, J. Ind. Bot. Soc.*, 25, 19, 1946.
290. **Newnham, R. A.**, *M.Sc. thesis, London University, London, 1962.*
291. **Moll, J. W.**, Observations on karyokinesis in *Spirogyra, Verb. d. Kon AK. V. Wetensch. te Amsterdam*, Abstr. Nat., 2, 1893.
292. **Van Wisselingh, C.**, Uber den nucleolus von *Spirogyra, Bot. Zeitg.*, Vol. 56, 1898.
293. **Allen, M. A.**, Ph.D. thesis, Indiana University, Bloomington, 1958.
294. **Berghs, J.**, Le noyau et la cinese *Spirogyra, Cellule*, 25, 53, 1906.
295. **Trondle, A.**, Uber die Reduktionsteilung in den Zygoten von *Spirogyra, Zeitschr. Bot.*, 3, 593, 1911.
296. **Sinha, J. P. and Noor, M. N.**, Chromosome numbers in some members of Chlorophyceae of Chotanagpur (India), *Phykos*, 6, 106, 1967.
297. **Prasad, B. N.**, Ph.D. thesis, London University, London, 1958.
298. **Patel, R. J. and Ashok Kumar, C. K.**, Morphological and cytological studies in *Zygnemopsis godwardense* sp. nov., *Phykos*, 10, 12, 1971.
299. **Chowdary, Y. B. K.**, Cytological and morphological observations on *Sirocladium vandalurense* in unialgal cultures, *Nucleus*, 11, 13, 1968.
300. **Nijam, J.**, Cytological studies in certain desmids, *Advancing Frontiers in Cytogenetics, A Collection of Papers in Honour of Prof. P. N. Mehra*, Hindustan Publishing, Delhi, 1974, 259.
301. **Karsten, G.**, Über die Tagesperiode der Kern und Zellteilung., *Z. Bot.*, 10, 1, 1918.
302. **Shashikala, J.**, Cultural and Cytological Studies on Eukaryotic Algae, Ph.D. thesis, Banaras Hindu University, Varanasi, India, 1979.
303. **Acton, E.**, Studies on nuclear division in desmids. I. *Hyalotheca dissiliens* (Sm.) Breb., *Ann. Bot.*, 30, 379, 1916.
304. **Potroff, H.**, Beitrage zur Kenntris de Conjugaten. I. Untersuchungen über die Desmidiaceae *Hyalotheca dissiliens* Breb. f. *minor, Planta*, 4, 261, 1927.
305. **Kopetzky-Rechtperg, O.**, Die Nucleolen in Kern der Desmidiaceae, *Beih. Bot. Centralbl.*, 49, 686, 1932.
306. **Maguitt, M.**, Karyokinese chez le *Penium., J. Soc. Bot. Reiss.*, 10, 415, 1925.
307. **Waris, H.**, Cytophysiological studies on *Micrasterias*. I. Nuclear and cell division, *Physiol. Plantarum*, 3, 1, 1950.
308. **Van Wisselingh, C.**, Über die Kernstructur und Kernteilung bei *Closterium, Beih. Bot. Centralbl.*, 29, 409, 1912.
309. **Schulze, K. L.**, Cytologische Untersuchungen aur *Acetabularia mediterranes* und *Acetabularia wettsteinii, Arch. Prostenkunde*, 92, 179, 1939.
310. **Werz, G.**, Über die Kernverhaltnisse der Dasycladaceen, besonders von *Cympolia barbata* (L.) Harv., *Arch. Prostenkunde*, 99, 198, 1953.

311. **Puiseux-Dao, S.**, Siphonales and Siphonocladales, in *The Chromosome of the Algae,* Godward, M. B. E., Ed., Edward Arnold, London, 1966.
312. **Chowdary, Y. B. K.** Cytological observations on two siphonaceous marine algae, *Bot. Mar.,* 13, 3, 1970.
313. **Chowdary, Y. B. K. and Singh, S. J.**, On the cytology of *Trichosolen mucronata* (Boergs.) Taylor (= *Pseudobryopsis mucronata* Boergs.), *Phykos,* 8, 28, 1969.
314. **Chowdary, Y. B. K., Singh, S. J., and Jose, G.**, On the cytology of *Bryopsis lamouroux* from Indian coast, *Phykos,* 21, 32, 1982.
315. **Singh, S. J. and Chowdary, Y. B. K.**, Some observations on the cytology of *Dichotomosiphon tuberosus* (A. Br.) Ernst, *Phykos,* 12, 8, 1973
316. **Singh, S. J. and Chowdary, Y. B. K.**, Some observations on the cytology of *Halicystis boergesenii* Iyengar and Ramanathan, *Phykos,* 13, 90, 1974.
317. **Singh, S. J. and Chowdary, Y. B. K.**, On the cytology of two species of *Dictyosphaeria* Decaisne, *Bot. Mar.,* 17, 161, 1974.
318. **Singh, S. J. and Chowdary, Y. B. K.**, Cytological observations on the material of *Boergesenia forbesii* (Harv.) Feldmann from the south and west coast of India, *Hydrobiologia,* 47, 469, 1975.
319. **Singh, S. J. and Chowdary, Y. B. K.**, Cytology of *Codium dwarkense* Boergs., *Phykos,* 21, 28, 1982.
320. **Singh, S. J.**, Cytology of Some Marine Siphonaceous Green Algae, Ph.D. thesis, Banaras Hindu University, Varanasi, India, 1973.
321. **Dao, S.**, Recherches caryologiques chez le *Neomeris annulata* Dickie, *Rev. Algol. (Nile Ser.),* 3, 192, 1958.
322. **Spring, H., Trendelenburg, M. F., Scheer, U., Franke, W. W., and Herth, W.**, Structural and biochemical studies of the primary nucleus of two green algal species, *Acetabularia mediterrania* and *Acetabularia major, Cytobiologie,* 10, 1, 1974.
323. **Spring, H., Scheer, U., Franke, W. W., and Trendelenburg, M. F.**, Lampbrush-type chromosomes in the primary nucleus of the green alga *Acetabularia mediterranea, Chromosoma,* 50, 25, 1975.
324. **Burr, F. A. and West, J. A.**, Light on electron microscope observation on the vegetative and reproductive structures of *Bryopsis hypnoides, Phycologia,* 9, 17, 1970.
325. **Burr, F. A. and West, J. A.**, Comparative ultrastructure of primary nucleus in *Bryopsis* and *Acetabularia, J. Phycol.,* 7, 108, 1971.
326. **Zinnecker, E.**, Reductionsteilung, Kernphasenwechsel and Geschlechtstimmung bei *Bryopsis plumosa* (Huds.) Ag., *Oesterr. Bot. Zeitsch.,* 84, 53, 1935.
327. **Borden, C. A. and Stein, J. R.**, Reproduction and early development in *Codium fragile* (Swingar) Hariot: Chlorophyceae, *Phycologia,* 8, 91, 1969.
328. **Williams, M. M.**, Cytology of the gametangia of *Codium tomentosum, Proc. Linn. Soc.,* 50, 91, 1925.
329. **Enomoto, S. and Hirose, H.**, On the life-history of *Anadyomene wrightii* with special reference to the reproduction, development and cytological sequences, *Bot. Mag.,* 83, 270, 1970.
330. **Godward, M. B. E., Beth, K., and Pacey, J.**, Nuclear division in the cyst and white spot nuclei preceding cyst formation in *Acetabularia wettsteinii, Protoplasma,* 101, 37, 1979.
331. **Davies, D. R. and Plaskitt, A.**, *John Innes Annual Report,* John Innes Horticultural Institution, Wisley, 1970, 64.
332. **Neumann, K.**, Der Ort der Meiosis und die Sporenbildung bei der siphonalen Grunalge *Derbesia marina, Naturwissenschaften,* 54, 121, 1967.
333. **Neumann, K.**, Beitrag zur Cytologie und Entwicklung der siphonalen Grunalge *Derbesia marina, Helgolander Wiss. Meeresunters,* 19, 355, 1969.
334. **Green, B. R.**, Evidence for the occurrence of meiosis before cyst formation in *Acetabularia mediterrania* (Chlorophyceae, Siphonales), *Phycologia,* 12, 233, 1973.
335. **Koop, H. U.**, The life-cycle of *Acetabularia* (Dasycladales, Chlorophyceae — a compilation of evidence for meiosis in the primary nucleus, *Protoplasma,* 100, 353, 1979.
336. **Franke, W. W. and Scheer, U.**, Structure and functions of nuclear envelope, in *The Cell Nucleus,* Vol. 1, Busch, H., Ed., Academic Press, New York, 1974.
337. **Scheer, U., Spring, H., and Trendenburg, M. F.**, Organisation of transcriptionally active chromatin of lamp-brush loops, in *The Cell Nucleus,* Vol. 7, Busch, H., Ed., Academic Press, New York, 1979.
338. **Liddle, L., Berger, S., and Schweiger, H.**, Ultrastructure during development of the nucleus of *Batophora oerstedii* (Chlorophyta, Dasycladaceae), *J. Phycol.* , 12, 261, 1976.
339. **Johnson, S. and Puiseux-Dao, S.**, Observations on morphologiques et caryologiques relatives a la reproduction chez le *Siphonocladus pusillus* (Kütz.) Hauck, Siphonocladacees, en culture, *C. R. Acad. Sci.,* 249, 1383, 1959.
340. **Steinmeyer, L. A. and Anderson, R. G.**, Effects of colchicine on *Protosiphon botryoides, J. Phycol. (Suppl.),* 8, 13, 1972.

341. **Bold, H. C.**, The life-history and cytology of *Protosiphon botryoides, Bull. Torrey Bot. Club*, 60, 241, 1933.
342. **Schüssnig, B.**, Der Generations und Phasenwechsel bei den Chlorophyceen. III, *Oesterr. Botan. Zeitschrift*, 18, 296, 1932.
343. **Schüssnig, B.**, Die Gametogenese von *Codium decorticatum* (Woods) Howe, *Svensk. Bot. Tidskr.*, 44, 55, 1950.
344. **Iyengar, M. O. P. and Ramanathan, K. R.**, On the life-history and cytology of *Microdictyon tenuis* (Ag.) Decsne., *New Phytol.*, 20, 157, 1941.
345. **Puiseux-Dao, S.**, Les Acetabulaires, materiel de laborateire. Les resultats obtenus avec ces Chlorophycees, *Annee Biol.*, 2(3—4), 100, 1963.
346. **Schüssnig, B.**, Der Kernphasenwechsel von *Valonia utricularis* (Roth) Ag., *Planta*, 28, 43, 1938.
347. **Schechner-Fries, M.**, Der Phasenwechsel von *Valonia utricularis* (Roth) Ag., *Oesterr. Botan. Zeitschrift*, 83, 241, 1934.
348. **Khan, M. and Sarma, Y. S. R. K.**, Studies on cytotaxonomy of Indian Charophyta. II. *Nitella, Phykos*, 6, 48, 1967.
349. **Khan, M. and Sarma, Y. S. R. K.**, Studies on cytotaxonomy of Indian Charophyta. I. *Chara, Phykos*, 6, 36, 1967.
350. **Khan, M. and Sarma, Y. S. R. K.**, Some observations on the cytology of Indian Charophyta, *Phykos*, 6, 62, 1967.
351. **Khan, M.**, Studies on Cytology and Cytotaxonomy of Indian Charophyta, Ph.D. thesis, Banaras Hindu University, Varanasi, 1966.
352. **Ramjee**, Contributions to the Systematics, Cytology and Cytotaxonomy of Indian Charophyta, Ph.D. thesis, Banaras Hindu University, Varanasi, 1968.
353. **Chatterjee, P.**, An Analysis of the Structure and Behaviour of Chromosomes as an Aid in the Taxonomy of Certain Higher and Lower Groups of Plants and the Cytological Effects of Certain Chemical and Physical Agents in the Latter, Ph.D. thesis, University of Calcutta, Calcutta, 1971.
354. **Kanahori, T.**, Cytotaxonomical research on the Characeae: karyotype on the section *Nitella* of the genus *Nitella, Bot. Mag. Tokyo*, 84, 327, 1971.
355. **Kasaki, H. and Kanahori, T.**, Karyotype of *Nitella inokasiraensis, Kromosoma*, 69—70, 2267, 1967.
356. **Kanahori, T.**, The karyotypes of two species of Characeae, *Chromosome Inf. Serv.*, 16, 29, 1974.
357. **Ramjee and Sarma, Y. S. R. K.**, Some observations on the morphology and cytology of Indian Charophyta, *Hydrobiologia*, 37, 367, 1971.
358. **Khan, M. and Sarma, Y. S. R. K.**, Cytogeographic study of Charophyta with particular reference to India, *Proc. Symp. Recent Adv. Crypto. Bot. Geophytol. Suppl.*, 55, 1981-82.
359. **Sarma, Y. S. R. K. and Ramjee**, Significance of chromosome numbers in Charophyta — a discussion, *Caryologia*, 24, 391, 1971.
360. **Noor, M. N. and Mukherjee, S.**, On the aneuploid chromosome numbers in *Chara hydropitys* Reich. from India, *Cytologia*, 40, 803, 1975.
361. **Noor, M. N. and Mukherjee, S.**, Some new records of chromosome numbers in Indian Charophyta, *Cytologia*, 42, 227, 1977.
362. **Tindall, D. R. and Sawa, T.**, Chromosomes of the Characeae of Woods Hole (Massachusetts), *Am. J. Bot.*, 51, 943, 1964.
363. **Ramjee and Bhatnagar, S. K.**, Significance of a new chromosome number in *Nitella mirabilis* Nordst. ex Grove. em RDW, *Hydrobiologia*, 57, 99, 1978.
364. **Sarma, Y. S. R. K., Khan, M., and Ramjee**, A cytological approach to phylogeny, interrelationships and evolution in Charophyta, *Indian Biol.*, 2, 11, 1970.
365. **Desikachary, T. V. and Sunderalingam, V. S.**, Affinities and interrelationships of the Characeae, *Phycologia*, 2, 9, 1962.
366. **Tuttle, A. H.**, The reproductive cycle of the Characeae, *Science*, 60, 412, 1924.
367. **Tuttle, A. H.**, The location of the reduction division in the Charophyte, *Univ. Calif. Publ. Bot.*, 13, 227, 1926.
368. **Oehlkers, F.**, Beitrage zur kenntnis der Kernteilungen bei de Charazeen, *Ber. Deutsch. Bot. Ges.*, 34, 222, 1916.
369. **Moutschen, J.**, Data on the nuclear structure of *Chara vulgaris* during morphogenesis and in relation to gene activity amplification phenomenon, *Rev. Cytol. Biol. Veg.*, 40, 125, 1977.
370. **Wood, R. D. and Imahori, K.**, *A Revision of the Characeae*, J. Cramer, Weinheim, Germany, 1965.
371. **Sarma, Y. S. R. K. and Khan, M.**, Dioecism and monoecism as taxonomic criteria in Charophyta, *Curr. Sci.*, 36, 245, 1967.
372. **Proctor, V. W.**, Taxonomic significance of monoecism and dioecism in the genus *Chara, Phycologia*, 10, 299, 1971.
373. **Sinha, J. P., Noor, M. N., and Srivastava, O. N.**, A review of karyological studies on chlorophycean members of Bihar, *Frontiers Plant Sci. (Prof. P. Parija felicitation volume)*, p. 123, 1977.

374. **Pickett-Heaps, J. D.**, The behaviour of the nucleolus during mitosis in plants, *Cytobios*, 6, 69, 1970.
375. **Telezynski, H.**, Garnitures des chromosome et synchronisme des divisions du genre *Chara* Vaill., *Acta. Soc. Bot. Poloniae*, 6, 230, 1929.
376. **Guerlesquin, M.**, Contribution a l'etude chromosomiques des charophycees d' Europe occidentale et d'Afrique du Nord. II, *Rev. Gen. Bot.*, 70, 355, 1963.
377. **Peshkov, M. A., Mirsaidov, T. I., and Tikhomirov, L. A.**, Etude caryologique de quelques especès d'algues du genre *Chara* de l'Uzbekistan, *Izvest. Akad. Nauk S.S.S.R. Ser. Biol.*, 5, 672, 1973.
378. **Rodrigues, M.**, Sobre o localizacao da meiose no ciclo da vida das Characeae, *Bot. Soc. Broteriana*, 19, 609, 1945.
379. **Geitler, L.**, Über die Teilungsrhythmen in den spermatogenen Faden der Characeen, *Osterr. Bot. Zeitschr.*, 95, 147, 1948.
380. **Karling, J. S.**, Nuclear and cell division in the antheridial filaments of the Characeae, *Bull. Torrey Bot. Club*, 55, 11, 1928.
381. **Lindenbein, W.**, Beitrag zur Cytologie der Charales, *Planta*, 4, 436, 1927.
382. **Corollion, R.**, Les Charophycees de France et d'Europe Occidentale These, Toulose et aussi, *Bull. Soc. Sci. Bret*, 32, 1, 1955.
383. **Guerlesquin, M.**, Contribution a l'etude chromosomique des charophycees d'Italic peninsulaire, *Rev. Gen. Bot.*, 71, 282, 1964.
384. **Guerlesquin, M.**, Observations chromosomiques sur les Characées due sud Tunisien, *Bull. Soc. Phycol. Fr.*, 22, 60, 1977.
385. **Mirsaidov, T. I.**, Nombre chromosomique de quelques especès de Characées d'ouzbekistan (in Russian), *Uzbekbiol. Zh.*, 15, 59, 1971.
386. **Peshkov, M. A.**, Mirsaidov, T. I., and Tikhomirov, L. A., Etude des nombres chromosomiques et des caryotypes approximatifs de quelques algues du genre *Chara* en l Uzbekistan, *Izvest. Akad. Nauk S.S.S.R. Ser. Biol.*, 13, 114, 1974.
387. **Sato, D.**, The protokaryotype and phylogeny in plants, *Sci. Pap. Coll. Gen. Ed. Um. Tokyo*, 9, 303, 1959.
388. **Moutschen, J. and Dahmen, M.**, Sur les modifications de la spermiogenese de *Chara vulgaris* L. induites par les rayons-X, *Rev. Cytol. Biol. Veg.*, 17, 433, 1956.
389. **Gillet, C.**, Nombres chromosomiques de plusieurs especes de charophycees (Generes *Nitella* et *Chara*), *Rev. Cytol. Biol. Veg.*, 20, 229, 1959.
390. **Sarma, Y. S. R. K. and Khan, M.**, A preliminary survey on the chromosome numbers of Indian Charophyta, *Nucleus*, 8, 33, 1965.
391. **Hotchkiss, A. T.**, Some chromosome numbers in Kentucky Characeae, *Trans. Ky. Acad. Sci.*, 19, 14, 1958.
392. **Williams, J. T. and Tindall, D. R.**, Chromosome numbers for species of Characeae from southern Illinois, *Am. Midl. Nat.*, 93, 330, 1975.
393. **Hotchkiss, A. T.**, *A Revision of the Characeae*, Vol. 1, Wood, R. D. and Imahori, K., Eds., J. Cramer, Weinheim, Germany, 1965, Vol. 1, 93.
394. **Schmuker, Th.**, Über Bildung sanomakien bei *Chara*, *Planta*, 4, 780, 1927.
395. **Hasitschka-Jenschke, G.**, Beitrag zur Karyology von Characeen, *Oesterr. Bot. Zeit.*, 107, 228, 1960.
396. **Guerlesquin, M.**, Recherches caryotypiques et cytotaxonomiques ches les Charophycees d'Europe occidentale et d'Afrique du Nord., *Bull. Soc. Sci. Bret.*, 41, 1, 1967.
397. **Ramjee and Bhatnagar, S. K.**, Studies on Charophyta from Rohilkhand division. I. Moradabad: taxonomic enumeration and chromosome counts, *Phykos*, 17, 87, 1978.
398. **Gonclaves da Cunha, G. A.**, Contribuicao para estudo des carofitos portugueses, *Trab. Inst. Bot. Lisboa*, 6, 136, 19-41—1942.
398a. **Mendes, E. J.**, Mitosis in the spermatogenous threads of *Chara vulgaris* var. *longibracteata* Kuetz., *Port. Acta Biol.*, 1, 251, 1946.
399. **Delay, C.**, Observations cytologiques sur les Characées. I. L'evolution du noyan pendant la spermiogenese de *Chara vulgaris* L., *Rev. Cytol. Biol. Veg.*, 11, 315, 1949.
400. **Guerlesquin, M.**, Contribution a' l'etude chromosomique des Charophycees d'Europe occidentale et d'Afrique du Nord-I, *Rev. Gen. Bot.*, 68, 360, 1961.
401. **Guerlesquin, M.**, Some cytological and cytotaxonomical observations of European and North African Charophyta, *3rd Ind. Geobot. Conf. Lucknow*, in press.
402. **Bhattacharya, S. S.**, Cytological studies on some Libyan charophytes, *Libyan J. Sci.*, 2, 1, 1972.
403. **Delay, C.**, Nombres chromosomiques chez les cryptogames, *Rev. Cytol. Biol. Veg.*, 14, 62, 1953.
404. **Ernst, A.**, *Bastardierung als Ursache der Apogamie in Pflanzen-reich.*, Jena, 1918.
405. **Strasburger, E.**, Einiges über Characeen und Amitoses, *Linsbauer Wiesner. Festschr. Vienna*, p. 24, 1908.
406. **Riker, A. J.**, Chondriosomes in *Chara*, *Bull. Torrey Bot. Club*, 48, 141, 1921.
407. **Debski, B.**, Beobachtungen über Kernteilung bei *Chara fragilis* Pringsh., *Jahrab Wissensch. Bot.*, 30, 227, 1897.

408. **Sasaki, M.**, Cytological studies in Charophyta with special reference to synchronous mitosis in antheridial filament, *St. Paul Rev. Sci. Nat. Sci.*, 8—9, 1, 1961.
409. **Chatterjee, P.**, Some additions to the charophytes of West Bengal, *Bull. Bot. Soc. Bengal India*, 29, 105, 1975.
410. **Sinha, J. P. and Noor, M. N.**, Studies on the karyology of some members of Charophyta, *Phykos*, 10, 112, 1971.
411. **Sarma, Y. S. R. K. and Khan, M.**, On two new charophytes from India, *Hydrobiologia*, 30, 405, 1967.
412. **Ahmad, S. and Sinha, J. P.**, Chromosome numbers in some taxa of *Chara*, *Proc. Ind. Sci. Congr. Assoc.*, p. 302, 1973.
413. **Sarma, Y. S. R. K. and Khan, M.**, Chromosome numbers in some Indian species of *Chara*, *Phycologia*, 4, 173, 1965.
414. **Sundaralingam, V. S.**, The cytology of spermatogenesis in *Chara zeylanica* Willd., *J. Ind. Bot.* Soc. (M.O.P. Iyengar Comm. Vol.) p. 289, 1946.
415. **Griffin, D. G., III and Proctor, V. W.**, A population study of *Chara zeylanica* in Texas, Oklahoma and New Mexico, *Am. J. Bot.*, 51, 120, 1964.
416. **Griffin, D. G., III**, The taxonomy of *Chara zeylanica* Klein ex Willd., *M.S. Dactylographie*, 1, 103, 1965.
417. **MacCracken, M. D., Proctor, V. W., and Hotchkiss, A. T.**, Attempted hybridization between monoecious and dioecious clones of Chara, *Am. J. Bot.*, 53, 937, 1966.
418. **Proctor, V. W. and Hotchkiss, A. T.**, Attempted hybridization between octo- and tetraseutate clones of monoecious-conjoined *Chara* (series *Gymnobasalis*), *M.S. Dactylographie*, 1, 1967.
419. **Guerlesquin, M.**, Recherches sur *Chara zeylanica* Klein ex Willd. Charophycées) d'Afrique Occidentale, *Rev. Algol.*, 10, 231, 1971.
420. **Sinha, J. P. and Verma, B. N.**, Cytological analysis of charophytes of Bihar, *Phykos*, 9, 92, 1970.
421. **MacDonald, M. B. and Hotchkiss, A. T.**, An estipulodic form of *Chara australis* R.Br. (= *Protochara australis*) Woms. & Ophel., *Proc. Linn. Soc. NSW*, 80, 274, 1956.
422. **Noor, M. N.**, A preliminary report on the chromosome numbers of some Indian Characeae, *Proc. Ind. Sci. Congr. Assoc.*, p. 387, 1969; *J. Ranchi Univ.*, 6—7, 233, 1969—70.
423. **Hotchkiss, A. T.**, A first report of chromosome number in the genus *Lychnothamnus* (Rupr.) Leonh. and its comparison with the other charophyte genera, *Proc. Linn. Soc. MSW*, 88, 268, 1963.
424. **Bharati, S. G. and Chennaveeraiah, M. S.**, Cytotaxonomic studies on *Chara zeylanica* complex, *Proc. Int. Symp. Taxo Algae*, University of Madras, 3, 1974.
425. **Sarma, Y. S. R. K. and Khan, M.**, Some new observations on the karyology of *Chara zeylanica* Klein ex Willd., *Curr. Sci.*, 34, 293, 1965.
426. **Sinha, J. P. and Ahmad, S.**, Karyological studies in *Chara vandalurensis* Sund. from India, *Proc. Int. Symp. Algol. Taxo.*, Abstr., 38, 1974.
427. **Corillion, R., Gillet, C., and Guerlesquin, M.**, Criteres cytologiques, anatomiques et ecologiques en faveur du maintien d'um genre *Charopsis* chez les Charophycées, *Bull. Soc. Sci. Bretag.*, 34, 65, 1959.
428. **Guerlesquin, M.**, Cytologie et nombres chromosomiques. In Recherches sur les Charophycées d'Afrique occidentale: systematique, phytogeographie, ecology et cytology (Corillion, R. and Guerlesquin, M.), *Bull. Soc. Sci. Bretag.*, 27, 125, 1972.
429. **Imahori, K. and Kato, T.**, Notes on chromosome numbers of charophytes in Fukui Perfecture, Japan. I., *Sci. Rep.*, 10, 39, 1961.
430. **Verma, S.**, A new chromosome count of *Chara braunii* Gmelin, *Proc. Ind. Sci. Congr. Assoc.*, Abstr., 366, 1967.
431. **Chatterjee, P.**, Cytotaxonomical studies of West Bengal Charophyta: karyotype analysis in *Chara braunii*, *Hydrobiologia*, 49, 171, 1976.
432. **Hotchkiss, A. T.**, Chromosome numbers in Characeae from the south Pacific, *Pac. Sci.*, 19, 31, 1965.
433. **Chatterjee, P.**, Karyological investigation of *Chara wallichii* A. Br., *Cell Chromosome Newsl.*, 2, 21, 1979.
434. **Chatterjee, P.**, Cytotaxonomic studies on *Chara socotrensis* f. *nuda* (Pal) RDW from West Bengal, India, *Cytologia*, 41, 650, 1976.
435. **Wood, R. D.**, Characeae of Australia, *Nova Hedwigia*, 22, 1, 1972.
436. **Sinha, J. P. and Noor, M. N.**, Studies on the karyology of some members of Charophyta, *Phykos*, 10, 112, 1971.
437. **Wood, R. D. and Manson, R.**, Characeae of New Zealand, *N. Z. J. Bot.*, 15, 87, 1977.
438. **Chennaveeraiah, M. S. and Bharati, S. G.**, Morphological and cytological observations on *Chara gymnopitys* A. Br., *Cytologia*, 39, 443, 1974.
439. **Chatterjee, P.**, *Chara fibrosa* var. *fibrosa* f. *longicorollata:* a new record for India and its cytology, *Curr. Sci.*, 48, 545, 1979.

440. **Hotchkiss, A. T.**, Chromosome numbers and relationships in *Chara leptopitys* A. Br., *Proc. Linn. Soc. NSW*, 98, 191, 1964.
441. **Tindall, D. R., Sawa, T., and Hotchkiss, A. T.**, *Nitellopsis bulbifera* in North America, *J. Phycol.*, 1, 147, 1965.
442. **Hotchkiss, A. T.**, Chromosome numbers in *Lamprothamnium*, *Proc. Linn. Soc. NSW*, 91, 118, 1966.
443. **Khan, M. and Sarma, Y. S. R. K.**, New chromosome numbers for Indian Charophyta, *Cell Chromosome Newsl.*, 3, 24, 1980.
444. **Sinha, B. D. and Srivastava, A. K.**, Record of *Lychnothamnus barbatus* (Meyen) Leonh. in Bihar, *Phykos*, 19, 175, 1980.
445. **Chatterjee, P.**, *Lychnothamnus barbatus* (Meyen) Leonh. in West Bengal and its chromosome number, *Bull. Bot. Soc. Bengal India*, 24, 131, 1970.
446. **Sundaralingam, V. S. and Bharatan, S.**, Occurrence of *Lychnothamnus barbatus* Leonh. near Madras, *Phykos*, 17, 123, 1978.
447. **Patel, R. J. and Jawale, A. K.**, On *Lychnothamnus barbatus* (Meyen) Leonh. from Gujarat, India, *Phykos*, 18, 51, 1979.
448. **Sarma, Y. S. R. K. and Khan, M.**, A new form of *Lychnothamnus barbatus* f. *iyengarii* forma novo, *Phykos*, 5, 195, 1966.
449. **Proctor, V. W., de Donterberg, C. C. C., Hotchkiss, A. T., and Imahori, K.**, Conspecificity of some charophytes, *J. Phycol.*, 3, 208, 1967.
450. **Sawa, T.**, Cytotaxonomy of the Characeae: karyotype analysis of *Nitella opaca* and *Nitella flexilis*, *Am. J. Bot.*, 52, 952, 1965.
451. **Sarma, Y. S. R. K. and Khan, M.**, Chromosome number in some Indian species of *Nitella*, *Chromosoma*, 15, 246, 1964.
452. **Tindall, D. R.**, A new species of *Nitella* (Characeae) belonging to *Nitella flexilis* species group in North America, *J. Phycol.*, 3, 229, 1967.
453. **Kasaki, H.**, The Charophyta from the lakes of Japan, *J. Hattori Bot. Lab. Bryol.*, 27, 217, 1964.
454. **Tindall, D. R.**, Observations on *Nitella acuminata* from south western United States and North Mexico, *J. Phycol.*, 6, 86, 1970.
455. **Mukherjee, S.**, Polyploid chromosome number in the genus *Nitella*, *Curr. Sci.*, 47, 386, 1978.
456. **Sinha, J. P. and Ahmad, S.**, Chromosome number of *Nitella* Ag. from Bihar, *Proc. Ind. Sci. Congr. Assoc.*, p. 319, 1973.
457. **Noor, M. N. and Mukherjee, S.**, An investigation into the karyology of *Nitella superba* from India, Proc. 2nd Int. Symp. Taxo. Algae (Madras Univ. India), p. 28, 1974.
458. **Sinha, J. P. and Ahmad, S.**, Karyological studies on three taxa of *Nitella* Agardh from Bihar, *J. Bihar Bot. Soc.*, 2, 22, 1973.
459. **Wood, R. D.**, Characeae in Samoa, *Bull. Torrey Bot. Club*, 90, 225, 1963.
460. **Noor, M. N.**, A new record of chromosome number in *Nitella mucronata* (A. Br.) Miquel from India, *Sci. Cult.*, 34, 214, 1968.
461. **Sarma, Y. S. R. K. and Ramjee**, Chromosome numbers in three taxa of Indian Charophyta, *J. Cytol. Genet.*, 4, 32, 1969.
462. **Ramjee and Bhatnagar, S. K.**, Cytotaxonomic studies on *Nitella furcata* subsp. *flagellifera* f. *patula* Gr. ex Allen and its comparison with other taxa of *N. furcata* complex, *Caryologia*, 31, 457, 1978.
463. **Sarma, Y. S. R. K. and Ramjee**, A new chromosome number in *Nitella tenuissima* Devs (Kütz.) RDW, *Caryologia*, 20, 61, 1967.
464. **Mukherjee, S. and Noor, M. N.**, A first report of karyological studies in *Nitella wattii* Groves from India, *Sci. Cult.*, 39, 459, 1973.
465. **Walther, E.**, Entwicklungsgeschichtliche und zytologische Untersuchungen an einigen Nitellen, *Arch. Julius Klaus-Stiftung*, 4, 23, 1929.
466. **Sinha, J. P. and Ahmad, S.**, Cytological analysis in two species of *Nitella* Ag. from Bihar, *Proc. 66th Ind. Sci. Congr. Assoc.*, p. 26, 1979.
467. **Stewart, L. M.**, Studies in the life history of *Nitella hyalina* Ag., *J. Elisha Mitchell Sci. Soc.*, 53, 173, 1937.
468. **Williams, M. B.**, A revision of *Nitella cristata* Br. (Characeae) and its allies. II, *Proc. Linn. Soc. NSW*, 54, 345, 1959.
469. **Hotchkiss, A. T.**, A new and revised base chromosome number for the genus *Tolypella*, *Bull. Torrey Bot. Club*, 93, 426, 1966.
470. **Corillion, R. and Guerlesquin, M.**, Premieres observations cytotaxonomiques sur le genre *Tolypella* (Charophycées), *Bull. Soc. Sci. Angers N.S.*, 2, 167, 1959.
471. **Corillion, R. and Guerlesquin, M.**, Sur une revision recente de la systematique chez les charophycées, *Bull. Mayenne Sci. Laval Paru en 1965*, p. 57, 1964.
472. **Bhatnagar, S. K. and Ramjee**, Karyological studies in *Tolypella glomerata* from India, *3rd All India Congr. Cytol. Genet. H.A.U. Hissar*, Abstr., 1978.

473. **Sawa, T.**, Chromosome numbers in the genus *Tolypella, J. Phycol.*, Suppl., 7, 1971.
474. **Sawa, T.**, New chromosome numbers for the genus *Tolypella* (Characeae), *Bull. Torrey Bot. Club*, 101, 21, 1974.
475. **Gross, S.**, The cytology of *Vaucheria, Bull. Torrey Bot. Club*, 64, 1, 1937.
476. **Hanatschek, H.**, Beiträge zur Entwicklungsgeschichte der Protophyten. X. Der Phasenwechsel bei der Gattung *Vaucheria, Arch. Protistenk*, 78, 497, 1932.
477. **Mundie, J. R.**, Cytology and life history of *Vaucheria geminata, Bot. Gaz.*, 87, 397, 1929.
478. **Ott, D. W. and Brown, R. M., Jr.**, Light and electron microscopical observations on mitosis in *Vaucheria litorea* Hofman ex C. Agardh, *Br. Phycol. J.*, 7, 361, 1972.
479. **Abbas, A. and Godward, M. B. E.**, Cytology of *Tribonema utriculosum* (Kuetz.) Hazen, *Phykos*, 2, 49, 1963.
480. **Lauterborn, R.**, *Untersuchungen über Ban, Kerniteilung und Bewegung der Diatomeen*, Engelmann, Leipzig, 1896.
481. **Klebahn, H.**, Beiträge zur Kenntnis der Auxosporebildung. I. *Ropalodia gibba* (Ehrenb.) O. Muller, *Jb. Wiss. Bot.*, 29, 595, 1896.
482. **Geitler, L.**, Die Reduktionsteilung und Copulation von *Cymbella lanceolata, Arch. Protistenk.*, 58, 465, 1927.
483. **Geitler, L.**, Somatische Teilung, Reduction Steilung, Copulation und Parthenogenese bei *Cocconies placenta, Arch. Protistenk.*, 59, 506, 1927.
484. **Geitler, L.**, Der Kernphasenwechsel der Diatomeen, *Beih. Bot. Centralf.*, 48, 1, 1931.
485. **Geitler, L.**, Reproduction and life history in Diatoms, *Botanical Rev.*, 1, 149, 1935.
486. **Geitler, L.**, Die Auxosporenbildung von *Synedra ulna, Ber. D. Deutsch. Bot. Ges.*, 57(9), 432, 1939.
487. **Geitler, L.**, Auxosporenbildung und Systematik bei pennaten Diatomeen und die Cytologie von *Cocconeis*— Sippen, *Oesterr. Bot. Z.*, 122, 299, 1973.
488. **von Cholnoky, B.**, Über die Auxosporenbildung der *Anomoeoneis sculpta* E. cl., *Arch. Protistenk.*, 63, 23, 1928.
489. **von Cholnoky, B.**, Beitrage zur Kenntnis der Auxosporenbildung, *Arch. Protistenk.*, 68, 471, 1929.
490. **von Cholnoky, B.**, Beitrage zur kenntnis der karyologie der Diatomeen, *Arch. Protistenk.*, 80, 321, 1933.
491. **von Cholnoky, B.**, Die kernteilung von *Melosira arenaria* einigen Bemerkungen über ihre Auxosporenbildung, *Zeitschr. Zellforsch. U. Mikr. Anat.*, 19, 698, 1933.
492. **Iyengar, M. O. P. and Subrahmanyan, R.**, On reduction division and auxospore-formation in *Cyclotella meneghiniana* Kütz., *J. Ind. Bot. Soc.*, 23, 125, 1944.
493. **Subramanyan, R.**, On somatic division, reduction division, auxospore-formation and sex-differentiation in *Navicula halophila* (Grun.) Cl., *Ind. Bot. Soc. (Prof. M.O.P. Iyengar Comm. Vol.)* p. 209, 1946.
494. **Subrahmanyan, R.**, On the cell division and mitosis of some south Indian Desmids, *Proc. Indian Acad. Sci.*, B22, 331, 1945.
495. **Manton, I., Kowallik, K., and von Stoch, H. A.**, Observations on the fine structure and development of spindle at mitosis and meiosis in a marine centric diatom *(Lithodesmium undulatum)*. I. Preliminary survey of mitosis in spermatogonia, *J. Microsc.*, 89, 295, 1969.
496. **Manton, I., Kowallik, K., and von Stoch, H. A.**, Observations on the fine structure and development of spindle at mitosis and meiosis in a marine centric diatom *(Lithodesmium undulatum)*. II. The early meiotic stages in male gametogenesis, *J. Cell Sci.*, 7, 271, 1969.
497. **Manton, I., Kowallik, K., and von Stoch, H. A.**, Observations on the fine structure and development of spindle at mitosis and meiosis in a marine centric diatom *(Lithodesmium undulatum)*. III. The latter stages of meiosis I in male gametogenesis, *J. Cell Sci.*, 6, 131, 1970.
498. **Manton, I., Kowallik, K., and von Stoch, H. A.**, Observations on the fine structure and development of spindle at mitosis and meiosis in a marine centric diatom *(Lithodesmium undulatum)*. IV. The second meiotic division and conclusion, *J. Cell Sci.*, 7, 407, 1970.
499. **Crawford, R. M.**, The protoplasmic ultrastructure of the vegetative cell of *Melosira varians* C. A. Agardh., *J. Phycol.*, 9, 50, 1973.
500. **Tippit, D. H. and Pickett-Heaps, J. D.**, Mitosis in the pennate diatom *Surirella ovalis, J. Cell Biol.*, 73, 705, 1977.
501. **Roy, P. S. and Sarma, Y. S. R. K.**, Report of chromosome number in a diatom, *Synedra ulna* (Nitzsch) Ehr., *Acta Botanica Indica*, 5, 178, 1977.
502. **Pickett-Heaps, J. D., Kent, L., McDonald, L., and Tippit, D. H.**, Cell division in the pennatediatom *Diatoma vulgare, Protoplasma*, 86, 205, 1975.
503. **Blochmann, F.**, Über die Kernteilung bei *Euglena viridis, Biol. Zbl.*, 14, 194, 1894.
504. **Keuten, J.**, Die Kernteilung von *Euglena viridis, Z. Wiss. Zool.*, 60, 215, 1895.
505. **Leedale, G. F.**, Mitosis and chromosome numbers in Euglenineae (Flagellata), *Nature (London)*, 181, 502, 1958.

506. **Leedale, G. F.**, Nuclear structure and mitosis in the Euglenineae, *Arch. Mikrobiol.*, 32, 32, 1958.
507. **Leedale, G. F.**, The time scale of mitosis in the Euglenineae, *Arch. Mikrobiol.*, 32, 352, 1959.
508. **Leedale, G. F.**, Amitosis in three species of *Euglena, Cytologia*, 24, 213, 1959.
509. **Leedale, G. F.**, The Euglenophyceae, in *The Chromosomes of the Algae*, Godward, M. B. E., Ed., Edward Arnold, London, 1966, 78.
510. **Leedale, G. F.**, Euglenoid Flagellates, Prentice-Hall, Englewood Cliffs, N.J., 1967.
511. **Leedale, G. F.**, The nucleus in *Euglena*, in *The Biology of Euglena*, Vol. 1, Beutow, D. E., Ed., Academic Press, New York, 1968.
512. **Singh, K. P.**, Cytological studies in *Trachelomonas grandis* Singh, *Agra Univ. J. Res. (Sci.)*, 7, 159, 1958.
513. **Baker, W. B.**, Studies on the life-history of *Euglena*. I. *Euglena agilis* Carter, *J. Biol. Bull.*, 51, 321, 1926.
514. **Hollande, A.**, Etude cytologique et biologique de quelques flagelles libres, *Arch. Zool. Exp. Gen.*, 83, 1, 1942.
515. **Pickett-Heaps, J. D.**, Cell division in Eukaryotic algae, *Biol. Sci.*, 26, 445, 1976.
516. **Leedale, G. F., Meeuse, B. J. D., and Pringsheim, E. G.**, Structure and physiology of *Euglena spirogyra* I and II, *Arch. Mikrobiol.*, 50, 68, 1965.
517. **Shashikala, Jayaraman, and Sarma, Y. S. R. K.**, Karyology of three species of *Euglena, J. Ind. Bot. Soc.*, 3, 196, 1980.
518. **Shahanara, S.**, Electron Microscope and Radiation Studies on *Euglena*, Ph.D. thesis, London University, London, 1977.
519. **Krichenbauer, H.**, Beitrag zur Kerntnis der Morphologie und Entwicklunggeschichte der Gattungen *Euglena* and *Phacus, Arch. Protistenk.*, 90, 88, 1937.
520. **Leedale, G. F.**, Studies on nuclear division on euglenoid flagellates, with special reference to a possible meiosis in *Hyalophacus ocellatus* Pringheim, *Proc. 1st Int. Conf. Protozool. (Prague)*, 265, 1961.
521. **O'Donnell, E. H. J.**, Nucleolus and chromosomes in *Euglena gracilis, Cytologia*, 30, 118, 1965.
522. **Dangeard, P. A.**, Recherches sur les Eugléniens, *Botaniste*, 8, 97, 1902.
523. **Dodge, J. D.**, The nucleus and nuclear division in the Dinophyceae, *Arch. Protistenk.*, 106, 442, 1963.
524. **Dodge, J. D.**, Chromosome structure in the Dinophyceae. I. The spiral chromonema, *Arch. Mikrobiol.*, 45, 46, 1963.
525. **Dodge, J. D.**, Chromosome numbers in some marine dinoflagellates, *Bot. Mar.*, 5, 121, 1963.
526. **Dodge, J. D.**, Chromosome structure in the Dinophyceae. II. Cytochemical studies, *Arch. Microbiol.*, 48, 66, 1964.
527. **Dodge, J. D.**, Nuclear division in the dinoflagellate *Gonyaulax tamarensis, J. Gen. Microbiol.*, 36, 269, 1964.
528. **Dodge, J. D.**, Cytochemical staining of sections from plastic-embedded flagellates, *Stain Techn.*, 39, 381, 1964.
529. **Dodge, J. D.**, The Dinophyceae, in *The Chromosomes of the Algae*, Godward, M. B. E., Ed., Edward Arnold, London, 1966, 96.
530. **Dodge, J. D.**, *The Fine Structure of Algal Cells*, Academic Press, New York, 1973.
531. **Sarma, Y. S. R. K. and Shyam, R.**, Cytology of Indian freshwater dinophyceae, Proc. 1st All India Congr. Cytol. Genet., *J. Cytol. Genet.*, Congr. Suppl., 72, 1971.
532. **Sarma, Y. S. R. K. and Shyam, R.**, On the morphology, reproduction and cytology of two new freshwater dinoflagellates from India, *Br. Phycol. J.*, 9, 21, 1974.
533. **Sarma, Y. S. R. K. and Shyam, R.**, Karyology of some Indian freshwater Dinophyceae, *12th Int. Bot. Congr. (Leningrad)*, 1, 44, 1975.
534. **Shyam, R. and Sarma, Y. S. R. K.**, An unusual dinophycean nucleus, *Nucleus*, 18, 86, 1975.
535. **Shyam, R. and Sarma, Y. S. R. K.**, *Woloszynskia stoschii* and *Gymnodinium indicum*, two new freshwater dinoflagellates from India: morphology reproduction and cytology, *Plant Syst. Evol.*, 124, 205, 1975.
536. **Shyam, R. and Sarma, Y. S. R. K.**, Cytology of Indian freshwater Dinophyceae, *Bot. J. Linn. Soc.*, 76, 145, 1978.
537. **Loeblich, A. R., III**, Dinoflagellate evolution, speculation and evidence, *J. Protozool.*, 23, 13, 1976.
538. **Ris, H. and Kubai, D. F.**, An unusual mitotic mechanism in the parasitic protozoan *Syndinium* sp., *J. Cell. Biol.*, 60, 702, 1974.
539. **Taylor, F. J. R.**, On dinoflagellate evolution, *Biosystems*, 13, 65, 1980.
540. **Rizzo, P. J. and Nooden, L. D.**, Chromosomal proteins in the dinoflagellate alga *Gyrodinium cohnii, Science*, 176, 796, 1972.
541. **Rizzo, P. J. and Nooden, L. D.**, Isolation and partial characterisation of dinoflagellate chromatin, *Biochem. Biophys. Acta*, 349, 402, 1974.

542. **Rizzo, P. J. and Nooden, L. D.**, Partial characterisation of dinoflagellate chromosomal proteins, *Biochem. Biophys. Acta,* 349, 415, 1974.
543. **Stosch, H. A.**, Zum chromosomen Formwechsel der Dinophyten sowie zur Mechanik und Terminologie von Schranben, *Arch. Protistenk.,* 103, 229, 1958.
544. **Skoczylas, O.**, Über die Mitose von *Ceratium cornutum* und einigen anderen *Peridinum, Arch. Protistenk,* 103, 229, 1958.
545. **Shyam, R.**, Studies on the Cytology and Systematics of Freshwater Algal Flagellates, Ph.D. thesis, Banaras Hindu University, Varanasi, 1974.
546. **Chaudhury, B. R. and Sarma, Y. S. R. K.**, On a freshwater *Gymnodinium* sp. with unusual nuclear organisation, *Cytologia,* 44, 915, 1979.
547. **Loeblich, A. R., III, Klotz, L. C., Roberts, T. M., Tuttle, R. C., and Allen, J. R.**, The dinoflagellate nucleus, *J. Phycol.,* 10 (Suppl.), 14, 1974.
548. **Allen, J. R., Roberts, T. M., Tuttle, R. C., Loeblich, A. R., III, and Klotz, L. C.**, Physicochemical characterisation of dinoflagellate DNA, *J. Phycol.,* 10 (Suppl.), 15, 1974.
549. **Roberts, T. M., Klotz, L. C., and Loeblich, A. R., III**, Measurement of molecular weight of dinoflagellate chromosomal DNA, *J. Phycol.,* 10 (Suppl.), 15, 1974.
550. **Zingmark, R. G.**, Sexual reproduction in the dinoflagellate *Noctiluca miliaris* Suriray, *J. Phycol.,* 6, 122, 1970.
551. **Dodge, J. D. and Crawford, R. M.**, The fine structure of *Gymnodinium fuscum* (Dinophyceae), *New Phytol.,* 68, 613, 1969.
552. **Dodge, J. D.**, A dinoflagellate with both a mesocaryotic and eucaryotic nucleus. I. Fine structure of the nuclei, *Protoplasma,* 75, 145, 1971.
553. **Tomas, R. N., Cox, E. R., and Steidinger, K. A.**, *Peridinium balticum* (Levander) Lemmermann, an unusual dinoflagellate with a mesokaryotic and eucaryotic nucleus, *J. Phycol.,* 9, 91, 1973.
554. **Leadbeater, B. S. C.**, An electron microscopic study of nuclear and cell division in a dinoflagellate, *Arch. Mikrobiol.,* 57, 239, 1967.
555. **Zubay, G. and Watson, M. R.**, The absence of histone in the bacterium *Escherichia coli.* I. Preparation and analysis of nucleoprotein extract, *J. Biophys. Biochem. Cytol.,* 5, 51, 1959.
556. **Rizzo, P. J. and Cox, E. R.**, Histone occurrence in chromatin from *Peridinium balticum,* a binucleate dinoflagellate, *Science,* 198, 1258, 1977.
557. **Rizzo, P. J.**, Histones in the marine *Olithodiscus luteus, Plant Physiol.,* 63 (Suppl.), 2, 1979.
558. **Rizzo, P. J. and Burghardt, R. C.**, Chromatin structure in unicellular algae *Olisthodiscus luteus, Crypthecodinium cohnii* and *Peridinium balticum, Chromosoma (Bull.),* 76, 91, 1980.
559. **Leadbeater, B. S. C. and Dodge, J. D.**, An electron microscope study of nuclear and cell division in dinoflagellate, *Arch. Mikrobiol.,* 57, 239, 1967.
560. **Kubai, D. F. and Ris, H.**, Division in the dinoflagellate *Gyrodinium cohnii* (Schiller). A new type of nuclear reproduction, *J. Cell Biol.,* 40, 508, 1969.
561. **Oakley, B. R. and Dodge, J. D.**, Kinetochores associated with the nuclear envelope in the mitosis of a dinoflagellate, *J. Cell Biol.,* 63, 322, 1974.
562. **von Stosch, H. A.**, Zur Problem der sexuallen Fortpflanzung in der Peridineengattung *Ceratium, Helgoland, Wiss. Meeresunters,* 10, 140, 1964.
563. **von Stosch, H. A.**, Sexualitat bei *Ceratium cornutum* (Dinophyta), *Naturwissenschaften,* 52, 112, 1965.
564. **von Stosch, H. A.**, La signification of cytologique de la 'Cyclose nucleaire' dans le cycle de vie des Dinoflagelles, *Soc. Bot. Fr. Mem.,* p. 201, 1972.
565. **von Stosch, H. A.**, Observations on vegetative reproduction and sexual life-cycles of two freshwater dinoflagellates, *Gymnodinium pseudopalustre* Schiller and *Woloszynskia apiculata* sp. nov., *Br. Phycol. J.,* 8, 105, 1973.
566. **Pfiester, L. A.**, Sexual reproduction of *Peridinium cinctum* f. (Dinophyceae), *J. Phycol.,* 11, 259, 1975.
567. **Grasse, P. P. and Dragesco, J.**, L'ultrastructure du chromosome des Peridinians et ses consequence genetiques, *C. R. Acad. Sci. Paris,* 245, 2447, 1957.
568. **Hall, R. P.**, Binary fission in *Oxyrrhis marina* Dujardin, *Univ. Calif. Publ. Zool.,* 26, 281, 1925.
569. **Entz, G.**, Uber die mitotische Teiling von *Ceratium hirundinella, Arch. Protistenk.,* 43, 415, 1921.
570. **Fine, K. E. and Loeblich, A. R.**, A comparison of *Crippsiella sweeneyae* (IUCC 1956) and *Peridinium trichoideum* (IUCC 1017), *J. Phycol.,* 10 (Suppl.), 13, 1974.
571. **Allen, J. R., Roberts, T. M., Loeblich, A. R., and Klotz, L. C.**, Characterisation of the DNA from the dinoflagellate *Crypthecodinium cohnii* and implications for nuclear organisation, *Cell,* 6, 161, 1975.
572. **Borgert, A.**, Kern und Zellteilung bei marine *Ceratium* — Arten, *Arch. Protistenk.,* 20, 1, 1910.
573. **Loeblich, A. R.**, A new marine dinoflagellate genus, *Cachonina,* in axenic culture from the Salton sea, California with remarks on the genus *Peridinium, Proc. Biol. Soc. Wash.,* 81, 91, 1968.

574. **Dodge, J. D. and Crawford, R. M.**, The morphology and fine structure of *Ceratium hirundinella* (Dinophyceae), *J. Phycol.*, 6, 137, 1970.
575. **Thakur, M.**, Ph.D. thesis, London University, London, 1965.
576. **Hollande, A.**, Etude cytologique et biologique de quelques flagelles libres, *Arch. Zool. Exp. Gen.*, 83, 1, 1942.
577. **Hollande, A.**, Classe des Cryptomonadineae, *Traite de Zoologie*, Grasse, P., Ed., 1, 285, 1954.
578. **Godward, M. B. E.**, The Cryptophyceae, in *The Chromosomes of the Algae*, Godward, M. B. E., Ed., Edward Arnold, London, 1966, 116.
579. **Okley, B. R. and Dodge, J. D.**, Mitosis in Cryptophyceae, *Nature*, 244, 521, 1973.
580. **Oakley, B. R. and Bisalputra, T.**, Mitosis and cell division in *Cryptomonas* (Cryptophyceae), *Can. J. Bot.*, 55, 2789, 1977.
581. **Howasse, R.**, Contribution a l'etude des chloromonadines: *Gonyostomum semen* Diesing, *Arch. Zool. Exp. Gen.*, 84, 239, 1945.
582. **Heywood, P.**, Mitosis and cytokinesis in the Chloromonadophycean alga *Gonyostomum semen*, *J. Phycol.*, 10, 355, 1974.
583. **Mignot, J. P.**, Structure et ultrastrutre de quelques Chloromonadines, *Protistologica*, 3, 5, 1967.
584. **Heywood, P. and Godward, M. B. E.**, Centromeric organisation in Chloromonadophycean alga *Vacuolaria virescens*, *Chromosoma*, 39, 333, 1972.
585. **Slankis, T. and Gibbs, S. P.**, The fine structure of mitosis and cell division in the Chrysophycean alga *Ochromonas danica*, *J. Phycol.*, 8, 243, 1972.
586. **Christensen, T.**, Alger, in *Botanik (Systematik Botanik)*, 1st ed., Böcher, T. W., Lange, M., and Sorensens, T., Eds., Munksgaard, Copenhagen, 1962.
587. **Manton, I.**, Observations with electron microscope on the division cycle in the flagellate *Prymnesium parvum* Carter, *J. R. Microsc. Soc.*, 83, 317, 1964.
588. **Stewart, K. D., Mattox, K. R., and Chandev, C. D.**, Mitosis and cytokinesis in *Platymonas subcordiformis*, a scaly green monad, *J. Phycol.*, 10, 65, 1974.
589. **Pearson, B. R. and Norris, R. E.**, Fine structure of cell division in *Pyramimonas parkeae* Norris and Pearson (Chlorophyta, Prasinophyceae), *J. Phycol.*, 11, 113, 1975.
590. **Pickett-Heaps, J. D. and Ott, D. W.**, Cell structure and division in *Pedinomonas*, *Cytobios*, 1974.
591. **Hibberd, D. J. and Leedale, G. F.**, A new algal class. The Eustigmatophyceae, *Taxon*, 20, 523, 1971.
592. **Leedale, G. F.**, How many are the kingdoms of organisms?, *Taxon*, 23, 261, 1974.
593. **Charles, H. P. and Knight, B. C. J.**, *Organisation and Control in Prokaryotic and Eukaryotic Cells*, Cambridge University Press, New York, 1970.
594. **Stainer, R. Y. and Cohen-Basire, G.**, Phototrophic prokaryotes: the cyanobacteria, *Ann. Rev. Microbiol.*, 31, 225, 1977.
595. **Klein, K. M. and Cronquest, A.**, A consideration of evolutionary and taxonomic significance of some biochemical, micromorphological and physiological characters in the Thallophytes, *Q. Rev. Biol.*, 42 (105), 296, 1967.
596. **Floyd, G. L., Stewart, K. D., and Mattox, K. R.**, Comparative cytology of *Ulothrix* and *Stigeoclonium*, *J. Phycol.*, 8, 68, 1972.
597. **Taylor, F. J. R.**, Flagellate phylogeny: a study in conflicts, *J. Protozool.*, 23, 28, 1976.
598. **Woodcock, C. L. F.**, The anchoring of nuclei by cytoplasmic microtubules in *Acetabularia*, *J. Cell Sci.*, 8, 611, 1971.
599. **McBride, G. E.**, Cytokinesis in the green alga *Fritschiella*, *Nature (London)*, 216, 939, 1967.
600. **McBride, G. E.**, Cytokinesis and ultrastructure in *Fritschiella tuberosa* Iyengar, *Arch. Protistenk.*, 112, 365, 1970.
601. **Stewart, K. D. and Mattox, K. R.**, Comparative cytology, evolution and classification of the green algae with some consideration of the origin of other organisms with Chlorophylls a and b, *Bot. Rev.*, 41, 104, 1975.
602. **Bold, H. C. and Wynne, M. J.**, *Introduction to the Algae — Structure and Reproduction*, Prentice-Hall, New Delhi, 1978.

Chapter 5

CHROMOSOMES IN THE EVOLUTION OF THE BRYOPHYTA

A. J. E. Smith

TABLE OF CONTENTS

I. INTRODUCTION

The Bryophyta is a relatively small division of the Plant Kingdom, within which most modern authorities recognize five classes: Hepaticopsida (liverworts or hepatics, 5500 to 6000 species), Anthocerotopsida (hornworts, 300 species), Sphagnopsida (bog mosses or sphagna, 150 species), Andreaeopsida (rock mosses, 115 species), and Bryopsida (true mosses, with between 7000 and 14,000 species depending upon the authority). Although long regarded as a natural group of plants derived from the green algae, current views are that the classes are unrelated — as also may be some of the subdivisions within them — and that they have been derived from different Psilophytalean ancestors by progressive simplification.[1-4]

With regard to the origin or the course of evolution within the various classes, the fossil record is fragmentary and, with the possible exception of *Sphagnum*, fossil remains resemble to a considerable degree present-day forms. Earliest known fossil liverworts are from the Devonian, hornworts from the Palaeocene or early Eocene, stone mosses from the Pleistocene, and true mosses from the Permian;[5] a report of a *Sphagnum* ancestral form from the Permian is very doubtful.[6]

Despite their small size the bryophytes have excited much interest, partly because of the dominance of the gametophyte generation, partly because of their apparent intermediate position between the green algae and vascular plants, and partly because of their relevance to speculation about the origin of the alternation of generations. Cytological observations were first reported upon at the turn of this century.[7-11] Many of the early observations were inaccurate, but since the 1930s there has been a steady accumulation of reliable chromosomal data. Cytological observations have been made upon about 9% of liverworts, 8% of hornworts, 14% of sphagna, and 13% of true mosses.[4] Sex chromosomes in plants were first reported from the liverwort *Sphaerocarpos donnellii*,[12,13] and heterochromatin was first described from the liverwort genus *Pellia*.[14,15] There are still, however, very little data based upon autoradiography, chromosome banding, and DNA measurement.

There have been a number of surveys of bryophyte chromosome numbers in recent years.[4,16-21] Several attempts have been made at phylogenetic speculation based on cytological observations, but there is no corroborative evidence whatsoever and chromosome data cannot be used in isolation.[22,23] A number of authors have adopted a system of nomenclature for bryophyte chromosomes based on length and centromere position. Similarities in the proportions of chromosome types have been taken to imply homology, not only within the liverworts[24-26] and mosses[27,28] but also between the two groups.[29,30] Such sweeping comparisons are of doubtful value[31] as they assume that the Bryophyta are monophyletic and that morphologically similar chromosomes are homologous. The first assumption is now highly unlikely and the second not necessarily true.[32]

Chromosomal, morphological, and anatomical studies at the species and generic levels are, however, more informative and it is possible in certain instances to suggest evolutionary trends. References to chromosome numbers in the Bryophyta are to the gametophyte number as, with only a few exceptions,[33,34] this has been customary in publications dealing with bryophyte chromosomes. Generalizations made about chromosome numbers within particular taxa are based upon data quoted by Fritsch.[18]

II. HEPATICOPSIDA

Almost all chromosome studies of liverworts have been made upon gametophyte mitoses, partly because of the paucity of or even absence of sporophytes in many species and partly because of the difficulty of obtaining sporophytes at the right stage for meiotic investigation.

A. Basic Chromosome Number

There are two schools of thought concerning the basic number in the liverworts: one considering that $x = 4$ or 5 and the other that $x = 9$ or possibly 8 or 10. With the exception of the only two species of *Takakia* and some clearly derived ones of *Radula*, all liverworts have $n = 8$, 9, or 10 or multiples thereof. The two species of *Takakia* have $n = 4$ and 5. A number of authorities[24,26,29,35,36] consider that *Takakia* is a primitive liverwort and that it, or something very similar, is ancestral to the rest of the class. It is argued that $x = 4$ and 5 gave rise to $n = 8$ and 10 and then to $n = 9$. Arguments put forward in support of this are that the karyotype of the only known diploid hornwort *Anthoceros husnotii*, considered to be a liverwort, is identical to that of many liverworts[37] and that in some members of the Metzgeriales the presence of two nucleolar organizers is indicative of diploidy.[36] Arguments against this are that *Anthoceros* is totally unrelated to the hepatics[3] and there is no evidence that *Takakia* is a liverwort[4,38] or, even if it is, that it is ancestral to the class. Third, there is no evidence from the distribution of constitutive heterochromatin that the complements of *Pellia, Riccardia,* and *Cryptothallus,* all members of the Metzgeriales, are diploid.[39] Fourth, if $x = 4$ and 5, then it would be expected that there would be two evolutionary lines within the liverworts, one with $n = 8$ and the other with $n = 10$. This is not so, genera with $n = 8$, 9, and 10 being distributed through the three major taxa of liverworts. This supports the contention that $x = 9$.[4,40] The genus *Riccia* has a very reduced sporophyte and is regarded as specialized[3] and has $x = 10$, suggesting that this is derived from the much commoner $x = 9$. Only a few genera (e.g., *Radula, Porella*) have $x = 8$, and these are regarded as specialized on morphological and ecological grounds,[3] suggesting that $x = 8$, like $x = 10$, is derived. This would indicate fairly conclusively that the original basic number in the Hepaticopsida is 9.

Karyotypes are very uniform throughout the Hepaticopsida and there is no suggestion that there are pairs of chromosomes as might be expected if polyploidy was involved. The liverworts are an ancient group, however, dating back to the Devonian, and any similarity between pairs of homologs may have been lost during the course of chromosome evolution.

B. Major Evolutionary Trends

The remarkably uniform karyotypes of liverworts provide no indication of the course of evolution, and there are no proven morphological trends, either. The major phyletic trends in the liverworts are a complete mystery despite arguments to the contrary.[3,19]

There has been a convention, initiated for both liverworts[25] and mosses,[41-43] of constructing karyotype formulas in which the chromosomes are labeled V, v, J, j, I, and i, depending upon size and position of centromere. Tatuno[25] also initiated the labeling of the largest and smallest chromosomes as heterochromatic (H- and h-chromosomes). On the basis of these formulas spurious homologies between karyotypes have been recognized. These formulas conceal subtle but, nonetheless, significant differences (see, for example, the situation in *Pellia,* Section C).

Similarly, the labeling of the largest and smallest members of the complement as heterochromatic is misleading. Proskauer[37] points out that he "cannot subscribe to Tatuno's[25,44] interpretation of the largest chromosome as being a heterochromatic chromosome and its purported interphase behaviour. I still believe that the interphase extranucleolar heterochromatin material ... is not a single chromosome." Newton[39] says with reference to the Metzgeriales, "The reaction of these five species to the Giemsa C-banding technique reveals a number of difficulties and lack of precision associated with the terms H- and h-chromosomes. It is therefore recommended that their use in liverwort cytology should be discontinued."

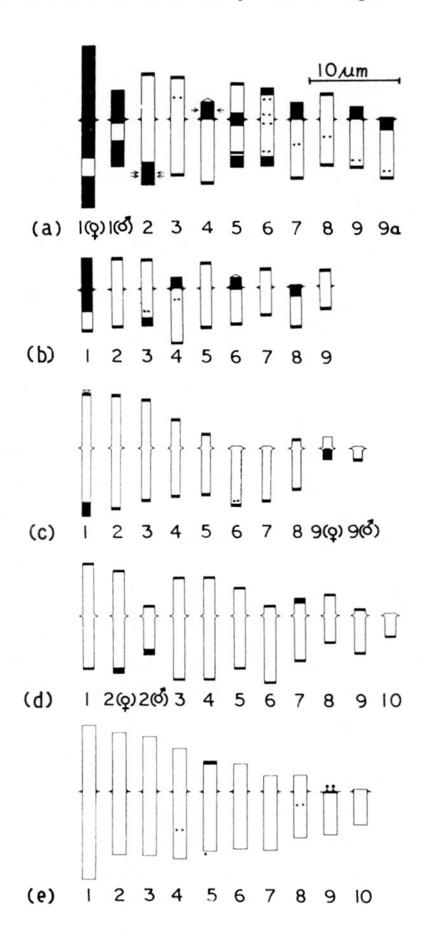

FIGURE 1. Giemsa C-banded idiograms of (a) *Pellia neesiana*, (b) *P. epiphylla*, (c) *P. endiviifolia*, (d) *Riccardia pinguis*, (e) *Cryptothallus mirabilis*. ♂ and ♀ represent alternatives from male and female plants; arrows indicate positions of secondary C-bands. In *P. neesiana* chromosomes 9 and 9a are the two types mentioned in Section II.C. (From Newton, M. E., *J. Bryol.*, 9, 327, 1977. With permission of Blackwell Scientific Publications Limited.)

Phylogenies[24,26,29] of both the Hepaticosida and the Bryophyta are as a whole based on such karyotype formulas and are unacceptable. It is not possible to relate either chromosome number or supposed homologies based on karyotype formulas to major evolutionary trends in the total absence of any supporting evidence.

C. Structural Changes

Because of cytological uniformity there is little information available about the evolutionary role of chromosomes, except in relation to polyploidy and aneuploidy at the generic and species levels in liverworts. There are, however, two examples that indicate the type of structural changes that may have taken place even though their direction is uncertain.

Karyotype idiograms[39] of the three British *Pellia* species, *P. neesiana, P. epiphylla,* and *P. endiviifolia*, as revealed by Giemsa C-banding, are shown in Figure 1. If these karyotypes are expressed in terms of the formulas used by Japanese cytologists they would be as follows:

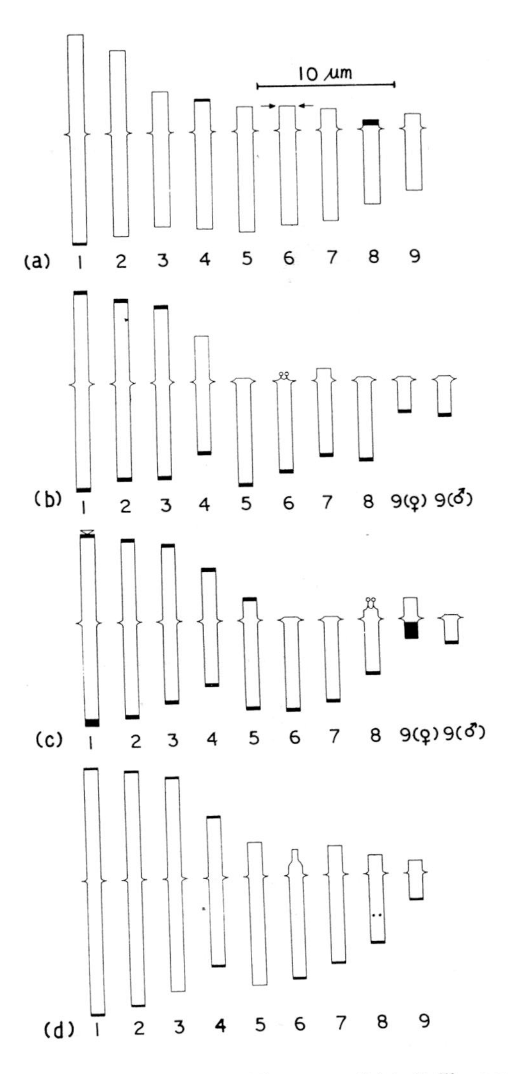

FIGURE 2. Giemsa C-banded idiograms of (a) *Pellia megaspora*, (b) *P. endiviifolia* from Japan, (c) *P. endiviifolia* from Britain, (d) *P. endiviifolia* from Canada. ♂ and ♀ represent alternatives from male and female plants; arrows indicate probable site of nucleolar organization. (From Newton, M. E., *J. Bryol.*, 11, 433, 1981. With permission of Blackwell Scientific Publications Limited.)

P. neesiana	V(H) + 5V + 3J
P. epiphylla	V(H) + 5V + 2J + I
P. endiviifolia	4V + 2J + 2I + v(h) or i(h)

It would appear from these formulas that the first two species have very similar karyotypes and the formula of *P. endiviifolia* suggests highly speculative homologies. In the latter species the nucleolar organizer is carried on a V-chromosome while in the other two it is on J-chromosomes.

Although constitutive heterochromatin as revealed by C-banding is associated with limited genetic coding, the differences between the three *Pellia* species is of evolutionary significance, *P. endiviifolia* showing the greatest difference with a more asymmetric karyotype and less heterochromatin. It is not possible to say in which direction evolution has proceeded, but clearly structural changes have been involved as has the degree of heterochromatization.

In two other species, *Riccardia pinguis* and *Cryptothallus mirabilis,* it is, from the karyotypes,[39] possible to suggest the direction in which change has occurred. The two species are thought to be closely related, with *Cryptothallus,* which is saprophytic, being the more advanced. In *Cryptothallus* there is less heterochromatin so evolution appears to have occurred by decrease in heterochromatin. It cannot, however, be assumed that the same applies to *Pellia.*

Idiograms of *P. megaspora* and *P. endiviifolia* from prepared Giemsa C-band squashes[45] show that the two differ in karyotype symmetry although on morphological grounds they are closely related (Figure 2). There are intraspecific differences within *P. endiviifolia* from Britain, Canada, and Japan suggesting that cytological changes have taken place resulting in different geographical cytotypes.

C-banding in *P. neesiana* provides the first and, as yet, definite evidence of a structural change in the Bryophyta. Seven populations from various parts of Britain exhibited the karyotype (Figure 1) with the smallest member of the complement, an acrocentric, being labeled 9. In two samples from an eighth population, the smallest chromosome, labeled 9a in the idiogram, was a telocentric. Comparison of 9 and 9a indicates that a pericentric inversion involving a block of heterochromatin has occurred.

A second example is provided by a detailed investigation of the large genus *Frullania.*[46] Within the genus there is interspecific variation in the length of the mitotic metaphase chromosomes, the degree of heteropycnosis at prophase and telophase, and the position of the centromere and of secondary constrictions in the longer chromosomes. Further, most monoecious species have $n = 9$ and dioecious ones $n = 9$ in both sexes or $n = 9$ in the female and $n = 8$ in the male.

There are two main karyotypes, one with a long and a short heteropycnotic chromosome and seven euchromatic chromosomes and a second with seven euchromatic chromosomes and two long heterochromatic ones in the female and one in the male. Divergent karyotypes are found mostly in monoecious species and monoecism may have arisen as the result of karyotype changes. The various karyotypes correlate with the morphological subdivisions of the genus. It is suggested that factors involved in karyotype evolution included loss of the smallest heteropycnotic chromosome, the doubling of a large heteropycnotic chromosome, translocations resulting in changes in chromosome symmetry, and decreases or increases in the amount of heterochromatin.

D. Polyploidy

Assuming the basic haploid number to be $x = 9$, the proportion of polyploid liverwort species is relatively small. In the Marchantiales it is 18%, in the Metzgeriales it is 12%, and in the Jungermanniales it is 6%.[4] These percentages also include a few species within which there is a polyploid series (e.g., *Dumortiera hirsuta* with $n = 9$, 18, and 27). The higher proportion of polyploids in the Marchantiales may be related to the occurrence of many members of the order, especially in the genus *Riccia,* in temporary or xeric habitats. It is likely that polyploids are autopolyploids and arose as the result of diplospory[4] (see later).

E. Aneuploidy

There are only three examples of interspecific aneuploidy based upon reliable count: in *Radula* there are some species with $n = 8$ and others with $n = 6$; in *Telaranea* there is one species with $n = 10$ and the remainder with $n = 9$; and in *Frullania* there is one species with $n = 10$.

F. Sex-Specific Chromosomes

Sex chromosomes have been reported from about 60 liverworts[47] and two types are recognized, morphological and structural, although in the complements of some dioecious species the chromosomes are completely homologous. Segawa[47-49] has suggested the evolution of sex chromosomes from completely undifferentiated X- and Y-chromosomes to those with differing amounts of heterochromatin (structural sex chromosomes) to morphologically different sex chromosomes. In the latter case, it is usually in the male that the sex chromosome is the smaller as in *Sphaerocarpos*[12,13] or even absent as in some *Frullania* species,[46] but in some Asian species of *Plagiochila* the female plant has $n = 8$ and the male $n = 9$.[50] It may be the largest pair of bivalents that is dimorphic as in *Pellia neesiana* or the smallest as in *P. endiviifolia.*

The recognition of structural and morphological sex chromosomes and their evolutionary relationships in both hepatics and mosses requires further investigation. In mosses there are several examples of dimorphic bivalents in monoecious species[34,51,52] as there are in the monoecious hornwort *Anthoceros husnotii.*[53] That two homologs are dimorphic does not of necessity mean that they are sex chromosomes. Further, with regard to structural sex chromosomes, Berrie[54] reports that while in the sporophyte of *Plagiochila praemorsa* there are no obvious differences between the two largest chromosomes, they do differ in the degree of heteropycnosis in the male and female gametophytes. He suggests that the differences in the amount of heterochromatin in the two sexes in the gametophyte generation may be a phenotypic response of the chromosomes to the male and female genomes.

Only from *S. donnellii* have sex chromosomes been unequivocally demonstrated in the Bryophyta.[12,13] Although chromosomes specific to one sex or the other have been reported from a number of mosses and liverworts, it has not been proved that they are sex chromosomes and they should better be referred to as sex-associated or sex-specific chromosomes.[4,39]

The origin of sex-specific chromosomes is unknown, but there seem to be marked differences within a species in different areas suggesting that they are highly polymorphic. In Britain the smallest bivalent in *Riccardia pirguis* is dimorphic and almost entirely euchromatic, with the larger chromosome occurring in the female gametophyte and the smaller, acrocentric one in the male.[39] In Japan the chromosomes are entirely heterochromatic and the male homolog is metacentric.[25,55] In material from North America[56] and central Europe[14,57,58] no sex-specific chromosomes were found. In central Europe, however, two races were found: one more or less without heterochromatin, the other with a block of heterochromatin on one of two chromosomes.[58]

G. Micro-Chromosomes

In the complements of many liverworts the smallest chromosome is usually 0.16 to 0.40 times the length of the largest chromosome.[36] Early workers[57,59] called the smallest chromosome an m-chromosome. On the other hand, later workers[25,60] specifically recognized them as not being m-chromosomes and designated them h-chromosomes. This latter term is dealt with elsewhere. Since there does not seem to be anything remarkable about hepatic m-chromosomes other than their small size, there seems little point in giving them a special designation.

H. Chromosome Structure and Replication

Apart from heterochromatic studies initiated by Heitz,[14,15] enlarged upon by Japanese bryo-cytologists, and refined by Newton,[39,45] relatively little is known of the structure and replication of liverwort chromosomes. In this group the position of the centromere is usually only detected when it is mechanically active at anaphase. In *Pellia epiphylla*, Lewis[61] suggests that at the onset of mitosis the centromere has divided and has the same cycle of spiralization as the euchromatin. He comments that the almost universal occurrence of a primary constriction in higher organisms with localized centromeres suggests that *Pellia* represents a low level of centromere organization. The centromeres are, however, localized and associated with small bands of heterochromatin[39] unlike the semilocalization reported in the moss *Pleurozium schreberi.*[62]

The data on DNA replication are conflicting but are, nevertheless, probably similar to those in other organisms. There are reports of late heterochromatin DNA replication in *Pellia endiviifolia* [63] and early replication in the sex-specific chromosomes of *P. neesiana*, using radioactive tracers.[64,65] It is not clear in the latter case, however, whether heterochromatin DNA replication began earlier or whether it merely ended earlier.

Ultrastructural studies of dividing cells in *Marchantia polymorpha* and *M. berteroana* [66] indicate that the behavior of the nuclear membrane, nucleolus, and chromosomes is typical of higher plants. Spindle formation, however, is somewhat different as the preprophase bands of microtubules characteristic of higher plants were not seen, although they have been reported from dividing cells in young *Sphagnum* leaves.[67]

III. ANTHOCEROTOPSIDA

A. Basic Number

A curious situation prevails in this class with respect to chromosome numbers. All recent reliable counts from five genera studied cytologically from Japan are $n = 6$. With the exception of one count of $n = 6$ [53] and another of $n = 9$,[37] all counts from the same five genera variously reported from Eurasia, Africa, the Americas, Oceania, and Australasia are $n = 5$. Thus, it is not possible to state whether $x = 5$ or 6 nor have experiments been carried out to elucidate homologies of the complements of Japanese and other populations.

B. Polyploidy

Polyploidy has played a negligible role in the evolution of the Anthocerotopsida, there being only a single record of a naturally occurring polyploid, *Anthoceros husnotii* with $n = 9$, from Portugal.[37]

In haploid material examined with $n = 5$, the fifth chromosome was very small and heterochromatic.[68] In the polyploid the ninth chromosome is very small and heterochromatic but it has no partner, and it is suggested that during evolution of the polyploid karyotype one of the small chromosomes was eliminated to give $n = 9$. The karyotype of the polyploid is very similar to that of many hepatics and has been used as evidence of a common origin of the two groups. Other evidence suggests, however, that any similarity is coincidental.[3,19]

C. Aneuploidy

While all Japanese counts are $n = 6$ and the majority of others $n = 5$, there is only one report of aneuploidy within a small area. A count of $n = 5$ was reported from two populations of *Anthoceros husnotii* from Wales[68] which were within 8 km of a population of the same species with $n = 6$.[53]

D. Accessory Chromosomes

Proskauer[37] presents convincing evidence of the existence of accessory chromosomes in *Phaeoceros laevis* ssp. *carolinianus.* Their number varies from population to population but is constant within a clone. Accessories are absent from the antheridia but present in the thallus and sporophyte tissue. It is suggested that they are eliminated during the formation of antheridial initials and are passed on via the egg cells. *P. laevis* ssp. *laevis* is dioecious and has no accessories while ssp. *carolinianus*, which has accessories, is monoecious. The connection between accessories and sex is obscure but it may be that accessories increase genetic variability in an inbreeding plant.

IV. SPHAGNOPSIDA

The basic number of the Sphagna is $x = 19 + 2m$, the great majority of reported counts being $n = 19 + 2m$ or $38 + 4m$. The few deviations involve the number of m-chromosomes and may well be due to misinterpretation of meiotic squashes, as m-chromosomes tend to disjoin prematurely and the subsequent m-univalents may be mistaken for bivalents.

It seems likely that $x = 19 + 2m$ arose at a very early stage in the evolution of the class from $x = 10$. Ramsay,[69] about the mitotic chromosomes of *Sphagnum capillifolium* (as *S. rubellum*), says "The 21 chromosomes show gradual decrease in length to the two 'm'-chromosomes which appear as dots. Eighteen of the chromosomes could be quite clearly divided into 9 pairs while the other three consisted of a small metacentric chromosome and two micro-chromosomes." The two m-chromosomes may be derived by centric fission of the homolog of the small metacentric chromosome, thus implying that $19 + 2m$ was derived by polyploidy and structural change from an original number of 10. As $x = 19 + 2m$ is universal throughout the Sphagnopsida, this suggests that the change occurred before evolution of the class commenced.

It has been reported that at meiosis in *Sphagnum* the spindle is quadripolar and the first division is equational and the second reductional.[70] However, these conclusions were based upon misinterpretations of meiotic squashes.[71]

Polyploidy has played a relatively minor role in evolution at the species level. Only 4 of the 22 species examined cytologically have $n = 38 + 4m$.

V. ANDREAEOPSIDA

Cytological data on this group are very sparse and provide no indication of relationships or phylogeny. Two numbers have been reported, $n = 10$ and 11, but there is no indication whether these are haploid numbers or polyploid derivatives.

VI. BRYOPSIDA

Except in Japan most studies of moss chromosomes have been made from meiotic preparations because of the ease with which these may be obtained compared with mitotic preparations. Recently, however, workers from elsewhere have started examining mitotic chromosomes and a considerable amount of information is now accumulating.

A. Basic Number

There are two groups within the Bryopsida based upon the structure of the peristome teeth of the sporophyte, the Nematodonteae with solid teeth, and the Arthrodonteae with articulated teeth. There are two subdivisions of the Arthodonteae, the Aplolepideae with a single whorl of teeth and the Diplolepideae with two concentric whorls. Attention was drawn to possible cytological differences between the various groups by Mehra and Khanna,[28] taken up later by Ramsay[69] and Smith.[4] In considering basic numbers these groups will be dealt with separately.

1. Nematodonteae

The only order that has received much attention is the Polytrichales. This order has a more elaborate internal structure than other mosses and circumstantial evidence suggests that evolution has proceeded in the Bryopsida by progressive simplification,[1,2,72] implying that the Polytrichales are the least evolved order. The basic number of the Polytrichales is possibly the same as that of the ancestral stock of the Bryopsida. Of the 234 recent counts from members of the order all but two are n = 7, 14, or 21, clearly indicating that x = 7.

2. Arthrodonteae-Aplolepideae

Although the majority of species have n = 12, 13, or 14, about 3% have n = 5, 6, or 7. Two reports suggest that the first set of numbers is polyploid. In two Japanese species of *Oncophorus* with n = 14 there are seven pairs of chromosomes suggesting a doubling of the complement.[73] In a second report about *Dicranella heteromalla* with n = 13 the complement consists of six pairs and a single chromosome in the gametophyte suggesting an origin from n = 14, one chromosome having been eliminated.[73] It is argued that in the Aplolepideae x = 7.[4,28]

3. Arthrodonteae-Diplolepideae

This group is subdivided into the Acrocarpeae with erect habit and sympodial growth and the Pleurocarpeae with prostrate habit and monopodial growth. There are marked chromosomal differences between the two which are, therefore, treated separately. The Orthotrichales, placed in the Acrocarpeae, would, on the basis of karyotype, be better placed in the Pleurocarpeae.

a. Diplolepideae-Acrocarpeae

There are two marked peaks in chromosome numbers, one at n = 5, 6, and 7 and the other at n = 10 and 11. In some orders (e.g., Funariales) the chromosome numbers are based on x = 7; in others (e.g., Bryales) they are based on 5 and 6. It is not possible to state which is the basic number of the group, although if the suggestion made earlier that the ancestral number was x = 7 is correct, then it is possible to read a series 7 → 6 → 5. This series cannot be related to any phyletic arrangement as the interrelationships of the group, even assuming it is monophyletic, are unknown.

b. Diplolepideae-Pleurocarpeae

About 76% of these mosses have n = 10 or 11 and only a few have a lower number. Many have n = 10 + 1m. It is suggested[4] that there is a series 11 → 10 + 1m → 10 resulting from an unequal reciprocal translocation and then elimination of the *m*-chromosome, supported by observations on the Brachytheciaceae.[74] It is likely that in this group x = 11,[4,74] and that its evolution commenced with a doubling of the chromosome complement in the ancestral form from 6 to 12 and reduction to 11 either by a nonreciprocal translocation or by Robertsonian fusion.

McAdam[74] suggests that the basic number for the whole of the Diplolepideae is x = 11. She points out that in the group n = 11 karyotypes are remarkably similar and suggests that centric fusion involving different chromosomes in different evolutionary lines took place, giving rise to the diversity of karyotypes in the Diplolepideae. However, only a few members of the Diplolepideae-Acrocarpeae have n = 11 (some Bryaceae and some Orthotrichales). In the remainder, the basic number appears to be 5, 6, or 7, there being too great a uniformity within various families and orders for these numbers to be derived. It is more likely that the taxonomy of the Diplolepideae is incorrect and those families with n = 11 and obvious derivatives thereof (such as the Orthotricales) should be removed from the Acrocarpeae and placed with the Pleurocarpeae. At present the Diplolepideae is a group that has been subdivided on an obvious but not necessarily phyletically important character.

B. Major Evolutionary Trends

A possible phylogeny of the Bryopsida based upon cytology and current views concerning morphological trends for which, however, there is no evidence, is presented in Figure 3.

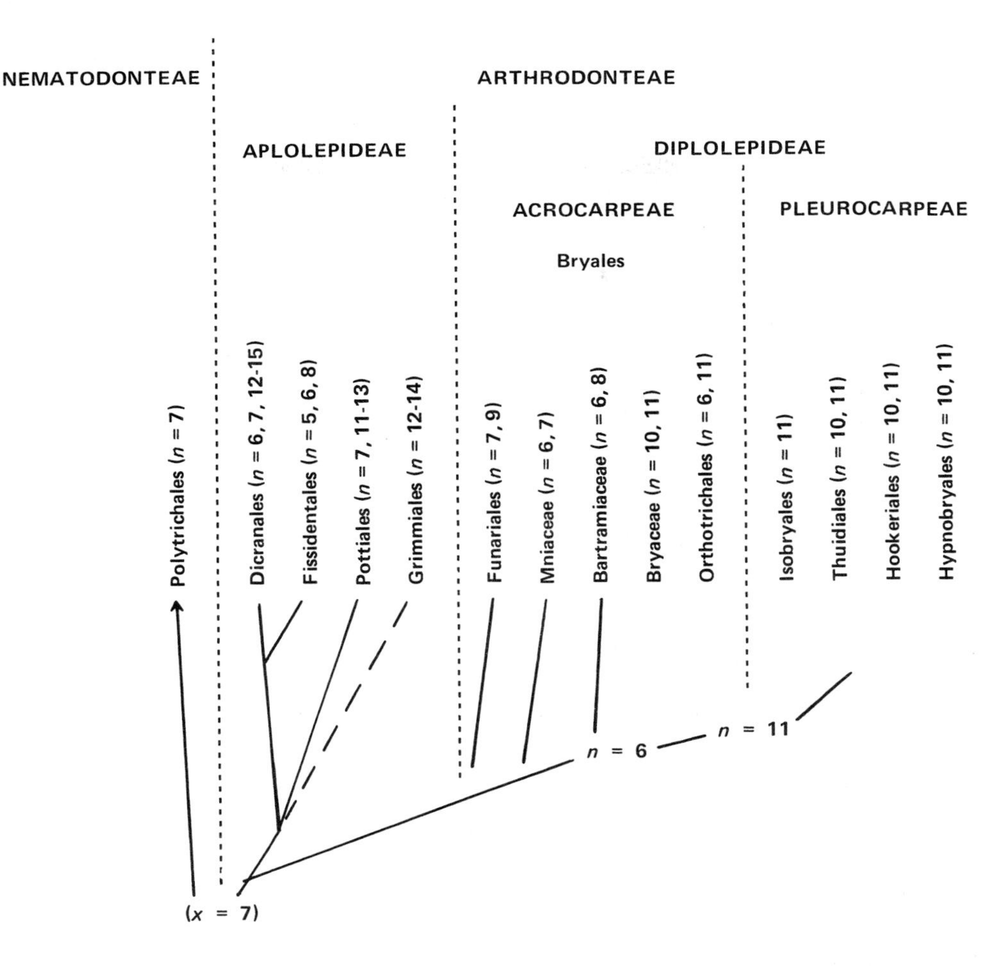

FIGURE 3. Outline phyletic scheme of the Bryopsida based on chromosome numbers and current views of morphological and anatomical evolutionary trends. (Modified with permission from Smith, A. J. E., *Adv. Bot. Res.*, 6, 207, 1978. Copyright: Academic Press Inc. [London] Ltd.)

C. Structural Changes

Due to the small size of the chromosomes of most moss species, studies of mitotic chromosomes have not been particularly informative because it has not usually been possible to locate centromeres. Most mitotic studies have been concerned with the location of heterochromatin from studies of heteropycnosis. Newton,[75] using Giemsa C-banding, has questioned the validity of the findings of such workers. Karyotype comparisons based on the doubtful distribution of heterochromatin and in the absence of centromere location are of dubious value. Recently techniques for locating centromeres have been developed[74] and the results obtained to date look promising.

Evidence of structural changes involved in speciation may be obtained from the study of meiosis in hybrids.[76] There are only three examples of such studies in mosses. In one, *Pleuridium acuminatum* × *Ditrichum pallidum,* the chromosomes failed to pair at meiosis. In *Weissia* (as *Atomum*) *ludoviciana* × *W. controversa* meiosis was perfectly normal although most of the resultant spores aborted as they did in the cross *Weissia* (as *Astomum*) *muhlenbergiana* × *W. controversa* where meiosis was again normal.[77] This suggests that in these species the isolating mechanisms leading to speciation may be genetic.

There are many reports of meiotic irregularities in nonhybrid sporophytes ascribed

Table 1
PERCENTAGES OF HAPLOIDS, PRIMARY POLYPLOIDS (DIPLOIDS), AND SECONDARY POLYPLOIDS IN VARIOUS TAXA[4]

Taxon	Probable basic no.	Percentages of		
		Haploids	Primary polyploids	Secondary polyploids
Nematodonteae	7	76	17	7
Arthrodonteae				
Aplolapideae	7	3	75	22
Diplolepideae-Acrocarpeae	5, 6, 7	33	44	23
Diplolepideae-Pleurocarpeae	11[a]	90	10	—

[a] This presumes that numbers less than n = 11 are derived and not haploid.

to structural heterozygosity. These abnormalities include bridges and fragments, nondisjunction, premature disjunction, asynapsis, lagging half-bivalents, and micronuclei. However, such irregularities may be due to genetic imbalance as well.

Further, apparent meiotic irregularities that suggest evidence of structural or genetic heterozygosity may be misleading. In diploid and triploid populations of *Atrichum undulatum*,[79] frequent meiotic irregularities were observed but spore formation was perfectly normal.

There is no conclusive evidence of structural changes in mosses, and attempts at using Giemsa C-banding have so far proved unsuccessful.[80]

There is, however, circumstantial evidence of structural changes during the course of evolution in the Brachytheciaceae.[74] As mentioned above, $n = 11$ is very common in pleurocarpous mosses and the karyotypes are very similar, suggesting an early common origin. In the Brachytheciaceae where there is a series from $n = 11 \rightarrow 10 \rightarrow 9 \rightarrow 7 \rightarrow 6$, the number of long arms in the various karyotypes remains the same (nine) but the number of long acrocentric (or telocentric) chromosomes decreases. This suggests that the reduction in number takes place by Robertsonian fusion. Thus, in Brachytheciaceae with $n = 11$ there are three long acrocentrics, six medium or short acrocentrics, and two short metacentrics. At the other end of the series, as in *Brachythecium rivulare* ($n = 6$), there are three long acrocentrics and three long metacentrics. In all instances the number of long arms remains nine. There is also increasing symmetry of the karyotype, going against the old but incorrect[22] theory that in plants karyotype evolution proceeds by increasing asymmetry.

There is also evidence of intraspecific cytological structural differences in different geographical areas. One such example is described[81] in *Pseudobryum cinclidioides.* There are marked differences in the relative lengths of the chromosomes in material from Japan,[82] the U.S.,[83] and Sweden.[81] In percentage terms the lengths of the longest and shortest chromosomes in the three areas are 18.8, 24.2, 24.6 and 15.2, 12.2, and 13.1, respectively. In the Japanese material the shortest chromosome is 81% the length of the longest, while in the American population it is 50%, and in the Swedish 53%. Structural changes are obviously involved but their precise nature is not known.

D. Polyploidy

1. Occurrence

Assuming that the basic numbers quoted are correct, then the proportion of poly-

Table 2
EXAMPLES OF INTERSPECIFIC POLYPLOIDY IN MOSSES — EACH PAIR CONSISTS OF TWO CLOSELY RELATED SPECIES

Haploid species	Chromosome number (*n*)	Polyploid species	Chromosome number (*n*)
Bryum capillare	10	*B. torquescens*	20
Distichium inclinatum	14	*D. hagenii*	42
Encalypta ciliata	13	*E. vulgaris*	26, 39
Fissidens cristatus	12	*F. adianthoides*	24
Plagiomnium affine	6	*P. medium*	12
Tortula intermedia	13	*T. princeps*	26
Weissia ludoviciana	13	*W. muhlenbergiana*	26

Table 3
EXAMPLES OF INTRASPECIFIC POLYPLOIDY

Species	Chromosome no.
Amblystegium riparium	20, 40
Atrichum undulatum	7, 14, 21
Distichium capillaceum	14, 28
Funaria brygrometrica	14, 28, 56
Hypopterygium rotulatum	9, 18, ca 27, 36
Physcomitrium pyriforme	9, 18, 27, 36
Pohlia nutans	11, 22, 33

ploid moss species is high. Table 1, based on chromosome numbers in Fritsch,[18] shows the percentages of haploids, primary polyploids, and secondary polyploids. It is clear that polyploidy has played an important role in the evolution, not only of species, but also of higher taxonomic categories. Interspecific and intraspecific polyploidy in mosses has been studied in some detail and is reviewed by Smith.[4]

In many instances polyploidy has led to evolution of species pairs, one haploid, the other diploid. The classical example of this is in the Mniaceae[83] (see Table 2). The differences between the species is usually small, suggesting autopolyploidy, and there is a series, as in flowering plants, from completely distinct cytotypes as listed in Table 2 to morphologically indistinguishable cytotypes as indicated in Table 3. With some cytotype pairs the only difference is sex. Thus, in *Mnium marginatum*, for example, var. *dioicum* has $n = 6$ and is dioecious and var. *marginatum* has $n = 12$ and is monoecious. This is a characteristic feature of many polyploid series in both mosses and liverworts, the haploid being dioecious and the diploid or higher ploid being monoecious.

Most reports on intraspecific polyploids being indistinguishable are based on the examination of a limited number of specimens, but a recent work[84] involved a study of a total of 472 gatherings of the three cytotypes — haploid, diploid, and triploid — of *Atrichum undulatum*. These proved completely indistinguishable both in the field and after cultivation under uniform conditions for 3 months.

There is experimental evidence[85] that new polyploids are larger than their haploid parents but that over a period of years they decrease in size, becoming indistinguishable from the haploids. In addition to the decrease in size there appears to be a decrease in the relative quantities of DNA, at least in *A. undulatum*.[84] In this species diploids have less than twice as much DNA per nucleus as haploids and triploids have proportionately less than three times as much.

2. Types of Polyploids

On the basis of morphological and cytological studies almost all polyploids studied are autopolyploids.[83,86,87] In new artificial polyploids there is a restoration of fertility over a period of time[85] and a mechanism is suggested to restore normal bivalent formation in new polyploids.[86] The situation appears similar in hepatics and it was found, for example, that by doubling the chromosome number of *Riccia fluitans* ($n = 8$), *R. rhenana* ($n = 16$) was produced.[88] The only known probable example of an allopolyploid is *Weissia exserta* ($n = 26$), which is intermediate between *W. crispa* and *W. controversa*, both with $n = 13$.[89]

Polyploidy also appears to have played a part in the initiation of the evolution of some genera, although in related genera haploid species are present. Examples are *Ditrichum* ($x = 13$), *Dicranella* ($x = 12$ to 16), *Dicranum* ($x = 11, 12$), *Grimmia* ($x = 13$), and *Ephemerum* ($x = 27$). Genera and families containing species that are weedy or occur in extreme environments tend to have much higher levels of polyploidy than usual. Examples include species of *Desmatodon* ($n = 25$ to 52), *Phascum* ($n = 21$ to 52), *Pottia* ($n = 15$ to 52), *Tortula* ($n = 7$ to 60), and Funariaceae ($n = 9$ to 56). In such genera diplospory probably occurs as a result of disturbances to meiosis from climatic fluctuations. The sporophyte, being small and unprotected in mosses, would, in exposed habitats, be very much affected by climatic fluctuations. In liverworts, when the sporophyte is protected by gametophyte structures until maturity, this is less likely and may account for the low frequency of polyploidy in that group.

3. Origin of Polyploidy

Polyploidy is considered to arise by apospory[6,83] because it is simple in the laboratory with mosses, though more difficult with liverworts, to induce the growth of gametophytes from damaged sporophyte tissue. This, however, only occurs under carefully controlled conditions that are unlikely to prevail in nature. It is much more likely that polyploidy arises by diplospory, as mentioned in the previous paragraph, involving the failure of one of the meiotic divisions. Evidence supporting this contention is provided by reports of spore dyads and syncytes in liverworts and mosses.[33,90,91] Polyploidy may also arise, at least in liverworts, by failure of mitosis, as there are reports of diploid shoots in otherwise haploid species.[25,92]

4. Polyploidy and Geographical Distribution

There is no evidence that polyploidy is related to geographical distribution either in the Northern[4,34] or in the Southern Hemisphere.[93] In Britain there is no difference in frequency between haploid and diploid species,[94] although species within which there is intraspecific polyploidy or aneuploidy are significantly ($p = 0.05$) more frequent than species with a uniform chromosome number.[94] Variation in chromosome number would appear to increase genetic variability and, hence, ecological amplitude.

The one example in which chromosome number is related to distribution is *Atrichum undulatum*.[84] The haploid cytotype is more frequent in southern and western Britain where the climate is atlantic (mild) than in the north and east when it is more extreme. In western U.S.S.R., where the climate is continental, the haploid is absent[95] and the triploid cytotype is much more common than in Britain.

E. Aneuploidy

There are numerous examples of inter- and intraspecific aneuploidy in mosses. Recently, however, reports of intraspecific aneuploidy have been called into question.

Table 4
COMPARISON OF UNIFORM COUNTS OBTAINED FROM GAMETOPHYTE MITOSIS BY McADAM[74] WITH COUNTS OBTAINED BY OTHER AUTHORS FROM METAPHASE I OF MEIOSIS[18]

Species	No. counts by McAdam	n (Found by McAdam)	n (Reported by other authors)
Brachythecium rutabulum	8	12	5, 10, 10 + 1B, 11, 12[a], 13, 20
Eurhynchium praelongum	71	11	8, 10, 11, 12
	3	11 + 1—4B	10 + 1B
E. schleicheri	4	11	7
E. striatum	18	11	6, 11, 12
E. swartzii	30	7	7, 10
	2	7 + 2B	

[a] Including 22 counts of $n = 12$ from British material by Smith and Newton.[52]

Nearly all such reports are based on meiotic studies, and interpretation of metaphase I is notoriously difficult, as it is often extremely hard to distinguish between bivalents and disjoined half-bivalents, especially when the bivalents tend to be sticky. McAdam,[74] in a very careful study of gametophyte mitosis in the Brachytheciaceae, found no examples of intraspecific aneuploidy, although in some instances varying numbers of accessories were found within species from which a variety of numbers have been reported by other authors. Examples of five species are given in Table 4. She also found $n = 11$ in 17 gatherings of 10 other species and $n = 10$ in 5 samples of 4 other species. The $n = 11$ karyotypes were all very similar as also were the $n = 10$ karyotypes. Yet, for the 15 species involved, other workers have reported a very wide range of numbers including $n = 6$, 8, 9, 10, 11, 12, and 18 with or without accessories. The implication is that some of these counts are based upon misinterpretations of meiotic spreads or possibly misidentification of the taxa concerned. These counts come from widely differing geographical locations and it is possible that populations in different countries, regarded as a single species, are not conspecific. It is likely, however, from McAdam's data that intraspecific aneuploidy is much less frequent than originally thought.

On the other hand, aneuploidy has evidently played a role in speciation in the Brachytheciaceae in which there is a range of numbers from $n = 6$ to 12,[74] and this doubtless applies to other moss families, although detailed karyotype analyses are required for confirmation. Karyotype studies suggest that *Brachythecium rutabulum* with $n = 12$ is a polyploid derivative of the $n = 6$ *B. rivulare*, indicating that a series of different chromosomal changes may be involved in speciation. Possible modes of origin of aneuploids are discussed in Section VI.D.3.

F. Sex-Specific Chromosomes

Dimorphic bivalents have been reported from dioecious mosses by a number of workers,[51,52,96] and there is circumstantial evidence that dimorphic chromosomes are sex specific.[96] These correspond to the morphological sex chromosomes of Japanese workers mentioned in Section II.F. However, as dimorphic bivalents also occur in monoecious mosses,[34,51,52] it does not necessarily follow that in dioecious species dimorphic chromosomes are sex chromosomes.

Ono[97] has also recorded structural sex chromosomes in mosses, but data about such chromosomes are conflicting. Thus, Yano[41-43] reported morphological sex chromosomes in some Polytrichaceae and Mniaceae while Ono[82,97,98] observed structural sex chromosomes from the same species. Tatuno and Yano[99] reported the X-chromosomes of *Mnium maximowiczii* as having more heterochromatin than the Y, while Tatuno and Segawa[55] recorded the reverse. Whether there are genuine differences between populations or whether the differences are artifacts resulting from different methods of squash preparation is not known. The situation with regard to sex-specific chromosomes in mosses requires further investigation, especially using banding techniques.

G. Accessory and m-Chromosomes

Micro-chromosomes were first reported from mosses by Heitz,[100] and the term is usually now applied to chromosomes that are less than one half or one third the size of the next smallest chromosome. They have been observed in a considerable number of moss species as well as from *Sphagnum*. Vaarama[101,102] reported accessory chromosomes from mosses, and since that time controversy has arisen as to the precise nature of very small members of the complement. There have been several reviews of the topic.[4,34,103-107] In some species, for example, those of *Sphagnum* and *Orthotrichum*, they are permanent members of the complement but in others, such as *Dicranum*, *Weissia*, and *Grimmia*, numbers vary, mostly from 0 to 4. Except in the hornwort *Anthoceros husnotii*[37] there is no evidence of the abnormal behavior characteristic of accessories, so there is no conclusive proof that accessories occur in mosses. That some minute chromosomes are permanent members of the complement while others apparently are not suggests that there may be two types of very small chromosome in mosses.

The origin of the two types of minute chromosome is unknown but it is possible that they arose as a consequence of nonreciprocal translocations.

Although the nature of these chromosomes is obscure they do appear to exert some genetic effect. In Britain monoecious species with accessory or m-chromosomes are significantly more frequent ($p = 0.025$ to 0.010) than are monoecious species without them.[94] The frequency of dioecious species is not significantly affected by their presence. It seems possible, therefore, that minute chromosomes increase genetic variability in inbreeding species, allowing greater ecological tolerance and hence greater frequency.

VII. CONCLUSION

The small size of many bryophyte chromosomes, the difficulty in obtaining Giemsa C-banding and of locating centromeres, and the lack of hybridization experiments are responsible for the very little direct evidence of the role of structural changes in chromosomal evolution in the Bryopsida. The little definite or circumstantial evidence available, however, suggests that structural changes have played an important role not only at the species level but also in the initiation of higher taxa. In the light of work currently in progress, it is very likely that this will be shown to be so.

Polyploidy and aneuploidy have played a much more important role in the evolution of mosses than in other groups. It is probable that modification of the present classification of mosses will, in the light of chromosome studies, provide the basis for a more convincing phylogeny.

As the dominant phase of the bryophyte life cycle is the haploid gametophyte, it means that recessive mutations and particular gene combinations cannot float in a population and, hence, are not available for selection. In the Hepaticopsida and Anthocertopsida this is largely true as in most instances the haploid number is the same as the basic one. In the Bryopsida in about 90% of species $n = 2x$ or nx, so the above argument does not apply. That mosses are much more frequent in terms of numbers of individuals and with a much greater ecological amplitude can be accounted for in terms of two or more sets of chromosomes in their complements. A similar argument applies to the Sphagnopsida, which appear to be of ancient polyploid origin, in that they account for 1 to 2% of the world total plant cover.[108] Thus, it may be argued that polyploidy has played a significant role in the evolution and dispersal of two classes of plant with a dominant gametophyte generation.

The appearance and behavior of bryophyte chromosomes is very similar to that of higher plants, lending support to the thesis that the Bryophyta have a common ancestry with vascular plants, although their course of evolution has been by simplification rather than by elaboration, and have no direct connection with the algae.

REFERENCES

1. **Crum, H.,** Mosses of the Great Lakes Forest, rev. ed., University of Michigan Press, Ann Arbor, 1976.
2. **Miller, H. A.,** The phylogeny and distribution of the musci, in *Bryophyte Systematics,* Clarke, G. C. S. and Duckett, J. G., Eds., Academic Press, New York, 1979, chap. 2.
3. **Schuster, R. M.,** The phylogeny of the Hepaticae, in *Bryophyte Systematics,* Clarke, C. G. S. and Duckett, J. G., Eds., Academic Press, New York, 1979, chap. 3.
4. **Smith, A. J. E.,** Cytogenetics, biosystematics and evolution in the Bryophyta, *Adv. Bot. Res.,* 6, 195, 1978.
5. **Miller, N. G.,** Fossil mosses of North America and their significance, in *The Mosses of North America,* Taylor, R. J. and Leviton, A. E., Eds., Pacific Division, American Association for the Advancement of Science, San Francisco, 1980, chap. 9.
6. **Crosby, M. S.,** The diversity and relationships of mosses, in *The Mosses of North America,* Taylor, R. J. and Leviton, A. E., Eds., Pacific Division, American Association for the Advancement of Science, San Francisco, 1980, 115.
7. **Beer, R.,** The chromosomes of *Funaria hygrometrica, New Phytol.,* 2, 164, 1903.
8. **Davies, B. M.,** Nuclear studies on *Pellia, Ann. Bot.,* 15, 147, 1901.
9. **Farmer, J. B.,** On spore-formation and nuclear division in the Hepaticae, *Ann. Bot.,* 9, 469, 1895.
10. **van Hook, J. M.,** Notes on the division of the cell and nucleus in liverworts, *Bot. Gaz.,* 30, 394, 1900.
11. **Moore, A. C.,** Sporogenesis in *Pallavicinia, Bot. Gaz.,* 40, 81, 1905.
12. **Allen, C. E.,** A chromosome difference correlated with sex differences in *Sphaerocarpos, Science (N.Y.),* 46, 466, 1917.
13. **Allen, C. E.,** The basis of sex inheritance in *Sphaerocarpos, Proc. Am. Phil. Soc.,* 88, 289, 1919.
14. **Heitz, E.,** Der bilaterale Bau der Geschlechtschromosomen und Autosomen bei *Pellia fabbraniana, P. epiphylla* und einigen anderen Jungermanniaceen, *Planta,* 5, 725, 1928.
15. **Heitz, E.,** Das Heterochromatin der Moose, *Jb. Wiss. Bot.,* 69, 762, 1928.
16. **Anderson, L. E.,** Chromosome numbers. Bryophytes, in *Growth,* Altman, P. L. and Dittmer, D. S., Eds., Federation of American Societies for Experimental Biology, Washington, D.C., 1962, 45.
17. **Berrie, G. K.,** The chromosome numbers of liverworts (Hepaticae and Anthocerotae), *Trans. Br. Bryol. Soc.,* 3, 688, 1960.
18. **Fritsch, R.,** Chromosomenzahlen der Bryophyten, eine Übersicht und Diskussion überes Aussagewertes für das System, *Wiss. Z. Friedriche Schiller Univ. Jena, Math.-Nat. Reihe,* 21, 839, 1972.
19. **Schuster, R. M.,** *The Hepaticae and Anthocerotae of North America East of the Hundredth Meridian,* Vol. 1, Columbia University Press, New York, 1966.
20. **Steere, W. C.,** Chromosome numbers in bryophytes, *J. Hattori Bot. Lab.,* 35, 99, 1972.
21. **Wylie, A. P.,** The chromosome numbers of mosses, *Trans. Br. Bryol. Soc.,* 3, 260, 1957.
22. **Jones, K.,** Chromosome changes in plant evolution, *Taxon,* 19, 172, 1970.
23. **Jones, K.,** Aspects of chromosome evolution in higher plants, *Adv. Bot. Res.,* 6, 119, 1978.
24. **Inoue, S.,** Karyological studies in *Takakia ceratophylla* and *T. lepidozioides, J. Hattori Bot. Lab.,* 37, 275, 1973.
25. **Tatuno, S.,** Zytologische Untersuchungen über die Lebermoose von Japan, *J. Sci. Hiroshima Univ.,* B4 (Div. 2), 73, 1941.
26. **Tatuno, S.,** Chromosomen von *Takakia lepidozioides* und eine Studie zur Evolution der Chromosomen der Bryophyten, *Cytologia,* 24, 138, 1959.

27. **Inoue, S. and Uchino, H.**, Karyological studies on mosses. VI. Karyotypes of fourteen species including some species with the intraspecific polyploid and aneuploid, *Bot. Mag. Tokyo,* 82, 359, 1969.
28. **Mehra, P. N. and Khanna, K. N.**, Recent cytological investigations in mosses, *Res. Bull. Panjab. Univ. N.S. Sci.,* 12, 1, 1961.
29. **Tatuno, S. and Nakano, M.**, Karyological studies on Japanese mosses. I, *Bot. Mag. Tokyo,* 83, 109, 1970.
30. **Yano, K.**, On the chromosomes in some mosses. VIII. Chromosomes in seven *Brotherella* species, *J. Hattori Bot. Lab.,* 23, 93, 1960.
31. **Newton, M. E.**, Chromosome morphology and bryophyte systematics, in *Bryophyte Systematics,* Clarke, G. C. S. and Duckett, J. G., Eds., Academic Press, New York, 1979, chap. 10.
32. **Rees, H.**, DNA in higher plants, in *Evolution of Genetic Systems,* Smith, H. H., Ed., Gordon and Breach, New York, 1972, 394.
33. **Allen, C. E.**, The genetics of bryophytes. II, *Bot. Rev.,* 11, 260, 1945.
34. **Steere, W. C.**, Chromosome number and behaviour in Arctic mosses, *Bot. Gaz.,* 116, 93, 1954.
35. **Berrie, G. K.**, The nucleolar chromosome in hepatics. I, *Trans. Br. Bryol. Soc.,* 3, 422, 1958.
36. **Berrie, G. K.**, Cytology and phylogeny of liverworts, *Evolution Lawrence Kan.,* 17, 347, 1963.
37. **Proskauer, J.**, Studies on Anthocerotales. V, *Phytomorphology,* 7, 113, 1957.
38. **Mizutani, M.**, Koke no Soseiki, *Shida Koke,* 71(2), 1, 1972.
39. **Newton, M. E.**, Heterochromatin as a cyto-taxonomic character in liverworts: *Pellia, Riccardia* and *Cryptothallus, J. Bryol.,* 9, 327, 1977.
40. **Fulford, M.**, Evolutionary trends and convergence in the Hepaticae, *Bryologist,* 6, 1, 1965.
41. **Yano, K.**, Cytological studies in Japanese mosses. I. Fissidentales, Dicranales, Grimminales, Eubryales, *Mem. Fac. Ed. Niigata Univ.,* 6, 1, 1957.
42. **Yano, K.**, Cytological studies in Japanese mosses. II. Hypnobryales, *Mem. Takada Branch Niigata Univ.,* 1, 85, 1957.
43. **Yano, K.**, Cytological studies in Japanese mosses. III. Isobryales, Polytrichales, *Mem. Takada Branch Niigata Univ.,* 1, 129, 1957.
44. **Tatuno, S.**, Über die Chromosomenzahlen bei drei Anthocerotaceen mit besonderer Rüchsicht auf ihre Heterochromosomen, *Bot. Mag. Tokyo,* 48, 54, 1934.
45. **Newton, M. E.**, Evolution and speciation in *Pellia* with special reference to the *Pellia megaspora-endiviifolia* complex (Metzgeriales). II. Cytology, *J. Bryol.,* 11, 433, 1981.
46. **Iverson, G. B.**, Karyotype evolution in the leafy liverwort genus *Frullania, J. Hattori Bot. Lab.,* 26, 119, 1963.
47. **Segawa, M.**, Karyological studies in liverworts with special reference to structural sex chromosomes. I, *J. Sci. Hiroshima Univ.,* B10 (Div. 2), 69, 1965.
48. **Segawa, M.**, Karyological studies in liverworts with special reference to structural sex chromosomes. II, *J. Sci. Hiroshima Univ.,* B10 (Div. 2), 81, 1965.
49. **Segawa, M.**, Karyological studies in liverworts with special reference to structural sex chromosomes. III, *J. Sci. Hiroshima Univ.,* B10 (Div 2), 149, 1965.
50. **Inoue, H.**, Some taxonomic problems in the genus *Plagiochila, J. Hattori Bot. Lab.,* 38, 105, 1974.
51. **Steere, W. C., Anderson, L. E., and Bryan, V. S.**, Chromosome studies on Californian mosses, *Mem. Torrey Bot. Club,* 20(4), 1, 1954.
52. **Smith, A. J. E. and Newton, M. E.**, Chromosome studies on some British and Irish mosses, *Trans. Br. Bryol. Soc.,* 5, 463, 1968.
53. **Newton, M. E.**, Chromosome studies in some British and Irish bryophytes, *Trans. Br. Bryol. Soc.,* 6, 244, 1971.
54. **Berrie, G. K.**, Sex chromosomes of *Plagiochila praemorsa* Stephani and the status of structural sex chromosomes in hepatics, *Bull. Soc. Bot. Fr.,* 121, 129, 1974.
55. **Tatuno, S. and Segawa, M.**, Über structurelle Geschlechtschromosomen bei *Mnium maximowicziia* und Nukleolinuschromosomen bei einigen Bryophyten, *J. Sci. Hiroshima Univ.,* B7 (Div. 2), 1, 1955.
56. **Showalter, A. M.**, The chromosomes of *Riccardia pinguis, Am. J. Bot.,* 10, 170, 1923.
57. **Lorbeer, G.**, Die Zytologie der Lebermoose mit besonderer Berüchstigung allgemeiner Chromosomenfragen. I, *Jb. Wiss Bot.,* 80, 567, 1934.
58. **Jachimsky, H.**, Beitrag nur Kenntnis von Geschlechtschromosomen und Heterochromatin bei Moosen, *Jb. Wiss Bot.,* 81, 208, 1935.
59. **Heitz, E.**, Geschlechtschromosomen bei *Pellia fabbroniana* und *Pellia epiphylla, Ber. Deut. Bot. Ges.,* 45, 607, 1927.
60. **Mehra, P. N.**, A study of the chromosome number of some Indian members of the family Codoniaceae, *Proc. Indian Acad. Sci.,* 8(B), 1, 1938.
61. **Lewis, K. R.**, Chromosome structure and organisation in *Pellia epiphylla, Phyton,* 11, 29, 1958.
62. **Vaarama, A.**, Cytological observations in *Pleurozium schreberi* with special reference to centromere evolution, *Suomal. Eluan-Ja Kasvit. Seur. Julk.,* 28, 1, 1954.

63. **Masubuchi, M.**, Early replicating DNA in heterochromatin of *Plagiochila ovalifolia* (liverworts), *Bot. Mag. Tokyo*, 87, 229, 1974.
64. **Masubuchi, M.**, Early DNA synthesis of heterochromatin and replication of Y-chromosomes in *Pellia neesiana, Bot. Mag. Tokyo*, 84, 24, 1971.
65. **Tatuno, S., Tanaka, R., and Masubuchi, M.**, Early DNA synthesis in the X-chromosome of *Pellia meesiana, Cytologia*, 35, 220, 1970.
66. **Fowke, L. C. and Pickett-Heaps, J. D.**, Electron microscope study of vegetative cell division in two species of *Marchantia, Can. J. Bot.*, 56, 467, 1978.
67. **Schnepf, E.**, Mikrotubulus-Anordnung und -Unordnung Wandbildung und Zellmorphogenese in jungen *Sphagnum*-blättchen, *Protoplasma*, 78, 145, 1973.
68. **Proskauer, J.**, Studies on the morphology of *Anthoceros*. I and II, *Ann. Bot.*, 12, 237 and 427, 1948.
69. **Ramsay, H. P.**, Cytological studies on some mosses from the British Isles, *Bot. J. Linn. Soc.*, 62, 85, 1969.
70. **Sorsa, V.**, The quadripolar spindle and the change of orientation of the chromosomes in meiosis of *Sphagnum, Suomol-Ugr. Seur. Aikak.*, Ser. A (IV, Biol.) 53, 1, 1956.
71. **Bryan, V. S.**, The question of post-reduction in *Sphagnum, Am. J. Bot.*, 53, 1012, 1967.
72. **Steere, W. C.**, A new look at evolution and phylogeny in the bryophytes, in *Current Topics in Plant Sciences*, Gunckel, J., Ed., Academic Press, New York, 1969, 135.
73. **Yano, K.**, On the chromosomes in some mosses, *Bot. Mag. Tokyo*, 67, 38, 1954.
74. **McAdam, S. V.**, Chromosome evolution in the Brachytheciaceae, *J. Bryol.*, 12, 233, 1982.
75. **Newton, M. E.**, Chromosome relationships of the heterochromatin bodies in a moss, *Dicranum taur-icum* Sapehin, *J. Bryol.*, 9, 557, 1977.
76. **Stebbins, G. L.**, *Chromosomal Evolution in Higher Plants*, Edward Arnold, London, 1971.
77. **Anderson, L. E.**, Cytology and reproductive biology of mosses, in *The Mosses of North America*, Taylor, R. J. and Levitan, A. E., Eds., Pacific Division, American Association for the Advancement of Science, San Francisco, 1980, 37.
78. **Lewis, K. R. and John, B.**, *Chromosome Marker*, Churchill Livingstone, London, 1963.
79. **Newton, M. E.**, personal communication, 1981.
80. **McAdam, S. V.**, personal communication, 1981.
81. **Wigh, K.**, Cytotaxonomical and modification studies in some Scandinavian mosses, *Lindbergia*, 1, 130, 1972.
82. **Ono, K.**, Karyological studies on Mniaceae and Polytrichaeceae with special reference to the structural sex chromosomes. I, *J. Sci. Hiroshima Univ.*, B13 (Div. 2), 91, 1970.
83. **Lowry, R. J.**, A cytotaxonomic study of the genus *Mnium, Mem. Torrey Bot. Club*, 20(2), 1, 1948.
84. **Abderrahman, S. and Smith, A. J. E.**, Studies on the cytotypes of *Atrichum undulatum*, I, *J. Bryol.*, 12, 265, 1982.
85. **Wettstein, F. von and Straub, J.**, Experimentelle Untersuchungen zum Artbildungs Problem, *Zeit. Indukt. Abst. Vererb.*, 80, 271, 1942.
86. **Moutschen, J. H.**, Etude de la meiose et induction de artificiopolyploides chez *Amblystegium riparium, Cellule*, 54, 353, 1952.
87. **Moutschen, J. H. and Fransson, M.**, Abnormal meiotic processes in the moss *Pogonatum aloides* P. Beauv., *Nucleus Calcutta*, 11, 89, 1973.
88. **Berrie, G. K.**, Experimental studies on polyploidy in liverworts. I. The *Riccia fluitans* complex, *Bryologist*, 67, 140, 1964.
89. **Khanna, K. R.**, Studies in natural hybridization in the genus *Weissia, Bryologist*, 63, 1, 1960.
90. **Newton, M. E.**, Cytology of British Bryophytes, Ph.D. thesis, University of Wales, 1968.
91. **Anderson, L. E. and Lemmon, B.**, Syndiploidy in a moss, *Am. J. Bot.*, 54, 641, 1967.
92. **Steel, D. T.**, The taxonomy of *Lophocolea bidentata* (L.) Dum. and *L. cuspidata* (Nees) Limpr., *J. Bryol.*, 10, 49, 1978.
93. **Newton, M. E.**, Chromosome studies in some Antarctic and sub-Antarctic bryophytes, *Br. Antarct. Surv. Bull.*, 50, 77, 1980.
94. **Smith, A. J. E., and Ramsay, H. P.**, Sex, cytology, and frequency of bryophytes in the British Isles, *J. Hattori Bot. Lab.*, 52, 275, 1982.
95. **Lazarenko, A. S. and Lesnyak, E. N.**, On chromosome races of the moss *Atrichum undulatum* (Hedw.) Brid. in the west of the U.S.S.R., *Ukr. Bot. Zh.*, 34, 383, 1977.
96. **Ramsay, H. P.**, Sex chromosomes in *Macromitrium, Bryologist*, 69, 293, 1966.
97. **Ono, K.**, Karyological studies of Mniaceae and Polytrichaceae with special reference to the structural sex chromosomes. II, *J. Sci. Hiroshima Univ.*, B13 (Div. 2), 107, 1970.
98. **Ono, K.**, Karyological studies of Mniaceae and Polytrichaceae with special reference to the structural sex chromosomes. III, *J. Sci. Hiroshima Univ.*, B13 (Div. 2), 167, 1970.
99. **Tatuno, S. and Yano, K.**, Geschlechtschromosomen bei vier Arten von Mniaceae, *Cytologia*, 18, 36, 1953.

100. **Heitz, E.**, Über die Beziehung zurischen Polyploidie und Gemischtgeschlechtlichkeit bei Moosen, *Arch. Julius Klans Stift Vererb. Sozialanth. Rassenhyg.*, 17, 444, 1942.
101. **Vaarama, A.**, Meiosis in moss species of the family Grimmiaceae, *Port. Acta Biol.*, A.R.B. Goldschmidt, Vol. 47 (Ser. A), 1949.
102. **Vaarama, A.**, Studies on chromosome numbers and certain meiotic features of several Finnish moss species, *Bot. Notiser*, 1950, 239, 1950.
103. **Anderson, L. E.**, Biosystematic evolutions in the Musci, *Phytomorphology*, 14, 27, 1964.
104. **Vaarama, A.**, Structurally and functionally deviating chromosome types in Bryophyta, *Nucleus Calcutta*, Suppl., 285, 1968.
105. **Inoue, S.**, B-chromosomes in two moss species, *Miscnea Bryol. Lichen. Nichinan*, 4, 167, 1968.
106. **Newton, M. E.**, Chromosome studies in some British and Irish bryophytes, *J. Bryol.*, 7, 379, 1973.
107. **Wigh, K.**, Accessory chromosomes in some mosses, *Hereditas*, 74, 211, 1973.
108. **Clymo, R.**, *Sphagnum*, in *Bryophyte Ecology*, Smith, A. J. E., Ed., Chapman and Hall, London, 1982, chap. 6.

INDEX

A

B

C

D

E

F

G

H

I

K

L

M

N

O

P

Q

R

S